Spring Batch 权威指南

[美] 迈克尔·T. 米内拉(Michael T. Minella) 著
张坤　张渊 译

清华大学出版社

北京

北京市版权局著作权合同登记号　图字：01-2020-2330

Michael T. Minella
The Definitive Guide to Spring Batch: Modern Finite Batch Processing in the Cloud, Second Edition
EISBN: 978-1-4842-3723-6
Original English language edition published by Apress Media. Copyright © 2019 by Apress Media. Simplified Chinese-Language edition copyright © 2021 by Tsinghua University Press. All rights reserved.

本书封面贴有清华大学出版社防伪标签，无标签者不得销售。
版权所有，侵权必究。举报：010-62782989，beiqinquan@tup.tsinghua.edu.cn。

图书在版编目(CIP)数据

Spring Batch 权威指南 / (美)迈克尔·T. 米内拉(Michael T. Minella)著；张坤，张渊译. —北京：清华大学出版社，2021.1
书名原文：The Definitive Guide to Spring Batch: Modern Finite Batch Processing in the Cloud, Second Edition
ISBN 978-7-302-56772-1

Ⅰ.①S… Ⅱ.①迈…②张…③张… Ⅲ.①数据处理 Ⅳ.①TP274

中国版本图书馆 CIP 数据核字(2020)第 218041 号

责任编辑：王　军
封面设计：孔祥峰
版式设计：思创景点
责任校对：成凤进
责任印制：丛怀宇

出版发行：清华大学出版社
网　　址：http://www.tup.com.cn，http://www.wqbook.com
地　　址：北京清华大学学研大厦 A 座　　　　邮　编：100084
社 总 机：010-62770175　　　　邮　购：010-62786544
投稿与读者服务：010-62776969，c-service@tup.tsinghua.edu.cn
质 量 反 馈：010-62772015，zhiliang@tup.tsinghua.edu.cn

印 装 者：大厂回族自治县彩虹印刷有限公司
经　　销：全国新华书店
开　　本：170mm×240mm　　　印　张：26.25　　　字　数：702 千字
版　　次：2021 年 1 月第 1 版　　　印　次：2021 年 1 月第 1 次印刷
定　　价：98.00 元

产品编号：086011-01

译 者 序

提到数据处理，很多人可能会想到大数据平台上的批处理和流式处理。为了使用 Hadoop 生态，应用开发者可能会在层出不穷的"大数据概念"中举步不前，在眼花缭乱的组件面前望而却步。假如要处理的数据集不是那么"大"，处理也不是很复杂，那么对于 Java 开发者而言，有没有简单易用的框架和工具呢？Spring Batch 就是可能的答案。对于数据工程师而言，Spring Batch 未免有些鸡肋；但是对于 Java 应用开发者，尤其是有一定经验的 Java 应用开发者而言，Spring Batch 未尝不是易于上手且能够解决实际问题的实用工具。

本书将理论与实际相结合，介绍了使用 Spring Batch 读取输入、加工处理以及输出时的各种考虑因素和不同实现选项，贯穿本书的是一个对账单示例：根据客户数据和客户交易记录文件，生成汇总的对账单文件。在开发过程中，本书使用了轻量级的敏捷开发方法：使用用户故事描述需求，进行多次迭代，不断地完善功能。如果跳过其中的理论讲解，你就会发现，其实并不需要编写太多代码，就可以完成端到端的数据处理任务。仔细回顾，你甚至会发现，如果转换逻辑不多，那么大部分代码仅仅用于作业、步骤等的配置。Spring Batch 提供了常见的输入输出工具，这样开发人员就可以更多地关注于编写数据转换逻辑。在实际场景中，批处理程序可能无法满足性能要求，所以本书还介绍了单机环境中的性能调优方法，以及分布式环境中的批处理。本书最后简单介绍了不同层次的自动化测试。

希望读者在深入阅读各个章节的过程中，也能思考本书的组织逻辑以及各个技术决策背后的原因。这样在实现功能的同时，就能建立体系化的思考方式。

清华大学出版社的编辑老师，一直关注本书的翻译工作，期间给予我们极大的耐心和许多建设性的反馈意见，在此特别感谢。本书的作者 Michael T. Minella 也是一位热心人士，回答了我们在翻译过程中的不解之处。在翻译本书期间，我们经历了新冠肺炎疫情在国内的爆发和有效控制，人们的生活和工作都受到了巨大影响。在很长一段时间里，人们都居家办公。本书能够成功翻译，离不开家人在这段时间内给予的有力支持，正因为他们，我们才能在安心的环境中投入时间进行翻译和校对。

本书涉及大量的专业术语和技术实践，我们也竭尽所能让这本译作简单易懂。但毕竟水平有限，错误和失误在所难免。如有任何意见和建议，请不吝指正，不胜感激！

译　者

2020 年 6 月

作者简介

Michael T. Minella 是一位软件工程师、作家和演说家,拥有超过18年的专业经验。Michael是Pivotal的软件工程主管,领导了Spring Batch和Spring Cloud Task项目,他也是JSR-352(Java Batch)的专家组成员。另外,他还是一名Java Champion和JavaOne Rockstar,曾在许多不同的Java国际会议上发表演讲。

在日常工作外,Michael还在播客OffHeap上扮演"脾气暴躁的人"。他个人对信息安全话题(InfoSec)也很感兴趣。Michael的爱好包括摄影和木工,他和爱人Erica、孩子Addison生活在一起。

技术审稿人简介

Wayne Lund 在为埃森哲(Accenture)的全球架构群组工作时，担任首席技术架构师，是 Spring Batch 最初的创造者之一，他在 JavaOne 2007 大会上将 Spring Batch 交付给 Java 社区。埃森哲的全球架构群组专注于 OSS 项目，帮助客户采用 Spring 作为抽象 JEE 的首选平台，并偏爱轻量级框架。在 Spring 被 VMWare 收购之后，Wayne 加入了 vFabric 群组(现在是 Pivotal 服务的一部分)，该群组在销售渠道中支持 SpringSource、RabbitMQ、Gemfire 和其他开源的轻量级框架。Wayne 目前是 Pivotal Data Services 的咨询平台架构师，负责帮助提供启用了 Spring 的数据产品解决方案，包括 Spring Cloud Data Flow(Spring Cloud Stream、Spring Cloud Task 和 Spring Batch)以及使用 RabbitMQ 和 Kafka 的消息处理。

Felipe Gutierrez 是解决方案软件架构师，拥有墨西哥蒙特雷大学蒙特雷分校计算机科学专业的学士和硕士学位。Felipe 有超过 20 年的 IT 经验，他为多个垂直行业的公司开发项目，比如零售、医疗保健、教育和银行等。Felipe 目前是 Pivotal 的平台和解决方案架构师，擅长 Cloud Foundry PAS 和 PKS、Spring 框架、Spring 云原生应用、Groovy 和 RabbitMQ 等技术。他曾在诺基亚、苹果、Redbox、高通等大公司担任解决方案架构师。Felipe 还是畅销书 *Spring Boot Messaging* 和 *Introducing Spring Framework* 的作者。

致　　谢

从我写第一本书到现在的这段时间里，我的生活发生了很大的变化，其中的很多变化都对我的职业生涯产生了实质性影响，这让我有机会撰写有关 Spring Batch 的第二本书。在此，我想利用本书的"致谢"部分，对一路帮助过我的各位人士表达我的感激之情。

首先，我要感谢 Dave Syer，不仅因为他是 Spring Batch 框架的创造者，让我在过去 6 年多的时间里有幸撰写了两本书，而且因为他是令我敬仰的开源实践者。当初，在我写完第一本书之后，我遇到了他并送了他一本。他向我推荐了一份使用 Java Community Process (JCP)围绕批处理(JSR-352)创建 Java Specification Request (JSR)的工作。这让我遇到了下一个我要感谢的人。

Wayne Lund 是 JSR-352 专家组的最初成员之一，也是本书的技术编辑之一，他对我取得今天的成就有巨大的影响。正是在 JSR-352 专家组里，我们会面并一起工作，根据我们在 Spring Batch 方面的经验来改进 Java 批处理规范的设计。在开发 JSR 期间，Wayne 告诉我 Spring 团队正在为 Spring Batch 项目寻找新的领军人，并问我是否有兴趣。我依然相信，我的大多数非技术背景的家人和朋友并不明白被邀请加入 Spring 团队意味着什么。我在 Spring 工程团队工作时，享受到前所未有的快乐!谢谢Wayne，感谢你最初对我的信任和一直以来提供的支持。

我还要感谢我在 Pivotal 工作时的经理 Brian Dussault。在我的职业生涯中，我有幸为许多出色的经理工作过，对于其中的许多人，我愿意为之工作并且十分乐意再次与之共事，但是没有人能像 Brian 那样给予我支持和信任。

我还要感谢另外两组人。首先是 Apress 团队。在漫长的写作过程中，Steve Anglin 和 Mark Powers 给予我充分的理解。我敢肯定，我不是最容易共事的作者，但我很幸运他们成为本书的编辑。没有他们对本书提供的持续支持，本书不可能顺利完成。我还要感谢技术编辑 Felipe Guiterrez，他的评论和鼓励对最终的结果产生了巨大的影响。

最后且最重要的是，我要感谢我的家人。任何写过书的人都知道，写作的过程需要付出很大的代价。作者们需要将大量的时间、精力和情感投入到写作过程中。感谢我的女儿 Addison，她每天都用她的热情、无尽的好奇心激励着我。我还要感谢我的妻子 Erica，是她支持我完成了这本书。如果没有她一直以来的鼓励和支持，我早就半途而废了。对我来说，你们就是我的整个世界，谢谢你们!

前　　言

Spring Batch 是我深度参与的第一个开源项目。在我的记忆中，有两件事拖延了 Spring Batch 1.0 的发布：一是确保服务质量(Quality of Service，QoS)特性在实际工作中发挥作用；二是在 API 的设计上需要投入大量的精力。无论如何，错误是不可避免的，但我认为，至少可以说我们尽了最大努力，让生活有了一个良好的开端。

如果回顾一下 Spring Batch 的发展历程，你就会发现它起源于批处理领域，并且诞生于世界各地的许多企业长时间不断重复发明的过程之中。我第一次看到这些代码是在 2006 年，当时是 Rob Harrop 在伦敦一家银行做咨询工作时写的一个很小的原型。在我们将 Spring Batch 的一些有用特性分离出来，以便在其他项目中共享之后，这些部分最终在 Spring Retry 中完成。Spring Batch 剩下的大部分，以及 Spring Batch 面向状态机的世界观，都来自与埃森哲的合作。从那时起，有太多的贡献者加入，在此无法一一列出，但值得特别提及的是 Lucas Ward，他是 Spring Batch 在早期的另一位家长和看护者。我还记得，Robert Kasanicky 和 Dan Garrette 为 Spring Batch 1.0 在 2008 年的成功发布做出过巨大贡献。

以上贡献者也为 Spring Batch 2.0 在 2010 年的成功发布发挥了重要作用，我们在 Spring Batch 2.0 中引入了 chunk 的概念以及一些特性，以支持分布式处理、并行处理和 Java 5 的新语言特性。chunk("块")是可以一起处理的一组数据项，这为提高效率和可伸缩性提供了可能。Spring Batch 2.0 在很长一段时间里代表最高的技术水平，并且在 JSR-352 规范启动时，成为其中的一部分。来自埃森哲的 Wayne Lund 在 Spring Batch 项目的早期就已加入其中，他也是 JSR-352 专家组的一员，现在是 Pivotal 的平台架构师。

那时候，Michael Minella 也是 JSR-352 专家组的成员，他在现实生活中大量地使用 Spring Batch 并撰写了一本书。当他在 2012 年加入 Spring 团队时，Spring 正好开始 Spring Batch 3.0 的准备和发布工作。在 Spring Batch 3.0 中，我们第一次看到了 @EnableBatchProcessing 注解，并且将重心从 XML 配置转到使用 Java 进行配置。Michael 很快就以项目负责人的角色接管了这个项目，领导 Spring Batch 的 3.x 系列版本一直到 4.0 版本。在 4.0 版本中，Java 8 成为基线，并且添加了一些新的流畅风格 (fluent-style)的配置构建器。与 Spring Cloud Data Flow 的连接以及分布式处理的工业化也在这一时期发生。在 2018 年年初，Mahmoud Ben Hassine 作为新的项目联合负责人加入进来，他一直在帮助 Michael 推动 Spring Batch，并仔细听取了许多用户的反馈。

所以，在本书写作期间，Spring Batch 已经发展了十年，期间也不断有优秀的贡献者加入进来。在未来几年里，Spring Batch 肯定还有很多事情要做，因为批处理似乎永远不会消失。这确实很有意思。

Dave Syer，Spring Batch 项目创始人
2019 年于伦敦

目 录

第 1 章 批处理和 Spring 1
1.1 批处理的历史 2
1.2 批处理面临的挑战 3
1.3 为什么使用 Java 进行
批处理 4
1.4 Spring Batch 的其他用途 5
1.5 Spring Batch 框架 7
    1.5.1 使用 Spring 定义作业 8
    1.5.2 管理作业 9
    1.5.3 本地和远程的并行化 9
    1.5.4 标准化 I/O 10
    1.5.5 Spring Batch 生态系统的
其他部分 10
    1.5.6 Spring 的所有特性 10
1.6 如何阅读本书 11
1.7 本章小结 11

第 2 章 Spring Batch 入门 13
2.1 批处理的架构 13
    2.1.1 深入讨论作业和步骤 14
    2.1.2 执行作业 15
    2.1.3 并行化 16
    2.1.4 文档 18
2.2 项目设置 19
    2.2.1 获取 Spring Batch 19
    2.2.2 IntelliJ IDEA 21
2.3 "Hello，World!" 示例
程序 22
2.4 运行作业 25
2.5 本章小结 26

第 3 章 示例作业 27
3.1 了解敏捷开发 27
    3.1.1 通过用户故事捕捉
需求 28
    3.1.2 使用测试驱动开发捕捉
设计 29
    3.1.3 使用版本控制系统 29
    3.1.4 在真正的开发环境中
工作 30
3.2 理解作业需求 30
3.3 设计批处理作业 34
    3.3.1 作业描述 35
    3.3.2 理解数据模型 36
3.4 本章小结 37

第 4 章 理解作业和步骤 39
4.1 作业介绍 39
4.2 配置作业 41
    4.2.1 基本的作业配置 41
    4.2.2 作业参数 43
    4.2.3 使用作业监听器 55
    4.2.4 执行上下文 58
    4.2.5 操作 ExecutionContext ... 58
4.3 使用步骤 62
    4.3.1 Tasklet 和基于块的
处理 62

- 4.3.2 步骤的配置 63
- 4.3.3 理解其他类型的Tasklet 65
- 4.3.4 步骤流 80
- 4.4 本章小结 95

第5章 作业存储库和元数据 ... 97
- 5.1 作业存储库是什么 97
  - 5.1.1 使用关系数据库 97
  - 5.1.2 使用内存存储库 101
- 5.2 配置批处理基础设施 101
  - 5.2.1 BatchConfigurer接口 ... 101
  - 5.2.2 自定义JobRepository ... 102
  - 5.2.3 自定义 TransactionManager 103
  - 5.2.4 自定义JobExplorer 104
  - 5.2.5 自定义JobLauncher ... 105
  - 5.2.6 配置数据库 106
- 5.3 使用元数据 106
- 5.4 本章小结 110

第6章 运行作业 111
- 6.1 使用Spring Boot启动作业 111
- 6.2 使用REST API启动作业 113
- 6.3 使用Quartz进行调度 118
- 6.4 停止作业 121
  - 6.4.1 自然结束 121
  - 6.4.2 以编程方式结束 122
  - 6.4.3 错误处理 134
- 6.5 控制作业的重启 136
  - 6.5.1 阻止作业再次执行 136
  - 6.5.2 配置重启次数 137
  - 6.5.3 重新运行一个完整的步骤 138

- 6.6 本章小结 139

第7章 ItemReader 141
- 7.1 ItemReader接口 141
- 7.2 文件输入 142
  - 7.2.1 平面文件 142
  - 7.2.2 XML文件 167
- 7.3 JSON 172
- 7.4 数据库输入 174
  - 7.4.1 JDBC 174
  - 7.4.2 Hibernate 180
  - 7.4.3 JPA 184
  - 7.4.4 存储过程 186
  - 7.4.5 Spring Data 187
- 7.5 现有的服务 191
- 7.6 自定义输入 194
- 7.7 错误处理 198
  - 7.7.1 跳过记录 199
  - 7.7.2 把无效的记录记入日志 200
  - 7.7.3 处理没有输入的情况 ... 202
- 7.8 本章小结 203

第8章 ItemProcessor 205
- 8.1 ItemProcessor概述 205
- 8.2 使用Spring Batch提供的ItemProcessor 206
  - 8.2.1 ValidatingItemProcessor ... 207
  - 8.2.2 输入校验 207
  - 8.2.3 ItemProcessorAdapter ... 213
  - 8.2.4 ScriptItemProcessor ... 215
  - 8.2.5 CompositeItemProcessor ... 216
- 8.3 编写自己的条目处理器 220
- 8.4 本章小结 222

目　录

第 9 章　ItemWriter ·············· 223
  9.1　ItemWriter 概述 ············ 224
  9.2　基于文件的 ItemWriter ······ 225
    9.2.1　FlatFileItemWriter ·········· 225
    9.2.2　StaxEventItemWriter ······· 235
  9.3　基于数据库的 ItemWriter ···· 239
    9.3.1　JdbcBatchItemWriter ······ 239
    9.3.2　HibernateItemWriter ······ 244
    9.3.3　JpaItemWriter ············· 249
  9.4　NoSQL ItemWriter ········· 252
    9.4.1　MongoDB ················ 252
    9.4.2　Noe4j ···················· 255
    9.4.3　Pivotal Gemfire 和
        Apache Geode ············ 259
    9.4.4　Repository 抽象 ·········· 263
  9.5　输出到其他目标的
      ItemWriter ················ 266
    9.5.1　ItemWriterAdapter ········ 266
    9.5.2　PropertyExtractingDelegating-
        ItemWriter ··············· 268
    9.5.3　JmsItemWriter ············ 271
    9.5.4　SimpleMailMessage-
        ItemWriter ··············· 275
  9.6　复合的 ItemWriter ········· 280
    9.6.1　MultiResource-
        ItemWriter ··············· 280
    9.6.2　CompositeItemWriter ····· 288
    9.6.3　ClassifierComposite-
        ItemWriter ··············· 291
  9.7　本章小结 ················· 294

第 10 章　示例应用 ············· 297
  10.1　回顾银行对账单作业 ······ 297
  10.2　配置新项目 ·············· 298

  10.3　导入客户数据 ············ 300
    10.3.1　验证客户 ID ·········· 306
    10.3.2　写入客户更新 ········ 308
  10.4　导入交易数据 ············ 311
    10.4.1　读取交易 ············ 313
    10.4.2　写入交易 ············ 314
  10.5　计算当前余额 ············ 315
    10.5.1　读取交易 ············ 316
    10.5.2　更新账户余额 ········ 316
  10.6　生成对账单 ·············· 317
    10.6.1　读取对账单数据 ······ 317
    10.6.2　为对账单添加账户
         信息 ················ 320
    10.6.3　写对账单 ············ 322
  10.7　本章小结 ················ 326

第 11 章　伸缩和调优 ············ 327
  11.1　分析批处理作业的
       性能 ···················· 327
    11.1.1　VisualVM 之旅 ······· 328
    11.1.2　分析 Spring Batch 应用的
         性能 ················ 331
  11.2　伸缩作业 ················ 337
    11.2.1　多线程步骤 ·········· 337
    11.2.2　并行步骤 ············ 339
    11.2.3　组合使用
         AsyncItemProcessor 和
         AsyncItemWriter ······ 344
    11.2.4　分区 ················ 346
    11.2.5　远程分块 ············ 360
  11.3　本章小结 ················ 365

第 12 章　云原生的批处理 ······· 367
  12.1　"12 要素应用" ············ 367

IX

	12.1.1	代码库 368
	12.1.2	依赖 368
	12.1.3	配置 368
	12.1.4	支持服务 368
	12.1.5	构建、发布、运行 369
	12.1.6	进程 369
	12.1.7	端口绑定 369
	12.1.8	并发 369
	12.1.9	可丢弃性 369
	12.1.10	开发环境与线上环境的等价 370
	12.1.11	日志 370
	12.1.12	管理进程 370
12.2	一个简单的批处理作业 370
12.3	断路器 376
12.4	外部化配置 379
	12.4.1	Spring Cloud Config 379
	12.4.2	通过 Eureka 进行服务绑定 381
12.5	批处理过程的编排 384
	12.5.1	Spring Cloud Data Flow 385
	12.5.2	Spring Cloud Task 386
	12.5.3	注册和运行任务 387
12.6	本章小结 390

第 13 章 批处理的测试 391

13.1	使用 JUnit 和 Mockito 进行单元测试 391
	13.1.1	JUnit 392
	13.1.2	mock 对象 394
	13.1.3	Mockito 395
13.2	使用 Spring 的实用工具进行集成测试 398
	13.2.1	使用 Spring 进行通用集成测试 398
	13.2.2	测试 Spring Batch 400
13.3	本章小结 408

第 1 章
批处理和 Spring

在最近的 IT 技术会议上，几乎没怎么出现有关批处理的话题。快速浏览一下最大型的 Java 会议，你会发现几乎没有专门针对这个主题的演讲。会议室里坐满了学习流式处理的听众，有关数据科学的讲座吸引了大批听众，专注于基于 Web 系统的云原生应用的博客斩获众多的访问流量。尽管如此，批处理依然存在。

在美国，你的个人银行报表和 401K 报表(401K 指的是美国在 1981 年创立的一种养老金计划，由于相关规定都在美国的《国内税收法》的条款 401 中，因此简称 401K)都是通过批处理生成的。你从喜爱的商城那里收到的折扣邮件也可能是通过批处理来发送的，甚至维修人员上门修理洗衣机的顺序也是由批处理决定的。在诸如 Amazon 的网站上，推荐相关产品的数据科学模型是通过批处理生成的。编排大数据的任务也是通过批处理实现的。在人们从 Twitter 获取新闻的时代里，Google 认为等待页面刷新会花费太长的时间以至于无法提供搜索结果，YouTube 可以让一个人一夜成名，那么为什么我们还需要批处理呢？

以下列举了几个合理的解释理由：

- 你并不总是能够立即获得所有的必要信息。对于给定的处理过程，批处理允许你在进行必要的处理前收集所需的信息。以每月的银行对账单为例，在每笔交易之后都打印采用一定文件格式的对账单是否有意义？更合理的做法是在月底重新查看经过审查的交易列表，然后从中构建对账单。
- 有时候批处理具有良好的业务意义。尽管大多数人都希望在网上购物时，当他们进行购买的那一刻，商品就被放到一辆送货卡车上，但对于零售商来说，这可能并不是最好的做法。假设顾客临时改变了主意，想要取消订单，这时如果商品还没有进行配送，那么取消订单的损失(否则就配送出去了)就会少一些。给顾客留出一些时间，并且批量地进行配送，能够为零售商节省很多成本。
- 批处理可以更好地利用资源。数据科学方面的示例在这里也是合理的。典型情况下，数据模型的处理被分为两个阶段。首先是模型的生成，此时需要对大量的数据进行密集的数学处理，这会花费很多时间；其次是对生成的模型进行评估，或者使用生成的模型对新数据进行评分，这一阶段非常快速。第一个阶段的意义在于，流式系统实时地使用了在流式用例之外通过批处理得到的结果(即数据模型)。

本书主要介绍如何使用 Spring Batch 框架进行批处理。本章将介绍批处理的历史，指出开发批量作业时面临的挑战，为使用 Java 和 Spring Batch 开发批处理提供案例，最后在较高层次上概述 Spring Batch 框架及其特性。

1.1 批处理的历史

对批处理历史的介绍实际上也就是对计算自身历史的介绍。

1951 年，UNIVAC 成为第一台商业化生产的计算机。在这之前，计算机都是为特定功能而设计的独特的定制机器。例如，1946 年美国军方委托制造了一台计算机来计算炮弹的弹道，名为 ENIAC，成本约为 500 万美元(按 2017 年美元价值计算)。UNIVAC 由 5200 个真空管组成，质量超过 14t，拥有 2.25 MHz 的惊人速度，并且能够运行从磁带驱动器里加载的程序。很快，UNIVAC 被认为是第一个商用的批处理器。

在深入研究历史之前，我们应该明确地定义什么是批处理。你开发的大部分应用都包含交互元素，无论是用户单击 Web 应用中的链接，在胖客户端将信息输入表单，还是通过某种中间件接收消息，抑或是在手机或平板电脑上单击应用。批处理与这些应用的类型刚好相反。批处理被定义为对有限数据进行处理，并且没有交互或中断。一旦启动了批处理过程，批处理就将以某种形式完成，而不需要任何干预。

在计算机和数据处理演进了四年之后，计算领域出现了巨大的变化——高级语言，它们最先由 IBM 704 上的 Lisp 和 Fortran 引入。但自那以后，面向业务的通用语言(Common Business Oriented Language, COBOL)成为批处理领域的重量级语言。COBOL 在 1959 年首次发布，并在 1968 年、1974 年、1985 年、2002 年和 2014 年进行了修订，现代商业中仍然运行着批处理。2012 年，《计算机世界》杂志所做的调查显示，超过 53%的受访企业使用 COBOL 进行新业务的开发。有趣的是，该调查还指出，他们的 COBOL 开发人员的平均年龄在 45~55 岁。

在过去的 20 多年里，COBOL 还没有出现过被广泛采用的重大修订。教授 COBOL 及相关技术的学校数量大幅减少，取而代之的是 Java 和.NET 等新技术。COBOL 的硬件十分昂贵，而资源变得稀缺。

大型机并不是进行批处理的唯一场所。前面提到的那些电子邮件可能是由不在大型机上运行的批处理发送的。我们从你最喜欢的快餐连锁店的销售终端下载的数据也是批量的。但是，大型机上的批处理过程与我们通常为其他环境(例如，C++和 UNIX)编写的批处理过程之间存在着显著差异。每个批处理过程都是定制开发的，它们几乎没有共用的地方。自从 COBOL 接管批处理以来，很少有新的工具或技术出现。虽然定时作业(Cron Job)已经能够在 UNIX 服务器上启动定制开发的进程，以及在 Microsoft Windows 服务器上启动调度任务，但是还没有出现被业界接受的用于执行批处理的新工具。

直到 Spring 出现后，在 2017 年，由 Accenture 的富大型机和批处理实践驱动，并由 Accenture 与 Interface21(Spring 框架的最初创造者，现在是 Pivotal 的一部分)合作，才为企业级批处理创建出一种全新的开源框架。受多年来被认为是 Accenture 架构的支柱这一理念的启发，这种协作产生了在 JVM 上进行批处理的实际标准。

Accenture 由于是首次正式进军开源领域，因此该公司选择将其在批处理方面的专长与 Spring 的流行度和特性集结合起来，以希望创建出健壮的、易于使用的框架。在 2008 年 3 月底，Spring Batch

1.0.0 正式向公众发布，它代表了在 Java 世界中首次基于标准进行批处理的方法。一年多以后，在 2009 年 4 月，Spring Batch 发布了 2.0.0 版本，其中添加了很多特性，例如，用 JDK 1.5+取代对 JDK 1.4 的支持、分块(Chunk-Based)处理、优化配置选项以及在框架内部大幅增加对伸缩性的支持选项。3.0.0 版本在 2014 年的春天出现，其中引入了新的 Java 批处理标准 JSR-352 的实现。4.0.0 版本包含了 Spring Boot 世界中的基于 Java 的配置。

1.2 批处理面临的挑战

毫无疑问，你非常熟悉基于 GUI 编程(胖客户端和 Web 应用等)的挑战：安全问题、数据校验以及实现对用户友好的错误处理。不可预测的使用模式会导致资源利用率大幅提升。所有这些都与同一件事情有关：用户与软件交互的能力。

然而，批处理与此不同。之前讲过，批处理不需要进行额外的交互就可以完成。正因为如此，许多 GUI 应用的问题不复存在。是的，这里需要考虑安全性，数据校验也是必要的，但是使用峰值和友好的错误处理是可预测的，也可能不适用于批处理。你可以预测批处理中的负载并进行相应的设计。只能根据可靠的日志记录和通知，才能快速而明显地表明批处理失败，因为技术资源可以解决任何问题。

批处理世界里的一切都是小菜一碟，没有挑战吗？很抱歉，事实不是这样，因为在许多常见的软件开发挑战中，批处理有自身独有的特点。软件架构通常包含许多非功能性需求：可维护性(maintainability)、易用性(usability)、伸缩性(scalability)等。这些都与批处理相关，只是方式不同。

易用性、可维护性和伸缩性是相关的。在批处理中，无须考虑用户界面，所以易用性并不是指好看的 GUI 和炫酷的动画。在批处理中，易用性与代码相关：包括错误处理和可维护性。你可以轻松地扩展公共组件以添加新特性吗？它们有不错的单元测试覆盖率，以便修改原有组件时，让你知道对整个系统的影响吗？在作业失败时，你能够知道是在何时、何地发生，并且在不花费大量时间进行调试就知道原因吗？以上是易用性影响批处理的所有方面。

接下来探讨一下伸缩性。是时候核实一下现状了：你上一次为一个每天有一百万访客的网站工作是什么时候？每天有十万访客呢？不过，企业内部开发的大多数 Web 站点都不会被浏览很多次。但是，需要在一晚上处理一百万个或更多个事务的批处理并不罕见。让我们把加载 Web 页面所需的 8 秒时间视为可靠的平均值。如果需要花费这么长的时间通过批处理来处理一个事务，那么处理十万个事务需要超过 9 天的时间(处理一百万个事务需要三个月以上)。在现代企业里，这样的情况对于任何系统都不实用。最重要的底线是，批处理过程需要能够处理的范围通常比过去开发的 Web 应用或胖客户端应用大一个或多个数量级。

我们还必须考虑可用性(Availability)。再说一次，这与你可能习惯的 Web 应用或胖客户端应用有所不同。通常批处理不是 24×7 运行的。事实上，它们通常有约定的运行时间。大多数企业都将作业安排在给定的时间运行，此时所需资源(硬件、数据等)已经可用。比如，以建立退休账户对账单为例。尽管可以在一天的任何时候运行作业，但是很可能最佳的进行时间是在闭市之后，这样就可以使用基金的收盘价来计算余额。你可以按需运行吗？你能在规定的时间内完成作业，以免影响其他系统吗？这些问题和其他问题都会影响批处理系统的可用性。

最后，我们必须考虑安全性。通常，在批处理世界中，安全性与侵入系统和破坏东西无关。在安全性方面，批处理所要扮演的角色是保证数据的安全。数据中的敏感字段是否已加密？是否无意中记录了个人信息？如何访问外部系统——是否需要登录凭证，是否能够以适当的方式确保这些登录凭证的安全性？数据验证也是安全性的一部分。一般情况下，虽然要处理的数据已被审查过，但是你仍然应该确保遵循一定的规则。

如你所见，在开发批处理时，会涉及大量的技术性挑战。从大多数系统的大规模伸缩到安全性，批处理都会遇到。这就是开发批处理的乐趣：你将更多地关注于解决技术问题，而不是调试最新的 JavaScript 前端框架。问题是，既然大型机上已经有了基础设施，而采用新平台也会有风险，那么为什么还要使用 Java 进行批处理呢？

1.3 为什么使用 Java 进行批处理

基于上面列出的种种挑战，为什么还要选择 Java 和开源工具(比如 Spring Batch)进行批处理呢？我能想到如下 6 个使用 Java 和开源工具进行批处理的原因：可维护性、灵活性、伸缩性、开发资源、社区支持和成本。

首先是可维护性。批处理必须考虑可维护性。批处理代码的生命周期通常比其他应用长得多。原因在于：没有人能看到批处理代码。与那些必须紧跟当前趋势和风格的 Web 应用和客户端应用不同，批处理程序用于处理数字并构建静态输出。只要还在运行，大部分人只需要享用批处理结果即可。正因为如此，我们需要以一种很容易但不会招致很大风险的方式修改代码。

下面进入 Spring 框架这一话题。Spring 框架是为下面这些可以利用的特性而设计的：可测试性和抽象。Spring 框架通过依赖注入进行对象的解耦，同时 Spring portfolio 提供了一些额外的测试工具，以允许你构建出健壮的测试套件，把维护风险降到最低。在不深入研究 Spring 和 Spring Batch 工作方式的情况下，Spring 提供了一些工具，可以用声明的方式对文件和数据库执行 I/O 操作等。你无须编写 JDBC 代码和 Java 中梦魇般的文件 I/O API。Spring Batch 给应用带来了注入事务和提交次数的功能，所以不必管理当前处于处理的什么位置，也不必管理故障发生时应该做什么。这些仅仅是 Spring Batch 和 Java 为你提供的可维护性方面的优势。

使用 Java 和 Spring Batch 的原因还包括灵活性。在大型机世界中，只能在大型机上运行 COBOL 或 CICS，就是这样。这最终会成为一种完全定制化的解决方案，因为没有被业界接受的批处理框架。大型机和 C++/UNIX 都没有提供用于部署的 JVM 和 Spring Batch 的特性集。想要使用 UNIX/Linux 或 Windows 在服务器、桌面或大型机上运行批处理流程？没关系。想要部署到应用服务器、Docker 容器或云环境中？选择适合需求的即可。想要使用瘦 WAR(Thin WAR)、胖 JAR(Fat Jar)或者其他任何新的热门技术？Spring Batch 都能应对。

然而，Spring Batch 的灵活性不仅来自 Java 的 "一次编写，随处运行" 特性，灵活性的另一方面是在系统之间共享代码的能力。对于 Web 应用中经测试和调试过的相同服务，批处理过程依然可以使用。事实上，访问曾被锁定在其他平台上的业务逻辑的能力是迁移到这个平台的最大胜利。通过使用 POJO 来实现业务逻辑，就可以在 Web 应用和批处理过程中——毫不夸张地说，可以在使用 Java 进行开发的任何地方——使用它们。

Spring Batch 的灵活性还包括扩展用 Java 编写的批处理的能力。让我们来看看用来对批处理过程进行伸缩的一些方案。

- 大型机：大型机在伸缩性方面的附加能力有限。为了并行地完成任务，唯一真正可行的方式是在单个硬件上并行地运行完整的程序。这种方式的局限性在于需要编写和维护代码来管理并行处理，比如错误处理和跨程序的状态管理。此外，处理能力受限于一台机器的资源。

- 定制化的处理：即使在 Java 中，从头开始也是让人望而却步的。在大量数据的基础上达到伸缩性和可靠性并非易事。你会再次面对编写负载均衡时的相同问题。当开始跨物理设备或虚拟机分发计算时，你还会面临巨大的基础设施方面的复杂性。你必须关注各个部分之间的通信是如何工作的。还有数据可靠性方面的问题。如果定制化的 Worker 宕机会发生什么？此外还有更多的问题。我并不是说这些无法实现。我的意思是，你最好把时间花在编写业务逻辑上，而不是重新发明轮子。

- Java 和 Spring Batch：尽管 Java 本身拥有处理上述大多数元素的工具，但是以可维护的方式将这些元素组合在一起则非常困难。Spring Batch 解决了这个问题。想在单独服务器上的单独 JVM 中运行批处理吗？没问题。随着业务持续增长，现在需要在五个不同的节点之间分配账单计算工作，以便在一夜之间完成所有计算。这也是可以的。每个月都有峰值，你能够在那一天使用云资源进行扩展吗？这当然可以。数据可靠性？只需要进行一些配置并记住一些关键原则，就可以完全处理事务回滚和提交计数。

你将看到，随着对 Spring Batch 框架及相关生态系统的深入，困扰前面的批处理方案的问题可以使用经过良好设计和测试的解决方案来缓解。到现在为止，本章已经讨论了选择 Java 和开源工具进行批处理的技术原因。然而，技术并不是做出此类决定的唯一原因。找到合格的开发资源来编写和维护系统是很重要的。前面提到过，批处理中的代码相比 Web 应用来说生命周期要长得多。正因为如此，找到理解相关技术的人与技术本身同样重要。Spring Batch 基于非常流行的 Spring 框架，遵循 Spring 中的惯例，并且使用 Spring 工具和其他基于 Spring 的应用。Spring Batch 是 Spring Boot 的一部分。所以，对于任何拥有 Spring 经验的开发人员来说，Spring Batch 的学习难度是最小的。但是，你能找到 Java，特别是 Spring 方面的资源吗？

在 Java 中，做很多事情的理由之一是社区支持。Spring 框架家族通过 GitHub、StackOverflow 和相关资源，在网上拥有一些非常活跃的大型社区。作为家族一员，Spring Batch 也有成熟的社区。再加上能够访问源代码的强大优势以及提供按需购买支持的能力，Spring Batch 已包含所有的支持基础。

最后是成本问题。任何软件项目都有很多成本：硬件、软件、许可费、薪酬、咨询费用、支持合同等。然而，Spring Batch 不仅仅最划算，而且也是最便宜的。在使用云资源和开源的框架后，唯一的经常性支出费用是开发薪酬——这相比与其他方案相关的定期许可费和硬件支持合同要少得多。

不言而喻。使用 Spring Batch 不仅仅是技术上最合理的方法，而且是最经济有效的方法。下面让我们开始了解 Spring Batch 到底是什么。

1.4 Spring Batch 的其他用途

到目前为止，你肯定想知道将大型机替换为 Spring Batch 是否有好处。当你持续地思考正在面对的项目时，你并不是每天都在提取 COBOL 代码。如果这就是 Spring Batch 框架所能带来的全部好处，那么用处并不大。然而，Spring Batch 框架可以帮助你处理许多其他用例。

Spring Batch 最常见的用例可能就是 ETL 处理了，也就是抽取(Extract)、转换(Transform)和加载(Load)。将数据从一种格式转换到另一种格式是企业数据处理的重要组成部分。Spring Batch 基于块的处理和极强伸缩能力，因而非常适合处理 ETL 工作负载。

Spring Batch 的另一个用例是数据迁移。在重写系统时，通常会将数据从一种形式迁移到另一种形式。风险在于，与通常的开发工作相比，你可能会编写缺乏测试的一次性解决方案，并且缺少对数据一致性方面的控制。然而，在考虑 Spring Batch 的特性时，这些似乎是顺其自然的。你无须编写大量代码就能获得并运行一个简单的批量任务，Spring Batch 还提供了许多数据迁移工具，它们应该但很少提供诸如提交次数和回滚的功能。

Spring Batch 的第三个用例是任何需要执行并行处理的过程。随着芯片制造商的生产工艺不断接近摩尔定律的极限，开发人员意识到继续提高应用性能的唯一方式并不是让单个操作变得更快，而是以并行方式执行更多的操作。人们最近发布了很多用来协助并行处理的框架。大部分的大数据平台，比如 Apache Spark、YARN、GridGain、Hazlecast 等，都是最近几年才出现的，它们的目的就是试图利用多核处理器和众多的云端服务器。然而，诸如 Apache Spark 的框架需要修改代码和数据，以便适应它们的算法或数据结构。Spring Batch 能够在多个核心或服务器上对处理进行伸缩(如图 1-1 所示的 Master/Worker 步骤配置)，并且能够使用与 Web 应用相同的对象和数据源。

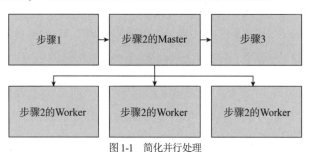

图 1-1　简化并行处理

编排工作负载是 Spring Batch 的另一个常见用例。通常企业级的批处理不止包含一个步骤，而是需要协调很多已解耦的步骤。你可能需要加载文件，对数据进行两种独立的处理，然后输出结果。Spring Batch 能够很好地编排这些任务。其中的一个示例是 Spring Cloud Data Flow 使用 Spring Batch 来处理"组合任务"(composed task)。这里，Spring Batch 调用 Spring Cloud Data Flow 来启动其他的功能，并且跟踪哪些已完成和哪些需要完成。图 1-2 展示了 Spring Cloud Data Flow 提供的用来构建"组合任务"的拖拉界面。

最后是持续处理，又称 24×7 处理。在许多用例中，系统接收到持续或接近持续的数据输入。尽管以数据到来的速率接收数据对于防止积压是必要的，但是，将数据按照块进行批量处理的话，可能会得到更好的性能(如图 1-3 所示)。Spring Batch 提供了一些工具，以便能够以可靠且可伸缩的方式进行这种类型的处理。使用 Spring Batch 框架的特性，可以从队列里读取消息，将每一批次作为一个块，然后在永不结束的循环中一起进行处理，从而能够在大数据量的情况下提高吞吐量，而不必了解从头开发此类解决方案的复杂细节。

第 1 章　批处理和 Spring

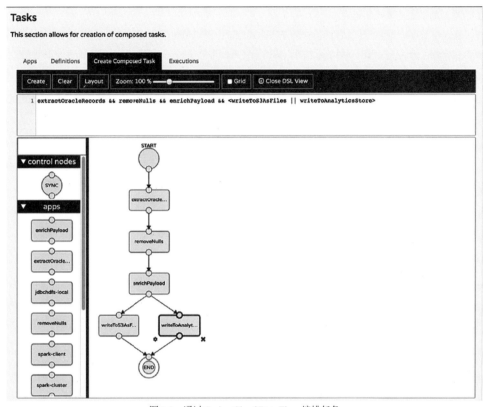

图 1-2　通过 Spring Cloud Data Flow 编排任务

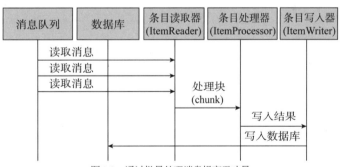

图 1-3　通过批量处理消息提高吞吐量

如你所见，尽管 Spring Batch 是为实现类似大型机的处理而设计的，但是作为框架，Spring Batch 可以用来简化各种各样的开发问题。在了解了有关什么是批处理以及为什么应该使用 Spring Batch 之后，让我们最终开始研究 Spring Batch 框架本身。

1.5　Spring Batch 框架

Spring Batch 是由埃森哲(Accenture)和 SpringSource 合作开发的框架，它以一种基于标准的方式来

7

实现常见的批处理模式和范式。

Spring Batch 实现的特性包括数据验证、格式化输出、以可重用的方式实现复杂逻辑以及处理巨大数据集的能力。当深入本书中的示例时，你就会发现如果熟悉 Spring，那么 Spring Batch 就很有意义。

Spring Batch 的架构如图 1-4 所示。

图 1-4　Spring Batch 的架构

Spring Batch 采用分层的配置并由三层组成。上面是应用层(application layer)，其中包括所有用来构建批处理的定制化代码和配置。业务逻辑、服务以及组织任务的配置等，都是应用层关心的内容。注意，应用层并不在其他两层之上，而是封装了其他两层。原因是，尽管你开发的大部分东西都由应用层(可与核心层一起协作)组成，但有时候你会编写定制化的基础设置部分，比如自定义的读取器和写入器。

应用层在大部分时间里都在与下一层——核心层——交互。核心层包含用于定义批处理域的所有部分。核心组件的元素包括作业(Job)和步骤(Step)接口，以及用来执行作业的如下两个接口：JobLauncher 和 JobParameters。

最下面是基础设施层。在处理任何东西前，都需要读取或写入文件、数据等。在任务失败后，必须能够进行重试。这些部分被认为是一些通用的基础设施，并且位于 Spring Batch 框架的基础设施组件中。

■ **注意**　人们对 Spring Batch 的常见误解是：认为 Spring Batch 是调度器，或者认为 Spring Batch 提供了调度器。事实并非如此。在 Spring Batch 框架中，没有办法进行作业调度，比如在给定时间或在给定的事件发生后执行。有很多方式可用来启动作业，从简单的定时作业脚本到 Quartz，甚至企业级的调度器(如 Control-M)都可以做到，但 Spring Batch 框架本身并不包括这些。第 4 章将介绍如何启动作业。

下面浏览一下 Spring Batch 的一些特性。

1.5.1　使用 Spring 定义作业

批处理中有不少领域特定的概念。作业(Job)表示处理过程，在从头到尾整个执行过程中，没有任何终端或交互。一个作业可能包括多个步骤(Step)。每个步骤可能有相关的输入输出。当步骤失败时，步骤可能是可重复的，也可能是不可重复的。作业流可能是有条件的(比如，仅仅在计算收入的步骤返回超过 1 000 000 美元时才执行计算奖金的步骤)。Spring Batch 提供了类、接口、XML 模式和 Java 配置实用程序，它们使用 Java 定义了这些概念以适当地划分关注点，并以 Spring 使用者熟悉的方式将它们连接在一起。例如，代码清单 1-1 展示了一个基本的 Spring Batch 作业，可通过 Java 语言进行配置。所以，你只需要对 Spring 有一定的基本了解，就可以非常快速地掌握 Spring Batch 框架。

代码清单 1-1　Spring Batch 作业配置示例

```
@Bean
public AccountTasklet accountTasklet() {
    return new AccountTasklet();
}

@Bean
public Job accountJob() {
    Step accountStep =
        this.stepBuilderFactory
            .get("accountStep")
            .tasklet(accountTasklet())
            .build();
    return this.jobBuilderFactory
            .get("accountJob")
            .start(accountStep)
            .build();
}
```

代码清单 1-1 创建了两个 Bean。第一个 Bean 是 AccountTasklet。AccountTasklet 是一个自定义的组件，里面包含了步骤中的业务逻辑。Spring Batch 将一遍又一遍地调用 execute 方法，每次调用都发生在新的事务中，直到 AccountTasklet 表明调用已经结束。

第二个 Bean 是真正的 Spring Batch 作业。在这个 Bean 中，我们从刚刚定义的 AcountTasklet 里创建了一个单独的步骤，并且使用了工厂(factory)提供的构建器(builder)。然后，我们使用工厂提供的构建器在这个步骤中创建了一个作业。在应用启动时，Spring Boot 会找到这个作业并且自动执行。

1.5.2　管理作业

编写能够一次性地处理一些数据，然后永远不再运行的 Java 程序确实是可行的。但是，关键型任务的处理需要更健壮的方法。一些特性，比如保存作业的状态以便重新执行、在作业失败时通过事务管理维护数据的一致性、将以往作业的执行性能指标保存为趋势等，都是企业级批处理系统期望的。Spring Batch 包含了这些特性，并且大部分默认都是开启的。在处理任务时，它们对性能和需求只有非常微小的影响。

1.5.3　本地和远程的并行化

正如前面所讨论的，批处理作业的规模以及能够扩展它们的需求，对于任何企业级批处理解决方案来说都是至关重要的。Spring Batch 提供了通过许多不同方法来解决这个问题的能力。从简单的基于线程的实现(每个提交间隔都在线程池自己的线程中处理)，到并行地运行所有步骤，再到配置通过分区从远程主机获得的工作单元的工作网格，Spring Batch 及相关的生态系统提供了不同选项的集合，包括并行块/步骤的处理以及远程块的处理和分区。

1.5.4 标准化 I/O

从具有复杂格式的平面文件、XML 文件(XML 是流，从不作为整体加载)、数据库或 NoSQL 存储，到以简单的配置写入文件或 XML，使用 Spring Batch 编写作业的可维护性表现在，可以通过代码抽象文件和数据库进行输入输出。

1.5.5 Spring Batch 生态系统的其他部分

与 Spring 产品组合中的大部分项目类似，Spring Batch 并不是孤立存在的。Spring Batch 只是生态系统的一部分，在 Spring Batch 生态系统中，一些其他项目对 Spring Batch 进行了扩展和补充，以提供更健壮的解决方案。Spring 产品组合中与 Spring Batch 一起工作的其他一些项目如下。

1. Spring Boot

Spring Boot 于 2014 年引入，它采用一种有主见的方式开发 Spring 应用。现在，Spring Boot 几乎是开发 Spring 应用的标准方式，它提供了易于打包、部署和启动包含批处理的所有 Spring 工作负载的设施和工具。Spring Boot 也是 Spring Cloud 提供的云原生方案的支柱之一。因此，Spring Boot 也将是本书开发批量应用的主要方法。

2. Spring Cloud Task

Spring Cloud Task 是 Spring Cloud 项目下的子项目，提供了在云环境中执行有限任务的设施。作为针对有限工作负载的框架，批处理是一种能够与 Spring Cloud Task 良好集成的处理风格。Spring Cloud Task 为 Spring Batch 提供了许多扩展，包括发布提示性消息(作业的开始/结束、步骤的开始/结束等)以及动态伸缩批处理作业的能力(而不是使用 Spring Batch 直接提供的各种静态方法)。

3. Spring Cloud Data Flow

自行编写批处理框架不仅仅意味着必须重新开发 Spring Batch 提供的开箱即用的性能、可伸缩性和可靠性等特性，还需要某种管理和编排工具集以完成作业的启动和停止，以及查看既往作业的运行数据等工作。然而，如果使用 Spring Batch，那么除了能够完成以上工作以外，你还可以使用 Spring Batch 提供的如下附加功能：Spring Cloud Data Flow。Spring Cloud Data Flow 是用来在云平台(CloudFoundry、Kubernetes 等)上编排微服务的工具。如果以微服务的方式开发批处理应用，就可以将它们以动态方式部署。

1.5.6 Spring 的所有特性

虽然 Spring Batch 包含了一系列令人印象深刻的特性，但最棒的特性在于 Spring Batch 构建于 Spring 之上，从而拥有了 Spring 为任何 Java 应用提供的详尽特性，包括依赖注入(Dependency Injection)、面向接口编程(Aspect-Oriented Programming，AOP)、事务管理以及用于处理大部分通用任务(JDBC、JMS、Email 等)的模板和助手，这些特性为你构建企业级的批处理过程提供了所需的几乎一切。

如你所见，Spring Batch 为开发人员带来了很多东西。Spring 框架的成熟开发模型以及可伸缩性和可靠性等特性，让你能够使用 Spring Batch 快速运行批处理任务。

1.6 如何阅读本书

在介绍了批处理和 Spring Batch 之后，我相信你一定迫不及待地想要深入研究一些代码，并了解使用 Spring Batch 框架构建批处理过程的全部内容。第 2 章将介绍批处理的架构，定义我们已开始使用的一些术语(如作业、步骤等)，并介绍如何设置第一个 Spring Batch 项目。

我写作本书的主要目的之一是，不仅让你深入了解 Spring Batch 框架是如何工作的，而且展示如何在现实案例中应用这些工具。第 3 章将提供第 10 章中示例应用的项目需求和技术架构。

本书的代码示例可以在 GitHub 上找到，下载网址为 https://github.com/Apress/def-guide-spring-batch，也可通过扫描封底的二维码下载。

1.7 本章小结

本章介绍了批处理的历史，其中涵盖了开发人员在进行批处理过程中面临的一些挑战，并且证明了使用 Java 和开源技术来克服这些挑战是合理的。最后，本章通过展示 Spring Batch 框架的高级组件和特性，让你对 Spring Batch 框架有了一定的了解。到目前为止，你应该已经很好地理解了自己将要面临的挑战，并了解了 Spring Batch 中用来应对这些挑战的工具。现在，你所要做的就是学习如何去做。让我们开始吧！

第 2 章

Spring Batch 入门

组装计算机是一项比较简单的任务，很多开发人员在他们的职业生涯中都实践过。但是，只有在理解了计算机的每个部分能做什么，以及如何在更大的系统中进行安装，才能做到真正简单。如果拿一包计算机零部件给一个不知道计算机为何物的人，并让他组装出一台计算机，那么事情可能就不会那么顺利了。

在企业级 Java 中，有许多领域知识十分易于传播。在大多数 Web 框架中，常见的 MVC 模式就是典型示例。一旦理解了一个 MVC 框架，那么使用另一个 MVC 框架时，只需要理解不同部分的语法即可。然而，现实中并没有太多的批处理框架。正因为如此，这些领域知识可能对你来说有点新。你可能不知道什么是作业、什么是步骤，也不知道 ItemReader 与 ItemWriter 的关系，以及 Tasklet 到底是什么。

本章将回答这些问题，并讨论以下话题。

- 批处理的架构：2.1 节将更深入地研究批处理过程的组成，并定义那些将贯穿本书其余部分的术语。
- 项目设置：你将通过实践进行学习。本书的组织方式是：通过示例展示 Spring Batch 框架如何工作，介绍如此工作的原因，并给予你独自编码的机会。本节将介绍如何搭建基于 Maven 的 Spring Batch 项目。
- "Hello, World!" 示例程序：热力学第一定律介绍的是能量守恒；第一运动定律研究的是静止的物体如何趋向于保持静止，除非受到外力的作用；计算机科学领域的第一定律似乎是无论你学什么新技术，都必须从编写 "Hello, World!" 示例程序开始，这里也将遵守这一定律。
- 运行作业：如何执行第一个作业可能不那么一目了然，因此，2.5 节将介绍作业是如何执行的，以及如何传入基本参数。

考虑到这些问题后，那么作业到底是什么？

2.1 批处理的架构

第 1 章花了一些时间讨论 Spring Batch 框架的三层架构：应用层、核心层和基础设施层。应用层代表你所开发的代码，它在大多数情况下与核心层交互。核心层包含组成批处理域的真正组件。最后，基础设施层包含条目读取器和条目写入器，以及处理可重启性等问题时所需的类和接口。

本节将首先深入 Spring Batch 的架构,并且定义第 1 章中引用的部分概念。然后引入一些可伸缩性方面的选项,它们对于批处理非常重要,并且让 Spring Batch 变得非常强大。最后,本节将讨论大纲管理选项(outline administration option),以及如何在文档中查找有关 Spring Batch 问题的答案。下面我们从批处理的架构开始,了解核心层的组件。

2.1.1 深入讨论作业和步骤

图 2-1 展示了作业的本质:通过 Java 或 XML 进行配置,作业(Job)是状态以及状态之间转换的集合。从本质上讲,Spring Batch 作业仅仅是状态机。由于步骤是 Spring Batch 使用的最常见的状态形式,因此我们现在仅关注它们。

以每天晚上处理用户的银行账户为例,步骤 1 可以是加载来自其他系统的交易文件。步骤 2 将所有的贷方记入账户。步骤 3 将所有借方记入账户。以上作业代表了将交易应用到用户账户的整个过程。

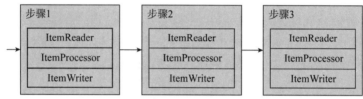

图 2-1 示例作业

当深入查看一个独立的步骤时,你会看到一个自包含的工作单元,它是作业的主构建块。有两类主要的步骤:基于 Tasklet 的步骤和基于块(chunk)的步骤。基于 Tasklet 的步骤相对简单一些。这种步骤使用 Tasklet 来实现,并且在事务范围内一遍又一遍地运行 execute(StepContribution contribution, ChunkContext chunkContext)方法,直到 execute 方法让步骤停止为止(每一次调用 execute 方法都将得到自己的事务)。这种步骤通常用于初始化、运行存储过程、发送通知等。

尽管基于块的步骤在结构上更严格一些,但它们更适用于基于条目的处理(item-based processing)。每个基于块的步骤最多有三个主要部分:一个 ItemReader(条目读取器)、一个 ItemProcessor(条目处理器)和一个 ItemWriter(条目写入器)。注意,刚才说的是最多有三个主要部分。步骤并不一定需要有 ItemProcessor。只包含 ItemReader 和 ItemWriter 的步骤也是可以的(例如,在数据迁移作业中通常就是这样)。表 2-1 列出了 Spring Batch 提供的代表这些概念的接口。

表 2-1 组成批处理作业的接口

接口	说明
org.springframework.batch.core.Job	这个接口代表作业,可在 ApplicationContext 中进行配置
org.springframework.batch.core.Step	与作业类似,这个接口代表配置的步骤
org.springframework.batch.core.step.tasklet.Tasklet	一个策略接口,提供了在一次事务的范围内执行逻辑的能力
org.springframework.batch.item.ItemReader<T>	一个策略接口,用来提供步骤的输入
org.springframework.batch.item.ItemProcessor<I,O>	这个接口用来将业务逻辑、验证等应用于提供的单个条目
org.springframework.batch.item.ItemWriter<T>	一个策略接口,用于在步骤中持久化条目

使用 Spring 构造作业的好处之一是能够把每一个步骤解耦到独立的处理器中。每个步骤负责自己的数据，把所需的业务逻辑应用到数据上，然后把数据写入合适的位置。这样的解耦方式提供了如下特性。

- 灵活性(Flexibility)：配置基于复杂逻辑的复杂工作流，是很难以可复用的方式实现的。然而，Spring Batch 提供了一组很好的构建器来完成这一任务。可使用 Spring Batch 的流畅 Java API(Fluent Java API)和传统的 XML 方式来配置批处理应用程序，这种功能是非常强大的。
- 可维护性(Maintainability)：将每个步骤的代码与前后的步骤解耦后，就可以在几乎不影响其他步骤的情况下，十分容易地对步骤进行单元测试、调试和更新。解耦后的步骤也使得在多个作业中复用步骤变得可能。在接下来的章节中可以看到，步骤仅仅是 Spring Bean，因此可以像 Spring 中的其他 Bean 一样来复用。
- 伸缩性(Scalability)：作业中的独立步骤提供了对作业进行伸缩的很多选项。多个步骤可以并行执行。步骤中的工作可以被切分到多个线程中，并且单个步骤中的代码可以并行执行。这些功能可以满足业务的伸缩需求，并且对代码的直接影响非常小。
- 可靠性(Reliability)：步骤中的不同阶段(比如通过 ItemReader 读取数据、通过 ItemProcessor 处理数据)都可以通过执行一些操作(比如重试操作或者在抛出异常时跳过条目)提供健壮的错误处理选项。

2.1.2 执行作业

执行作业后，许多组件会进行交互，以达到提供韧性(resiliency)的目的。下面从较高层次介绍这些组件以及它们之间如何交互。

首先介绍架构中的主要共享部分——JobRepository 组件。如图 2-2 所示，JobRepository 组件负责维护作业的状态和各种各样的处理指标(开始时间、结束时间、状态、读/写次数等)。JobRepository 组件通常由关系数据库支持，并且由 Spring Batch 的几乎所有主要组件共享。

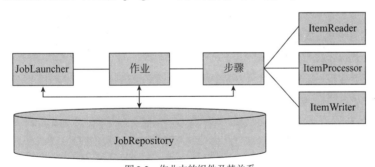

图 2-2 作业中的组件及其关系

接下来介绍 JobLauncher 组件，该组件负责作业的执行。JobLauncher 组件的任务不仅仅是调用 Job.execute 方法，还包括执行一些其他的工作，例如验证重启作业是否有效(并不是所有的作业都是可重启的)、决定如何执行作业(在当前线程中、使用线程池等)、验证参数等。然而，所有这些任务都与具体的实现无关。在 Spring Boot 的世界里，JobLauncher 是一个通常不需要直接使用的组件，因为 Spring Boot 提供了开箱即用的作业启动设施。

启动作业后，作业就会执行每个步骤。在每个步骤执行时，JobRepository 会使用最新的状态进行更新。执行的步骤，作业的当前状态，读取、处理和写入了多少条目等，这些都存储在 JobRepository 中。

作业会遍历配置的步骤列表，并执行每一个步骤。在处理了步骤中某一条目的一个块(chunk)之后，Spring Batch 将使用当前状态更新 JobRepository 中的 JobExecution 或 StepExecution。步骤使用 ItemReader 读取条目的列表。在步骤处理每一个条目的块时，JobRepository 中的 StepExecution 将根据自身在步骤中的位置进行更新。当前的提交次数、开始和结束次数以及其他信息都保存在 JobRepository 中。当作业或步骤完成时，相关执行的最终状态也会在 JobRepository 中进行更新。

上面已经多次提到 JobExecution 和 StepExecution，所以现在花点时间理解一下它们是什么以及它们之间的关系。图 2-3 描述了这种关系中的各个组件。JobInstance 是 Spring Batch 作业的一次逻辑执行，可由作业名称以及为这次逻辑执行提供的唯一参数集来标识。例如，假设有一个名为 statementGenerator 的对账单生成作业，那么当每次以相同的参数启动 statementGenerator 作业时，都会生成新的 JobInstance。在这种情况下，使用 2017 年 3 月 7 日作为参数运行 statementGenerator 作业时，就会生成新的 JobInstance。

图 2-3　JobInstance、JobExecution 和 StepExecution 的关系

JobExecution 代表 Spring Batch 作业的一次物理执行。每次启动作业时，都会得到新的 JobExecution，但可能不会得到新的 JobInstance。显而易见的示例就是重启失败的作业。当第一次运行作业时，你会得到新的 JobInstance 和 JobExecution。如果执行失败，那么在重启时，就不会得到新的 JobInstance，因为还在使用相同的逻辑运行作业(使用相同的识别参数)。然而，你会得到新的 JobExecution 来跟踪第二次物理运行。所以，一个 JobInstance 可以有多个 JobExecution。

最后，StepExecution 是步骤的一次物理运行。Spring Batch 中并没有 StepInstance 的概念。一个 JobExecution 可以有(并且通常有)多个 StepExecution 与之关联。

2.1.3　并行化

简单的批处理架构往往包括单线程结构的进程，用于从头到尾地执行作业中的步骤。然而，Spring Batch 提供了许多并行化选项，在深入 Spring Batch 的过程中你应该注意这些选项(第 11 章将详细介绍这些选项)。这里有 5 种不同的方式可用来进行并行处理：多线程的步骤、并行地执行步骤、异步的 ItemProcessor/ItemWriter、远程分块和分区。

1. 多线程的步骤

实现并行化的第一种方式是通过多线程的步骤划分工作。在 Spring Batch 中，作业被配置为以块来处理工作，每个块都被封装在自己的事务中。通常，每个块都是按顺序处理的。如果有 10 000 条记录，并且提交计数被设置为 50 条记录，那么作业将处理记录 1～记录 50，然后提交，接着继续处理记录 51~记录 100,以此类推,直到处理完 10 000 条记录。Spring Batch 允许通过并行地执行分成块的工作来改善性能。有了 3 个线程,就可以将理论吞吐量增加三倍,如图 2-4 所示。

图 2-4 多线程的步骤

2. 并行地执行步骤

下一种可用的并行化方式是并行地执行步骤,如图 2-5 所示。假设有两个步骤,每个步骤会把输入文件加载到数据库中,但是这两个步骤之间没有任何关系。在加载下一个文件之前,必须等待加载完上一个文件。

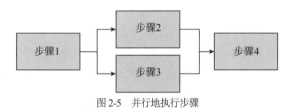

图 2-5 并行地执行步骤

3. 异步的 ItemProcessor/ItemWriter

在一些用例中,步骤中的 ItemProcessor 可能成为瓶颈。例如,你可能需要完成复杂的数学计算或者调用远程服务才能丰富 ItemReader 提供的数据。将步骤的这一部分并行化可能很有用。AsynchonousItemProcessor 是 ItemProcessor 实现的修饰器,前者可在自己的线程里执行对 ItemProcessor 的每一次调用。与返回 ItemProcessor 调用的结果不同,AsynchronousItemProcessor 对每个调用返回 java.util.concurrent.Future,在当前块中返回的 Future 列表被传给 AsynchronousItemWriter。AsynchronousItemWriter(步骤中真正需要使用的 ItemWriter 的修饰器)将打开 Future,并把实际结果传递给委托的 ItemWriter。

4. 远程分块

最后将要介绍的两种并行化方式允许跨多个 JVM 进行处理。在前面所有的示例中,处理都是在单个 JVM 中进行的,这会严重妨碍伸缩性。当能够对处理的任何部分进行跨 JVM 的水平扩展时,你的满足大量需求的能力就会增强。

第一种远程处理方式是远程分块(remote chunking)。在这种方式下,输入是在主节点的标准 ItemReader 中完成的,然后通过持久通信的形式把输入发送到远程的工作者 ItemProcessor,后者则被配置为消息驱动的 POJO。当处理完成时,工作者要么把更新过的条目发送给主节点进行写入,要么自行写入。由于要从主节点读取数据,由工作者处理后再把数据发回主节点,因此务必注意,这可能会占用大量的网络资源。这种方式适用于 I/O 成本比实际处理成本低的场景。

17

5. 分区

在 Spring Batch 中，进行并行化处理的最后一种方式是分区，如图 2-6 所示。Spring Batch 支持远程分区(主节点和远程工作者)和本地分区(使用工作者线程)。远程分区和远程分块的主要区别是：使用远程分区时不需要持久通信，主节点只是作为一组工作者步骤的控制器。在这种情况下，每一个工作者步骤都是自包含的，并且使用相同的配置，就像它们部署在本地一样。唯一的区别是，工作者步骤从主节点而不是从作业本身接收工作。当所有的工作者完成它们的工作后，主步骤就完成了。这种配置不需要持久通信，因为 JobRepository 能确保不出现重复的工作，并且所有工作都会完成——这与远程分块不同，在远程分块中，JobRepository 对分布式工作的状态一无所知。

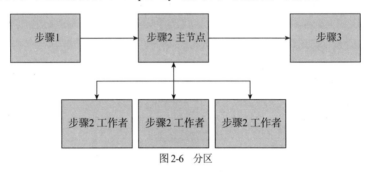

图 2-6　分区

2.1.4　文档

开源项目的好坏取决于项目的文档。我们努力创建的不仅仅是一套完整的文档，我们还将创建一套完整的示例，用以演示如何使用 Spring Batch 框架中的许多概念来执行批处理作业。表 2-2 提供了一些批处理作业示例。

表 2-2　批处理作业示例

批处理作业	描述
adhocLoopJob	一个无限循环，用来演示如何通过 JMX 来暴露元素，并且在后台线程中(而不是在主 JobLauncher 线程中)运行作业
amqpExampleJob	一个展示了如何把 AMQP 作为数据输入输出的作业
beanWrapperMapperSampleJob	一个包括两个步骤的作业，展示了文件字段到领域对象的映射以及基于文件输入的验证
compositeItemWriterSampleJob	一个步骤只能有一个读取器和一个写入器。CompositeWriter 就是用来解决这个问题的。这个示例作业对此进行了演示
customerFilterJob	使用 ItemProcessor 过滤掉无效的客户，这个作业还更新了步骤执行的筛选器计数字段
delegatingJob	使用 ItemReaderAdapter，把对输入的读取委托给 POJO 中的配置方法
footballJob	一个足球统计作业。加载两个输入文件(其中一个文件中是球员数据，另一个文件中是比赛数据)后，这个作业将生成球员和游戏的汇总统计信息，并将它们写入日志文件
groovyJob	使用 Groovy(一种动态的 JVM 语言)编写用来解压和压缩文件的脚本
headerFooterSample	使用回调函数，在输出中渲染文件头和文件尾
hibernateJob	Spring Batch 的读取器和写入器默认不使用 Hibernate。这个作业展示了如何把 Hibernate 集成到作业中

(续表)

批处理作业	描述
infiniteLoopJob	一个包含无限循环的作业，用来展示停止和重启场景
ioSampleJob	这个作业提供了许多不同的 I/O 选项示例，包括定界和定宽文件、多行记录、XML 和 JDBC 集成
jobSampleJob	演示从一个作业到另一个作业的执行
loopFlowSample	使用决策标签，展示如何以编程的方式控制执行流程
mailJob	使用 SimpleMailMessageItemWriter 发送 Email，作为每个条目的输出形式
multilineJob	将一组文件记录视为表示单个条目的列表
multilineOrder	扩展了多行输入的概念，使用自定义的读取器，从文件中读取多行的嵌套记录。输出也是多行的，使用了标准的写入器
parallelJob	把记录读入临时表，然后处理它们
partitionFileJob	使用 MultiResourcePartitioner 并行处理文件集合
partitionJdbcJob	与查找多个文件，然后并行处理每个文件不同，这个作业会对数据库中的记录数进行切分，以便进行并行处理
restartSampleJob	当处理开始时，抛出假的异常，演示如何重新启动有错误的作业并从中断处重新开始
retrySample	使用一些有趣的逻辑展示了 Spring Batch 能够在放弃执行和抛出错误之前，尝试对条目进行多次处理
skipSampleJob	这个作业基于 tradeJob，有一条记录验证失败并被跳过
taskletJob	Spring Batch 最基本的用途是 Tasklet。这个作业展示了如何通过 MethodInvokingTaskletAdapter 将任何现有的方法用作 Tasklet
tradeJob	对真实世界中的场景进行建模。这个三步作业会将贸易信息导入数据库，更新客户账户并且生成报告

2.2 项目设置

到目前为止，你已经了解了为什么要使用 Spring Batch，并考查了这个框架的各个组件。不过，仅仅查看图表和学习新的术语只会让你止步于此。到了一定的时间，你就需要深入代码之中：所以现在就打开编辑器，开始钻研吧！

本节将构建第一个批处理作业。我们将首先设置一个 Spring Batch 项目，包括从 Spring 中获取必要的文件。然后配置作业并且编写 Spring Batch 版本的 "Hello, World！" 示例程序。最后，我们将展示如何从命令行中启动批处理作业。

2.2.1 获取 Spring Batch

在开始进行批处理之前，需要获取 Spring Batch 框架。获取的方式有很多，包括从 GitHub 中获取代码、使用 Maven 或 Gradle，等等。然而，由于本书主要关注的是基于 Spring Boot 的批处理作业，因此我们将从 Spring Initializr 开始进行介绍。Spring Initailizr 是 Spring 团队提供的一个服务，它允许你使用一组经过验证的依赖项来生成项目的框架。

为了使用 Spring Initializr，你需要如下两样东西中的一样：支持与 Initializr 直接集成的 IDE(例如 Spring Tool Suite 或 IntelliJ)或 Web 浏览器。接下来，本书中的每个示例都假定你已经有了一个干净的基于 Spring Boot 的项目。

Spring Initializr 站点

在浏览器中打开 https://start.spring.io 即可访问 Spring Initializr 站点，如图 2-7 所示。你将看到一个 UI。

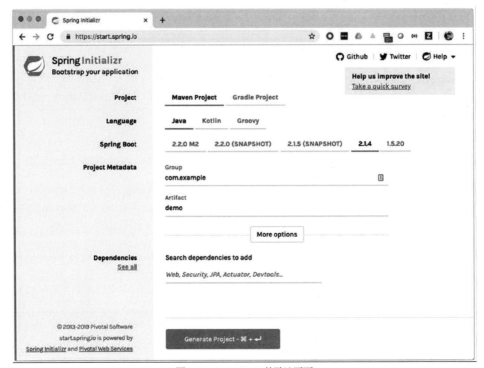

图 2-7　start.spring.io 的默认页面

可在这个 UI 中定义关于项目的以下一些基本参数。

- 构建系统(Build system)：在编写本书时，Maven 和 Gradle 都已得到支持。
- 语言(Language)：Spring 对 Java、Groovy 和 Kotlin 提供了健壮的支持，所以 Spring Initializr 允许你选择它们之一。
- Spring Boot 的版本(Spring Boot version)：Spring Boot 的有些特性因版本而异，因此 Spring Initializr 允许你选择版本。
- 构件坐标(Artifact coordinates)：允许 POM 或 gradle.build 生成带有正确的预填充坐标的构建文件。
- 依赖项(Dependencies)：允许指定项目中要包含的 Spring Boot 启动器。

这个 UI 提供了两个选项来添加依赖项。如果你很清楚需要哪些依赖项，那么可以把它们输入右边的搜索框里。如果不清楚，那么可以单击链接 Switch to the full version，页面将为 Spring Initializr 中的每一个可用选项提供一个复选框。对于本书中创建的每一个项目，至少都会选择 Spring Boot 启动器。

一旦输入相关的数据，就单击 Generate Project 按钮。之后会下载一个压缩文件，其中包含完整的项目框架，以便添加新代码。然后，可将这个项目直接导入 IDE 中，以便在接下来的开发过程使用。

Spring Tool Suite

Spring Tool Suite (STS)是基于 Eclipse 的 IDE，由 Spring 团队维护。STS 提供了与 Spring 框架和开发微服务相关的一些特性。STS 可以作为独立的 IDE 进行下载，也可以作为插件添加到已经安装的 Eclipse 中。STS 可以通过 Spring 站点 https://spring.io/tools 免费获取。

在安装了 STS 之后，选择 File | New | Spring Starter Project，你将看到如图 2-8 所示的对话框。

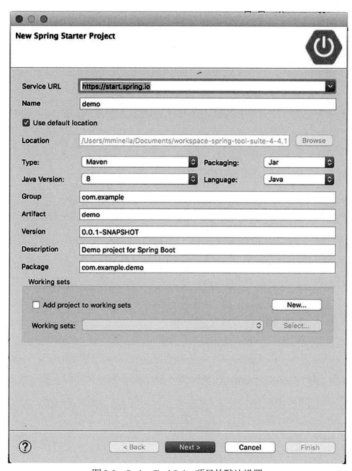

图 2-8　Spring Tool Suite 项目的默认设置

此处我们仍然使用 https://start.spring.io/作为服务的地址，在配置完图 2-8 所示对话框中的内容之后，单击底部的 Next 按钮。此时会打开新的界面，在上面可以选择项目中要包含的 Spring Boot 启动器。在选择了期望的依赖项之后，单击 Finish 按钮，STS 会下载并导入新的项目。

2.2.2　IntelliJ IDEA

IntelliJ IDEA 是 Java 开发中另一个十分流行的 IDE，它实现了与 Spring 功能的良好集成，包含与使用 Spring Initializr 相似的体验。安装了 IntelliJ IDEA 之后，选择 File | New | Project…，在打开的对话框的左边，你将能够看到 Spring Initializr。选择 Spring Initializr 之后，你将会看到如图 2-9 所示的界面。

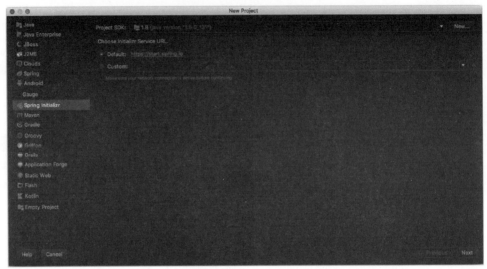

图2-9　IntelliJ IDEA 项目的默认设置

我们可以围绕图 2-9 所示界面中使用的默认值进行讨论。从现在开始，单击 Next 按钮，接下来的界面将会展示站点上可用的选项。填写完毕后单击 Next 按钮，下一个界面将会展示你正在查看的 Spring Intiailizr 实例的可用 Spring Boot 启动器。选择期望的选项，单击 Next 按钮，最后一个界面将允许你设置项目的名称以及保存位置。输入对应的值，单击 Finish 按钮。IntelliJ IDEA 将下载合适的项目并导入。

2.3　"Hello，World！" 示例程序

在开始 Spring Batch 的 "Hello, World！" 示例程序之前，需要创建新的项目。可使用任何一种偏爱的方式创建项目，命名为 hello-word。在使用任何 Spring Initializr 向导时，都需要进行以下设置。

- Group ID：io.spring.batch
- Artifact ID：hello-world
- Build System：Maven
- Language：Java 8+[1]
- Packaging：Jar
- Version：0.0.1-SNAPSHOT
- Spring Boot Version：2.1.2[2]
- Dependencies：Batch、H2 和 JDBC

对于依赖项，批处理是显而易见的。H2 会被作业存储库用作内存数据库。最后添加 JDBC 用于

---

[1] Java 8 及以上版本都应该能运行本书中的所有示例。

[2] 在编写本书时，Spring Boot 2.1.2 是最新版本。然而，任何 2.1 以上版本的 Spring Boot 都应该能够运行本书中的示例。

支持数据库(DataSource 等)。在把项目导入 IDE 后，你应该会得到如图 2-10 所示的项目结构[1]。

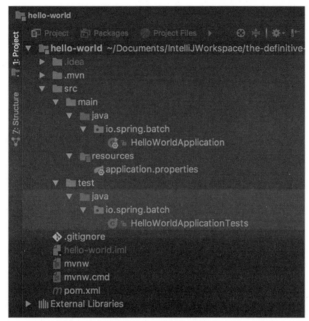

图 2-10　导入的项目结构

导入项目后，我们可以看看项目的不同部分。实际上，Spring Initializr 项目只包含如下 4 个主要部分：
1) 构建系统(本书使用的是 Maven，但你也可以选择使用其他的构建系统)。
2) 一个用来启动 Spring Boot 的源文件(位于/src/main/java/your/package/name/HelloWorldApplication.java)。
3) 一个测试文件，仅仅用来启动 Spring 上下文，以便确认能够正常工作(位于/src/test/java/your/package/name/HelloWorldApplicationTests.java)。
4) 一个 application.properties 文件，用来配置应用(位于/src/main/resources)。

对于 hello-word 项目而言，我们将聚焦于 HelloWorldApplication 类。代码清单 2-1 展示了所有需要添加到第一个批处理作业的代码(除了 import 语句)。

代码清单 2-1　"Hello, World!" 作业

```
@EnableBatchProcessing
@SpringBootApplication
public class HelloWorldApplication {

    @Autowired
    private JobBuilderFactory jobBuilderFactory;

    @Autowired
    private StepBuilderFactory stepBuilderFactory;

    @Bean
```

---

[1] 这个项目使用 Maven 作为构建系统。如果选择 Gradle，那么布局将略有不同。

```java
    public Step step() {
        return this.stepBuilderFactory.get("step1")
                        .tasklet(new Tasklet() {
                            @Override
                            public RepeatStatus execute(StepContribution
                                            contribution,
                                            ChunkContext
                                            chunkContext) {
                                System.out.println("Hello, World!");
                                return RepeatStatus.FINISHED;
                            }
                        }).build();
    }

    @Bean
    public Job job() {
        return this.jobBuilderFactory.get("job")
                        .start(step())
                        .build();
    }

    public static void main(String[] args) {
        SpringApplication.run(HelloWorldApplication.class, args);
    }
}
```

在仔细阅读代码清单 2-1 之后，你会从较高的层次上感觉陌生。这里实际上只是用 Java 呈现了本章前面 2.1.1 节中讨论的内容。然而，如果将其分解，那么里面实际上包含三个主要部分，我们将它们添加到了生成的 HelloWorldApplication 类中：@EnableBatchProcessing 注解，JobBuilderFactory 和 StepBuilderFactory 的注入，以及步骤和作业的定义。

通过自上而下的方式，作业从@EnableBatchPocessing 注解开始。这个注解由 Spring Batch 提供，用来启动批处理的基础设施。由于已经提供了大部分批处理基础设施的 Spring Bean 定义，因此不必自行提供。其中包括

- JobRepository：记录作业运行时的状态。
- JobLauncher：启动作业。
- JobExplorer：使用 JobRepository 执行只读操作。
- JobRegistry：在使用特定的启动器实现时找到作业。
- PlatformTransactionManager：在工作过程中处理事务。
- JobBuilderFactory：一个流畅的作业构建器。
- StepBuilderFactory：一个流畅的步骤构建器。

如你所见，@EnableBatchProcessing 注解做了很多事情，但你需要为之提供数据源（DataSource）。JobRepository 和 PlatformTransactionManager 可根据需要使用它。Spring Boot 则通过类路径中的 HSQLDB 来处理这个问题。在检测到 HSQLDB 之后，就会在启动时创建嵌入式 DataSource。

接下来是 HelloWorldApplication 类的@SpringBootApplication 注解，它实际上是一个结合了@ComponentScan 和@EnableAutoConfiguration 注解的元注解。这个元注解能触发自动配置(autoconfiguration)，从而创建 DataSource 和实现其他有意义的基于 Spring Boot 的自动配置。

第 2 章　Spring Batch 入门

　　在 HelloWorldApplication 类定义之后，我们装配(autowire)了 Spring Batch 提供的两个构建器，分别用来构建作业(jobBuilderFactory)和步骤(stepBuilderFactory)。@EnableBatchProcessing 注解会自动提供这两个构建器，所以你只需要让 Spring 把它们注入配置类中即可。

　　接下来我们创建了步骤。因为这个作业仅包含一个步骤，所以我们简单地将其命名为 step。在这个简单的示例中，step 被配置为 Spring Bean，并且仅仅需要两个元素：名称和 Tasklet。Tasklet 在此处是通过内联方式实现的，以完成作业中真正的处理工作。在这个示例中，我们仅仅调用了 System.out.println("Hello, World!")，然后返回 RepeatStatus.FINISHED。

　　把 RepeatStatus.FINISHED 返回给 Spring Batch 意味着已经完成了 Tasklet。其他的选项会返回 RepeatStatus.CONTINUABLE。在这个示例中，Spring Batch 会再次调用 Tasklet。如果在此处返回 RepeatStatus.CONTINUABLE，结果将是一个无限循环。

　　步骤配置完之后，就可以用来创建作业。前面提到过，作业由一个或多个步骤组成。在本例中，只有一个步骤。通过 JobBuilderFactory，可使用与步骤同样的范式来配置作业。我们配置了作业的名称以及开始步骤。在这个简单的示例中，要做的工作仅此而已。

　　在完成所有的配置之后，就意味着定义了第一个 Spring Batch 作业！下面我们看看作业在实战中是如何运作的。

2.4　运行作业

　　现在尝试构建和运行作业。假定正在使用 Maven，那么在项目的根目录下应该运行 mvn clean package 以进行编译。构建成功后，运行作业。默认情况下，Spring Boot 将在启动时运行你在已配置过的 ApplicationContext 中找到的所有作业。根据需求，可通过属性来配置这种行为。由于默认情况就是我们所期望的，因此只需要进入项目的 target 目录，执行 java -jar hello-world-0.0.1-SNAPSHOT.jar 即可。Spring Boot 将完成其余工作以运行作业。

　　运行作业之后，请注意，在传统的 Spring Boot 风格中，即使是简单的"Hello, World!"示例程序也有相当多的输出。如果仔细查看(在第 23 行左右)，你会看到如下信息：

```
2010-12-01 23:15:42,442 DEBUG 2017-05-22 21:15:04.274 INFO 2829 --- [           main]
o.s.batch.core.job.SimpleStepHandler     : Executing step: [step1]
Hello, World!
2017-05-22 21:15:04.293 INFO 2829 ---  [           main] o.s.b.c.l.support.
SimpleJobLauncher         : Job: [SimpleJob: [name=job]] completed with the following
parameters: [{}] and the following status: [COMPLETED]
```

　　祝贺你！你刚刚运行了第一个 Spring Batch 作业。但是，到底发生了什么？Spring Batch 有一个名为 JobLauncherCommandLineRunner 的组件。当 Spring Batch 在类路径(环境变量)中找到这个组件时，就会在启动时加载它。它使用 JobLauncher 来运行 Spring Batch 在 ApplicationContext 中找到的所有作业。所以，当你在 main 方法中启动 Spring Boot 并创建 ApplicationContext 时，就会执行 JobLauncherCommandLineRunner，开始运行作业。

　　作业执行第一个步骤，于是又开始一个事务，执行 Tasklet 并把结果更新到 JobRepository。本书在后面将详细介绍 JobRepository 最终的结果。

2.5　本章小结

在本章，你第一次接触到了 Spring Batch，浏览了批处理域，包括作业和步骤是什么，以及它们如何通过作业存储库进行交互。你还学习了 Spring Batch 框架的不同特性，包括在 Java 代码中映射批处理、健壮的并行化选项和正式文档。

此外，你编写了 Spring Batch 版本的"Hello, World!"示例程序，学习了使用 Spring Initializr 来上手 Spring Batch 的不同方式。在设置好项目后，你使用 Java 创建并执行了作业。

需要指出的是，你几乎还没有看到 Spring Batch 能做什么。第 3 章将介绍一个示例应用是如何设计的(本书接下来的章节会构建这个示例应用)，其中概述了 Spring Batch 如何处理那些原本必须由你自己处理的问题。

第 3 章

示例作业

本书将不仅解释 Spring Batch 的特性如何工作，还会对它们进行详细的演示。每一章都包括很多示例，以展示每个特性的工作方式。然而，用于展现各种不同技术的示例可能并不适合在真实环境中演示这些技术如何协同工作。所以，本书第 10 章将创建一个模拟真实世界场景的示例应用。

这里选择的场景已经过简化：一个易于理解的领域，但却提供了足够的复杂性，以便让 Spring Batch 体现出其价值。银行对账单是批处理的常见示例。处理过程通常在夜间进行，基于上个月的交易生成对账单。即将创建的批处理过程将把交易应用于现有的一组账户集合，然后为每个账户生成对账单。本章将介绍一些重要的批处理概念。

- 多种输入输出选项：Spring Batch 最重要的特性之一是对各种来源的读写进行了良好的抽象。银行对账单将从平面文件和数据库获取输入。在输出端，可以写入数据库和平面文件。这里利用了各种各样的读取器和写入器。
- 错误处理：维护批处理过程最糟糕的情形是在它们出错时，通常会在凌晨 2 点发生，而你正是那个接到电话要去解决问题的人。正因为如此，健壮的错误处理是不可或缺的。生成银行对账单的示例涵盖了许多不同的场景，包括记录日志、跳过有错误的记录以及重试逻辑。
- 可伸缩性：在真实世界中，批处理需要能够容纳大量的数据。本书在后面章节中将使用 Spring Batch 的可伸缩性来调整批处理过程，使之能够应对数百万的客户。

为了构建批处理作业，必须有一组需求。我们将使用用户故事来定义需求，因此我们首先从整体上了解一下敏捷开发过程。

3.1 了解敏捷开发

在深入介绍第 10 章将要开发的批处理过程的各个需求之前，让我们先花一点时间温习一下开发方法。业内已经有不少关于各种敏捷开发过程的讨论。所以，与其依赖于之前从这门学科学到的知识，还不如从建立敏捷和开发过程的基础开始。

敏捷开发过程包含 12 条原则，几乎所有的变体都是这样规定的，这些原则是：

- 客户的满意度来自可工作软件的快速交付。

- 无论处于哪个开发阶段，变化都是受欢迎的。
- 频繁地交付可工作的软件。
- 业务人员和开发人员必须每天携手并进。
- 与积极的团队一起建立项目。给他们工具，相信他们能完成工作。
- 面对面沟通是最有效的沟通形式。
- 可工作的软件是衡量成功的第一标准。
- 努力实现可持续发展。团队的所有成员都应该能够无限期地保持开发速度。
- 不断追求卓越的技术和优秀的设计。
- 可通过减少不必要的工作来减少浪费。
- 自组织团队，产生最佳的需求、架构和设计。
- 每隔一段时间，让团队反思，决定如何改进。

无论是使用极限编程(Extreme Programming，XP)、Scrum，还是使用当前流行的其他任何变体，这些都不重要。关键是这十几条原则仍然适用。

注意，并不是所有规定都适用于不同的具体情况。我们很难通过一本书进行面对面的工作。你可能需要独自地完成这些示例，所以团队激励的各个方面对你并不适用。但是，其中一些也许是适用的。例如，快速地交付可工作的软件这条规则将引导你构建应用的一个个小部分，使用单元测试验证它们是否可以工作，然后将它们添加到应用程序中。

虽然有例外，但这些敏捷开发原则为开发任何项目提供了坚实的框架，本书会尽可能多地应用它们。下面介绍如何应用它们，方法是检查示例作业的需求记录方式：用户故事。

3.1.1 通过用户故事捕捉需求

用户故事是用于记录需求的敏捷方法。用户故事记录了客户对应用应该做什么的看法，目的是表达用户如何与系统进行交互，并且记录交互的可测试结果。用户故事包含以下三部分。

- 标题(title)：用户故事应该有简洁明了的标题。例如，"加载交易文件""生成打印文件"等都是不错的标题。注意这些标题并不是专门用于描述 GUI 的。没有 GUI 并不意味着用户之间不能进行交互。在这种情况下，用户是你正在编写的批处理过程，或是任何你要与之进行交互的外部系统。
- 叙述(narrative)：从用户角度对交互进行的简短描述。典型情况下，格式类似于"给定情况 Y，X 做了一些事情，然后发生了其他事情"。在接下来的章节中，你会看到如何为批处理过程编写用户故事(假定它们在本质上都是纯技术)。
- 验收条件(acceptance criteria)：验收条件是可测试的需求，用来鉴定用户故事的完整性。只有能够以某种方式进行验证的验收条件才是有用的。这些不是主观要求，而是开发人员可以用来说"是"或"不是"的硬性规定。

下面以一个为通用遥控器编写的用户故事为例。

- 标题：打开电视机。
- 叙述：作为用户，在关闭电视、接收器和有线电视盒的情况下，按下通用遥控器上的电源按钮，就可以同时打开电视、接收器和有线电视盒的电源，并且打开一个电视节目。

- 验收条件：
  - 通用遥控器上有电源按钮。
- 当用户按下电源按钮时，会发生如下情况：
  - 电视将接通电源。
  - AV 接收器将接通电源。
  - 有线电视盒将接通电源。
  - 有线电视盒将被设置为 187 频道。
  - AV 接收器将被设置为 SAT 输入。
  - 电视将被设置为 Video 1 输入。

以上用户故事的标题为"打开电视机"，这一标题简短且具有描述性。然后是叙述信息。在本例中，叙述部分描述了用户按下电源按钮时发生的事情。最后，验收条件部分列出了对于开发人员和 QA 而言的可测试需求。注意，每个标准都是开发人员可以轻松检查的东西：他们可以查看开发的产品并回答是或否，进而判断产品是否符合验收条件。

用户故事与用例

用例(User Case)是另一种大家十分熟悉的需求文档形式。与用户故事类似，用例是以行动者为中心的。用例是统一软件开发过程(Rational Unified Process，RUP)选择的文档形式。它们旨在记录行动者和系统之间交互的每个方面。正因为它们主要关注文档且格式较为臃肿，所以用例在敏捷开发中已经不再受欢迎，取而代之的是用户故事。

用户故事标志着软件开发周期的开始。下面继续介绍你在开发周期的其余部分将要用到的一些工具。

3.1.2 使用测试驱动开发捕捉设计

测试驱动开发(Test-Drivien Development，TDD)是另一种敏捷实践。在使用 TDD 时，开发人员首先要编写失败的测试，然后通过实现代码让其通过。TDD(也称为 Test-First Development)被设计为让开发人员在编码之前思考将要编写的内容，事实已证明，TDD 能够提高开发人员的工作效率，在减少调试的同时编写出更加简洁的代码。

TDD 的另一个优势在于测试能够作为可执行的文档。由于缺乏维护，用户故事或其他形式的文档会变得陈旧，而 TDD 与之不同，自动化的测试始终可以作为能够持续维护的代码的一部分进行更新。如果想理解一段代码是如何运作的，可以通过查看单元测试得到蕴含代码意图的完整场景。

尽管 TDD 有很多积极的优点，但本书并不会大量使用。TDD 尽管十分优秀，但它并不是解释工作原理的最佳方式。第 13 章将介绍从单元测试到功能测试的所有测试类型，并介绍如何使用开源工具(比如 JUnite、Mockito 和 Spring)的附加功能等。

3.1.3 使用版本控制系统

无论如何，这都不是必需的，但是强烈建议对所有的开发工作使用源码控制系统。不论是使用 Git 和 GitHub，还是使用其他形式的版本控制系统，源码控制对于高效编程都是不可或缺的。

你可能会问:"我将来很可能会丢掉学习过程中编写的代码,为什么还要进行源码控制呢?"有了版本控制系统,你就可以给自己编织一张进行各种尝试的安全网:提交可工作的代码并尝试一些可能无法工作的东西,如果确实能够工作,那么提交新的修订;否则,回滚到上一次修订,但不会有任何影响。通过使用版本控制系统,你可以避免在可控的环境中犯错误。

3.1.4 在真正的开发环境中工作

在敏捷开发环境中还有很多其他的内容。一定要确保拥有良好的 IDE。由于本书有意地与 IDE 无关,因此我们不会讨论每个 IDE 的优点和缺点。然而,拥有合适的 IDE 是十分重要的,你要好好学习如何使用,包括快捷键。

尽管在学习一项给定的技术时,花费大量时间来搭建持续集成环境并不是很有意义,但一般来说,为了个人发展搭建持续集成环境是值得的。你永远不知道正在开发的小部件会在什么时候出现大问题。当事情开始变得无比糟糕时,才不得不回去设置源码控制和持续集成(Continuous Integration,CI),这种情景相信你一定不希望遇到。在使用 GitHub 这样的服务时,一种选择是使用云服务 Travis CI (https://travis-ci.org/)。Travis 通过 Webhook 实现了与 GitHub 的无缝集成,所以你只需要进行一些简单的配置,就能在项目中启用 CI。

3.2 理解作业需求

现在,你已经看到了在学习 Spring Batch 时,我们鼓励使用的开发过程的各个部分,下面介绍本书将要探讨的内容。图 3-1 展示了你在银行看到的每个月的对账单。尽管很多人习惯接收在线的对账单,但我们还是使用打印的对账单作为批处理作业的示例。

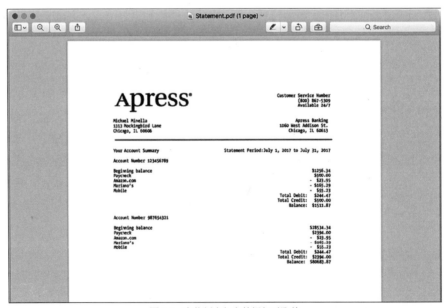

图 3-1　在信纸上打印的银行对账单

如果分析一下银行对账单是如何创建的，就会发现它实际上包含两部分。首先，它只不过是一张信纸，上面印着第二部分——文本信息。本书将创建第二部分，如图 3-2 所示。

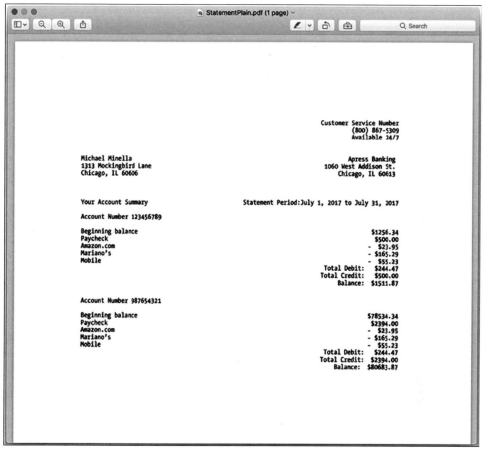

图 3-2　纯文本的银行对账单

通常，银行对账单可通过如下方式创建。除文本外，批处理过程创建的打印文件还包括其他一些内容。然后，打印文件会被发送到打印机，接下来将文本打印到装饰纸上，生成最终的银行对账单。以上过程中的打印文件就是使用 Spring Batch 创建的部分。批处理过程将执行下面的操作：

- 基于提供的文件更新客户信息。
- 在数据库中导入所有客户的交易数据。
- 使用账户余额更新账户余额。
- 打印用于上个月的银行账户的文件。

让我们来看看为完成以上操作所需的内容。批处理作业会接收一个平面文件(Flat File)，其中包含现有的待更新客户的有关信息。例如，客户的地址可能会发生变化。这个平面文件提供了地址变化的细节。批处理作业将读取该文件，并更新数据库中现有客户的数据。

批处理作业所需的下一个功能是导入所有客户的交易数据。这些数据以 XML 文件形式存在，我们将它们导入现有的数据库中。

一旦导入交易数据，就能够更新账户表。账户表中保存了客户的当前余额。有了账户表，就不需要每次都从头对所有的交易进行重新计算。

在所有数据库都更新完之后，就能够抽取包含了客户信息、交易列表和账户概要的打印文件。

介绍以上内容只是为了在真实世界中展示 Spring Batch 的完整用法。通过阅读本书，你将掌握 Spring Batch 提供的用来帮助开发批处理过程的各种特性，以便在类似场景下进行开发。第 10 章将实现使用如下用户故事描述的批处理作业。

更新客户信息：作为批处理过程，将导入客户信息，并且使用它们更新现有的客户记录。验收条件如下。

- 批处理作业将读取基于 CSV 格式的客户更新文件。
- 更新将基于类型被应用于客户记录(每一种类型都有自己的格式)。
  - 类型为 1 的记录表明名称发生变化。
  - 类型为 2 的记录表明邮件地址发生变化。
  - 类型为 3 的记录表明联系人信息发生变化。

记录的格式如表 3-1~表 3-3 所示。

表 3-1　记录类型 1

名称	是否必需	格式
Record Type ID	True	\d
Customer ID	True	\d{9}
Customer First Name	False	\w+
Customer Middle Name	False	\w+
Customer Last Name	False	\w+

类型为 1 的记录如下所示：

```
1,123456789,John,Middle,Doe
```

表 3-2　记录类型 2

名称	是否必需	格式
Record Type Id	True	\d
Customer Id	True	\d{9}
Address1	False	\w+
Address2	False	\w+
City	False	\w+
State	False	\w{2}
Postal Code	False	\d{5}

类型为 2 的记录如下所示：

```
2,123456789,123 4th Street,Unit 5,Chicago,IL,60606
```

表 3-3　记录类型 3

名称	是否必需	格式
Record Type ID	True	\d
Customer ID	True	\d{9}
Email Address	False	\w+
Home Phone	False	\d{3}-\d{3}-\d{4}
Cell Phone	False	\d{3}-\d{3}-\d{4}
Work Phone	False	\d{3}-\d{3}-\d{4}
Notification Preference	False	\d

类型为 3 的记录如下所示：

3,123456789,foo@bar.com,123-456-7890,123-456-7890,123-456-7890,2

验证出错的记录应该写入错误文件，以便将来进行验证和重新处理。

导入交易数据。作为批处理过程，我将导入通过 XML 输入文件提供的所有新的交易数据。验收条件如下：

- 批处理过程将读取交易数据的 XML 文件。
- 每个交易都会在交易表中创建新的记录。
- 文件中的每个交易都包含表 3-4 所示的记录类型。

表 3-4　记录类型 4

名称	是否必需	格式
Transaction ID	True	\d{9}
Account Id	True	\d{9}
Credit	False	\d+\.\d{2}
Debit	False	\d+\.\d{2}
Timestamp	True	yyyy-mm-dd hh:mm:ss.ssss

交易文件中包含的数据示例如下(每条记录都要填写借方或贷方)。

```
<transactions>
   <transaction>
      <transactionId>123456789</ transactionId>
      <accountId>987654321</accountId>
      <description>Paycheck</description>
      <credit>500.00</credit>
      <debit/>
      <timestamp>2017-07-20 15:38:57.480</timestamp>
   </transaction>
   ...
</transactions>
```

用交易数据更新账户表：作为批处理过程，将使用最新的余额更新账户表。验收条件如下：

- 账户表提供了余额字段，要求使用你在最近的导入中插入的交易数据对余额字段进行更新。

打印对账单页头：作为批处理过程，将在每一页的顶部打印页头，从而提供关于客户和银行的基本信息。验收条件如下：

- 除了客户的地址以外，页头都是静态文本。

如下页头示例中的 Michael Minella 和地址信息提供了客户的名称和联系地址：

```
                                        Customer Service Number
                                             (800) 867-5309
                                             Available 24/7
Michael Minella
1313 Mockingbird Lane                   Apress Banking
Chicago, IL 60606                       1060 West Addison St.
```

打印账户概要：作为批处理过程，在完成所有计算后，将打印每个客户的概要。概要提供了客户账户的整体情况以及构成账目总值的明细。验收条件如下：

- 批处理过程将为每个客户生成一个文件。
- 概要用一行陈述了以下完全合理的内容。

```
Your Account Summary        Statement Period:<BEGIN_DATE> to <END_DATE>
```

其中，BEGIN_DATE 是账户表中自上一次对账单日期之后的第一个日历日期，而 END_DATE 是作业的运行日期。在概要标题之后，将显示客户拥有的每个账户的标题。

- 在账户标题之后，是对应账户下的交易列表。
- 在交易列表之后，会有两行分别展示当前账单周期里的贷方总额和借方总额。
- 最后一行是账户的当前余额。

账户示例如下：

```
Your Account Summary        Statement Period: 07/20/2017 to 08/20/2017

Account Number 123456789

Beginning balance                                              $1256.34
Paycheck                                                        $500.00
Amazon.com                                                      - $23.95
Mariano's                                                      - $165.29
Mobile                                                          - $55.23
                                                   Total Debit: $244.47
                                                  Total Credit: $500.00
                                                      Balance: $1511.87
```

以上就是开发需求。如果现在不知所措，没关系。3.3 节将开始概述如何使用 Spring Batch 处理银行对账单。

3.3　设计批处理作业

如前所述，这个项目的目标是利用真实世界中的示例以及 Spring Batch 提供的特性，构建健壮的、可伸缩的且可维护的解决方案。为了完成这个目标，该例包含的元素现在看来可能有一点点复杂，例如标题、多种文件格式的输入以及包含子标题的复杂输出。这么做的原因是 Spring Batch 已经为这些特性提供了一些基础设施。下面列出银行对账单作业的大纲，并描述其中的步骤，从而深入研究如何构造批处理过程。

3.3.1 作业描述

为了生成银行对账单,我们需要构建一个包含 4 个步骤的作业,如图 3-3 所示,随后的章节将描述这些步骤。

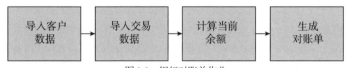

图 3-3 银行对账单作业

1. 导入客户数据

首先需要导入客户数据。前面提到过,这些数据保存在平面文件中,有复杂的格式,包含三种类型。Spring Batch 提供了在单个文件中处理多种记录格式的功能,我们将在这个步骤中应用该功能。一旦读取了数据,就使用 ItemProcessor 进行校验,以便在数据写入阶段减少错误。在此基础之上,基于记录的类型,使用合适的 ItemWriter 实现来完成更新。代码清单 3-1 展示了将在这个步骤中使用的输入示例。

代码清单 3-1 客户输入文件

```
2,3,"P.O. Box 554, 6423 Integer Street",,Provo,UT,10886
2,65,"2374 Aliquet, Street", ,Bellevue,WA,83841
3,73,Nullam@fames.net,,1-611-704-0026,1-119-888-1484,4
2,26,985 Malesuada. Avenue,P.O. Box 585,Aurora,IL,73863
2,23,686-1088 Porttitor Avenue,,Stamford,CT,89593
1,36,Zia,,Strong
2,60,313-8010 Commodo St.,,West Jordan,UT,26634
2,17,"P.O. Box 519, 3778 Vel Rd.",,Birmingham,AL,36907
```

2. 导入交易数据

在导入客户数据之后,下一步是导入交易数据。同样,Spring Batch 提供了一组健壮的 ItemReader 和 ItemWriter 实现,所以在这个步骤中,可以使用这些实现来读取 XML,并且写入数据库。在这个步骤中,我们将再次校验输入,并且把每条已经转换过的记录插入数据库中。代码清单 3-2 展示了使用 XML 格式的交易输入文件的示例。

代码清单 3-2 交易输入文件

```xml
<?xml version="1.0" encoding="UTF-8" ?>
<transactions>
    <transaction>
        <transactionId>1</transactionId>
        <accountId>15</accountId>
        <credit>5.62</credit>
        <debit>1.95</debit>
        <timestamp>2017-07-12 12:05:21</timestamp>
    </transaction>
    <transaction>
        <transactionId>2</transactionId>
```

```xml
        <accountId>68</accountId>
        <credit>5.27</credit>
        <debit>6.26</debit>
        <timestamp>2017-07-23 16:28:37</timestamp>
    </transaction>
    ...
</transactions>
```

3. 计算当前余额

导入交易数据后，需要在账户表中更新余额。对于在线账户来说，这是预先计算的，但我们也需要使用余额来生成对账单。为此，使用驱动查询模式来迭代每个账户并计算当前交易对当前余额的影响，然后在账户表中更新对应的余额。

4. 生成对账单

最后一个步骤最复杂。在这个步骤中，将为每个账户生成包含客户对账单的打印文件。与计算当前余额的方式类似，这个步骤也使用了驱动查询模式。我们将使用 ItemReader 从数据库中读取客户信息，把客户信息发送到 ItemProcessor 以添加每个对账单所需的数据，然后把这些数据发送给一个基于 ItemWriter 的文件，从而使用最少的定制代码满足生成对账单的所有需求。

以上这些在理论上听起来很好，但还是遗留了很多问题。在本书的其余部分，你将了解这些特性是如何在处理过程中实现的，以及如何检查异常处理和重启/重试逻辑、解决可伸缩性问题等。但在继续之前，你还应该熟悉数据模型，这将有助于澄清系统的结构。下面介绍数据模型。

3.3.2 理解数据模型

数据是任何应用的命脉所在，研究将要使用的数据模型是了解系统如何工作的好方法。本节将介绍示例应用使用的数据模型。

图 3-4 列出了特定于示例应用的数据表。需要澄清的是，其中并不包含运行银行对账单这个批处理作业所需的所有数据表。我们将在后续章节中介绍它们。但是，包括这些数据表在内的所有数据表都将存在于数据库中。

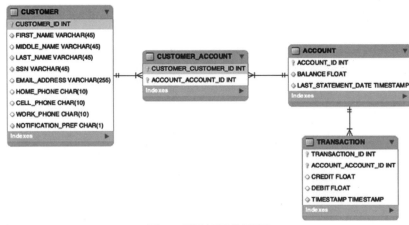

图 3-4　示例应用的数据模型

我们的示例应用包含如下 4 个数据表：CUSTOMER、ACCOUNT、TRANSACTION 和 CUSTOMER_ACCOUNT。如果查看数据表中的数据，就会注意到并没有保存用于生成对账单的所有字段。一些字段(例如，借方总额和贷方总额)会在计算的过程中生成。除此之外，数据模型应该看起来比较简单。

- CUSTOMER：其中包含所有特定于客户的信息，包括姓名和联系人信息。
- ACCOUNT：每个客户都有一个账户。为此，每个账户都有编号和当前余额。
- CUSTOMER_ACCOUNT：这是一个连接表。由于一个账户可能关联了很多客户，并且一个客户可能有很多个账户，因此账户和客户之间是多对多的关系。
- TRANSACTION：其中包含对于给定账户发生的所有交易。

3.4　本章小结

本章讨论了敏捷开发过程以及如何将其应用到批处理开发中。然后沿着这条思路，通过用户故事为本书中的示例应用定义了需求。从现在开始，本书的关注点将从"Spring Batch 是什么""为什么使用 Spring Batch"切换到"应该怎么做"。

第 4 章将深入探究 Spring Batch 中的作业和步骤的概念，并且讲解其他一些具体示例。

第 4 章
理解作业和步骤

在第 2 章，你已经创建了第一个作业，并逐步地配置作业和步骤、执行作业、配置数据库以保存作业存储库。在 "Hello, World!" 示例程序中，你开始了解 Spring Batch 中的作业和步骤是什么。本章将更深入地探究作业和步骤。我们首先从了解与 Spring Batch 框架相关的作业和步骤开始。

在此基础上，本章将深入讲解在执行作业和步骤时发生的细节，包括加载和校验它们是否有效，直到它们执行完毕。然后，详细讲解一些代码，看看作业和步骤的可配置的各个部分，并在此过程中学习最佳实践。最终，你将看到批处理难题的不同部分如何通过 Spring Batch 批处理过程中涉及的各个作用域相互传递数据。

本章尽管深入研究步骤，但并不涉及步骤中最主要的部分：条目读取器(ItemReader)和条目写入器(ItemWriter)。第 7 和 9 章将介绍 Spring Batch 中可用的输入输出功能。本章会让每个步骤中的 I/O 部分尽可能简单，以便重点介绍作业中各个步骤的复杂性。

4.1 作业介绍

在企业内部，现代开发人员会看到许多不同类型的工作负载。Web 应用、集成应用、大数据等，它们都具有一些通用范式。其中的一种思路是：完成一段业务流程，使应用提供业务价值。例如，在 Web 应用中，可能有购物车流程。在这个流程中，用户将商品添加到购物车中，提供配送地址、支付信息，最终确认订单。在集成风格的应用中，一条消息可能会经过许多的转换器、过滤器等，才能完成业务流程。

作业与这些流程类似。我们将作业定义为唯一的、有序的步骤列表，能够从头到尾独立执行。下面拆分一下上述定义，以便能够更好地理解。

- 唯一：Spring Batch 中的作业可通过 Java 或 XML 进行配置，与使用核心的 Spring 框架配置 Bean 的方式相同，因此可以重用它们。可以使用相同的配置，按需多次执行作业。因此，多次定义相同的作业是没有意义的。
- 有序的步骤列表：以结账流程为例，步骤的顺序是有意义的。如果没有提供配送地址，就无法确认订单。如果购物车是空的，就无法执行结账流程。作业中步骤的顺序非常重要。在将交易数据导入系统之前，无法生成客户的对账单。在将交易数据应用到余额之前，无法计算账户的余额。因此，在构建作业时，要让步骤在序列中按照逻辑顺序执行。
- 能够从头到尾执行：第 1 章将批处理定义为能够在没有额外交互的情况下，以某种形式完成处理的过程。作业中包含一系列的步骤，这些步骤能够在没有外部交互的情况下执行。如果不构建作业，步骤就将保持等待状态，直到某个文件被发送到要处理的目录中。你需要在文件到达时就开始执行作业。
- 独立：每个批处理作业都应该能够在不受外部依赖影响的情况下执行，这并不意味着作业不能有依赖。不过，实际中很少有作业是没有外部依赖的。然而，作业应该能够管理这些依赖。如果文件不存在，那么你应该能够优雅地处理这个错误，而不是一直等待文件的到来(这是调度器要完成的任务)。作业能够处理批处理过程中定义的所有元素。

如图 4-1 所示，当所有的输入可用时，才会执行批处理过程。这里没有用户交互。执行完一个步骤后，再执行下一个步骤。在深入研究如何在 Spring Batch 中配置作业的各个特性之前，下面先讨论一下作业的生命周期。

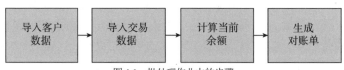

图 4-1　批处理作业中的步骤

跟踪作业的生命周期

作业在执行时会经历相应的生命周期。在构建作业和理解作业如何运行时，对生命周期的理解很重要。在定义作业时，真正要做的事情是提供作业运行时的蓝图(blueprint)。就像通过编写 Java 类来为 JVM 创建实例的蓝图一样，作业的定义用于为 Spring Batch 创建作业实例的蓝图。

作业可通过作业运行器(JobRunner)来执行。作业运行器通过名称和传递的参数来执行作业。Spring Batch 提供了如下两个作业运行器。

- CommandLineJobRunner：可以通过脚本来使用或直接在命令行中使用。使用时，CommandLineJobRunner 会启动 Spring，并且使用传递的参数来执行请求的作业。
- JobRegistryBackgroundJobRunner：在使用诸如 Quartz 或 JMX Hook 的调度器执行作业时，Spring 通常在作业执行前已经启动了，并且已经运行了 Java 进程。在这种情况下，当包含可运行作业的 Spring 启动时，会创建 JobRegistry(作业注册表)。JobRegistryBackgroundJobRunner 可用来创建 JobRegistry。

CommandLineJobRunner 和 JobRegistryBackgroundJobRunner(两者都位于 org.springframework.batch.core.launch.support 包中)是 Spring Batch 提供的两个作业运行器。Spring Batch 还提供了另一种通过 JobLauncherCommandLineRunner 启动作业的方式。JobLauncherCommandLineRunner 的实现会在 ApplicaitonContext(应用上下文)中查找所有类型为 Job 的 Bean，并且在启动时执行它们(除非另有配置)。在本书中，我们将使用这种机制启动所有的作业。

尽管作业运行器旨在与 Spring Batch 进行交互，但它并不是 Spring Batch 框架的标准部分。Spring Batch 框架中并没有 JobRunner 接口，因为每个场景下所需的实现是不同的(尽管 Spring Batch 提供的两个作业运行器都使用 main 方法启动)。Spring Batch 框架真正的执行入口是 org.springframework.batch.core.launch.JobLauncher 接口的一个实现。

Spring Batch 提供了一个 JobLauncher，也就是 org.springframework.batch.core.launch.support.SimpleJobLauncher 类。CommandLineJobRunner 和 JobLauncherCommandLineRunner 在内部使用了这个类，该类使用 Spring 核心的 TaskExecutor 接口来执行请求的作业。你需要稍微了解一下 TaskExecutor 接口的配置方式，但请特别注意，在 Spring 中有多种配置 org.springframework.core.task.TaskExecutor 接口的方式。如果使用了 org.springframework.core.task.SyncTaskExecutor，那么作业将在与 JobLauncher 相同的线程中执行。任何其他选项都会在自己的线程中执行作业。

当批处理作业运行时，会创建 org.springframework.batch.core.JobInstance。JobInstance(作业实例)代表作业的一次逻辑运行，可通过作业名称和传递给作业的识别性参数来标识。作业的运行与执行作业的尝试是不一样的。如果有一个每天都要运行的作业，那么你只需要对它配置一次。每天都会有一次新的运行或者一个新的 JobInstance，因为每次都会传给作业一组新的参数(例如日期)。当一个 JobInstance 有了一次尝试或 JobExecution 成功完成时，就认为这个 JobInstance 已经完成。图 4-2 展示了 Job、JobInstance 和 JobExecution 之间的关系。

图 4-2　Job、JobInstance 和 JobExecution 之间的关系

■ **注意**　JobInstance 只有一次能成功执行。因为 JobInstance 通过作业名称和传入的识别性参数来标识，所以这意味着使用相同的识别性参数只能运行作业一次。

你可能想知道 Spring Batch 如何从一次又一次的尝试中知道 JobInstance 的状态，这会在第 4 章进行深入研究。不过，在 JobRepository 使用的数据库中有一个 BATCH_JOB_INSTANCE 表，它是派生其他所有数据表的基础。BATCH_JOB_INSTANCE 和 BATCH_JOB_EXECUTION_PARAMS 表可用来标识 JobInstance(BATCH_JOB_INSTANCE.JOB_KEY 实际上是作业名称和识别性参数的哈希值)。

JobExecution 是真正地运行作业的一次尝试。如果作业在第一次时从头到尾执行完毕，那么对于给定的 JobInstance，只有一个 JobExecution 与之关联。如果作业在第一次运行后发生错误，那么每次尝试运行 JobInstance 时(传递相同的识别性参数给作业)，都会创建新的 JobExecution。Spring Batch 在为作业创建每一个 JobExecution 时，都会在 BATCH_JOB_EXECUTION 表中创建新的记录。当 JobExecution 执行时，作业的状态也会在 BATCH_JOB_EXECUTION_CONTEXT 表中得到维护。这样 Spring Batch 就可以在发生错误后，从正确的地方重启作业。

4.2　配置作业

对相关理论的介绍已经足够了，下面介绍代码。本节将深入介绍配置作业的各种方式。第 2 章提到过，就像 Spring 一样，Spring Batch 的配置可通过 XML 或 Java 来完成。我们的示例应用将使用 Java 进行配置。

4.2.1　基本的作业配置

在开始之前，我们先看一个简单的 Spring Batch 作业。从功能上说，这个作业将与第 2 章的示例

Spring Batch 权威指南

作业相同。然而，我们会做一些简化，将所有的代码放在单个类中。我们不会为作业存储库使用内存数据库。原因在于我们将查看一些跨执行的特性，因此需要持久化数据。这里将使用 MySQL 来完成示例应用，但任何基于 JDBC 的数据库也都可以做到。

下面从 Spring Initializr 创建一个新的项目，使用所有的默认值并选择如下依赖项：Batch、JDBC、MySQL。将这个新项目暂时命名为 Chatper04。在把新的项目导入 IDE 后，还需要更新两个地方。第一个地方是 Spring Boot 提供的 application.properties 文件中的数据库连接属性。代码清单 4-1 展示了我们需要的属性。

代码清单 4-1　application.properties 文件

```
spring.datasource.driverClassName=com.mysql.cj.jdbc.Driver
spring.datasource.url=jdbc:mysql://localhost:3306/spring_batch
spring.datasource.username=root
spring.datasource.password=p@ssw0rd
spring.batch.initialize-schema=always
```

在代码清单 4-1 中，将 Spring Boot 配置为使用 MySQL 驱动来创建数据源(DataSource)，然后将数据源指向本地的 MySQL 实例，并使用合适的登录凭据。此处还将应用配置为在批处理的数据库模式(batch schema)不存在时，自动进行创建[1]。

一旦配置了数据库，就可以创建应用的代码了。代码清单 4-2 展示了一个简单的 HelloWorld 作业，它将用于接下来的讨论。

代码清单 4-2　HelloWorld.java

```
...
@EnableBatchProcessing
@SpringBootApplication
public class HelloWorldJob {

    @Autowired
    private JobBuilderFactory jobBuilderFactory;

    @Autowired
    private StepBuilderFactory stepBuilderFactory;
    @Bean
    public Job job() {
        return this.jobBuilderFactory.get("basicJob")
                .start(step1())
                .build();
    }

    @Bean
    public Step step1() {
        return this.stepBuilderFactory.get("step1")
                .tasklet((contribution, chunkContext) -> {
                    System.out.println("Hello, world!");
                    return RepeatStatus.FINISHED;
```

---

[1] 以这种方式配置时，Spring Boot 将总是尝试创建批处理的数据库模式。如果已经创建过这种模式，那么这个脚本就会失败。默认情况下，在后续的运行中，上述失败将被忽略。

```
            }).build();
    }

    public static void main(String[] args) {
        SpringApplication.run(HelloWorldJob.class, args);
    }
}
```

配置的第一部分是@EnableBatchProcessing 注解。在应用中只需要使用该注解一次，就可以提供第 2 章讨论过的用来运行批处理作业的基础设施。然后是@SpringBootApplication 注解，用于启用 Spring Boot 魔法。默认情况下，因为 Spring Boot 会将 HelloWorldJob 作为配置类，所以我们不需要显式地添加@Configuration 注解。使用 Spring Boot 后，在大多数情况下，可以通过扫描类路径获得 HelloWorldJob 类。

在 HelloWorldJob 类的声明之后，我们自动装入(autowire)一个 JobBuilderFactory 和一个 StepBuilderFactory。这两个工厂用于创建 JobBuilder 和 StepBuilder 实例，我们将使用这些实例来创建真正的 Spring Batch 作业和步骤。在装入这些工厂后，我们定义了一个类型为 Job、名为 job 的 Bean。这个 Bean 将返回配置完整的 Spring Batch 作业。作业是通过 Spring Batch 提供的构建器(builder)构造的。把作业的名称传给 JobBuilderFactory.get(String name)调用，就会得到 JobBuilder，可以用来配置作业。我们指定了开始步骤，由于这是一个单步骤的作业，因此调用 JobBuilder.build()以生成真正的作业。

配置中的最后一个 Bean 定义是步骤。我们创建了一个返回类型为 Step 的方法，并使用 StepBuilderFactory 来构造它。传入步骤名称并调用 get()方法会返回一个 StepBuilder，我们接下来将使用它定义步骤。这个步骤将使用一个 Tasklet，因此我们传入一个 lambda 来代表 Tasklet 的实现。在这个示例中，我们仅仅调用了 System.out.println，并返回 Tasklet 已经成功的消息。我们最后调用了 build 方法。HelloWorldJob 类中的剩余代码与任何其他的 Spring Boot 应用相同，都是用来启动 Spring 的 main 方法。

就配置而言，作业配置的 90%涉及的是步骤的配置以及从一个步骤到下一个步骤的转换，本章后面将对此进行讨论。

4.2.2　作业参数

前面已经提到很多次，可通过传给作业的名称和识别性参数来标识 JobInstance。正因为如此，对于相同的作业，不能再次使用相同的识别性参数来执行，否则就会得到 org.springframework.batch.core.launch.JobInstanceAlreadyCompleteException 异常，这意味着如何再次执行作业，就需要修改参数。在构建应用之后，执行命令 java-jar target/Chapter04-0.0.1-SNAPSHOT.jar foo=bar，就可以看到代码清单 4-3 所示的正常日志。

代码清单 4-3　使用参数首次执行作业

```
...
2019-01-16 11:09:29.562 INFO 74578 --- [    main] o.s.b.a.b.
JobLauncherCommandLine Runner :Running default command line with: [foo=bar]
2019-01-16 11:09:29.669 INFO 74578 --- [    main] o.s.b.c.l.support.SimpleJobLauncher :
Job: [SimpleJob: [name=basicJob]] launched with the following parameters:[{foo=bar}]
2019-01-16 11:09:29.714 INFO 74578 --- [    main] o.s.batch.core.job.SimpleStepHandler :
```

```
Executing step: [step1]
Hello, world!
2019-01-16 11:09:29.793 INFO 74578 --- [ main] o.s.b.c.l.support.SimpleJobLauncher :
Job: [SimpleJob: [name=basicJob]] completed with the following parameters:
[{foo=bar}] and the following status: [COMPLETED]
...
```

现在,如果再次执行相同的命令,就会得到完全不同的结果,如代码清单4-4所示。

代码清单4-4 使用相同的参数,尝试再次执行作业

```
...
2019-01-16 11:09:34.250 INFO 74588 ---[ main] o.s.b.a.b.
JobLauncherCommandLineRunner: Running default command line with: [foo=bar]
2019-01-16 11:09:34.436 INFO 74588 --- [ main]
ConditionEvaluationReportLoggingListener :

Error starting ApplicationContext. To display the conditions report re-run your
application with 'debug' enabled.
2019-01-16 11:09:34.447 ERROR 74588 --- [ main] o.s.boot.SpringApplication :
Application run failed

java.lang.IllegalStateException: Failed to execute CommandLineRunner
    at org.springframework.boot.SpringApplication.callRunner(SpringApplication.
    java:816)
    [spring-boot-2.1.2.RELEASE.jar!/:2.1.2.RELEASE]
    ...
Caused by: org.springframework.batch.core.repository.JobInstanceAlreadyCompleteException:
A job instance already exists and is complete for parameters={foo=bar}. If you
want to runthis job again, change the parameters.
    at org.springframework.batch.core.repository.support.SimpleJobRepository.
    createJob Execution(SimpleJobRepository.java:132)
    ~[spring-batch-core-4.1.1.RELEASE.jar!/:
    4.1.1.RELEASE]
...
```

那么,如何把参数传给作业呢?在Spring Batch中,不仅可以把参数传给作业,而且可以自动让参数递增,或者在运行作业之前进行验证。我们首先看看如何将参数传给作业[1]。

传递参数给作业的方式取决于调用作业的方式。作业运行器的功能之一是创建一个org.springframework.batch.core.JobParameters实例,然后将其传给JobLauncher并执行。这是有意义的,因为从命令行启动作业与从 Quartz 调度器启动作业时传递参数的方式是不同的。到目前为止,由于一直在使用 Spring Boot 的 JobLauncherCommandLineRunner,因此下面从它开始进行介绍。

向JobLauncherCommandLineRunner 传递参数的方式,与前面示例中演示的在命令行上传递 key=value 一样简单。代码清单4-5 展示了如何使用调用作业的方式将参数传递给作业。

代码清单4-5 向 CommandLineJobRunner 传递参数

```
java -jar demo.jar name=Michael
```

---

[1] 把自增的参数作为 JobInstance 的参数可能更有意义。例如,如果作业运行的日期是 JobInstance 的参数之一,那么可以通过参数自增器(Parameter Incrementer)自动解决。

代码清单 4-5 传递了参数 name。在把参数传给批处理作业时，作业运行器会创建一个 JobParameters 实例，用以充当接收到的所有作业的参数容器。

> **注意** Spring Batch 的 JobParameters 与通过 Spring Batch 的命令行配置属性不同。因此，作业的参数不是通过--前缀传递的。Spring Batch 的 JobParameters 与系统属性也不同，不能在命令行上通过 -D 参数传给批处理应用。

JobParameters 只不过是 java.util.Map<String,JobParameter>对象的封装器而已。注意，尽管本例中传递的是字符串，但 Map 的值是一个 org.springframework.batch.core.JobParameters 实例。其中的原因就是类型。Spring Batch 提供了参数的类型转换功能，并在 JobParameters 类中提供了类型特定的访问器。如果指定参数的类型为 long，就会成为 java.lang.Long.String，Double 和 java.util.Date 都是可以直接用来进行类型转换的。为利用这些转换，需要在参数名后边的括号中指定参数的类型，如代码清单 4-6 所示。注意，这些类型必须小写。

代码清单 4-6 指定参数的类型

```
java -jar demo.jar executionDate(date)=2017/11/28
```

可以通过作业存储库查看传递给作业的参数。在作业存储库的数据库模式中，有一个数据表 BATCH_JOB_EXECUTION_PARAMS。在执行了代码清单 4-6 中的命令之后，如果查看这个数据表，就会看到表 4-1 所示的内容(为简便起见，此处删除了 LONG_VAL 和 DOUBLE_VAL 字段)。

表 4-1 BATCH_JOB_EXECUTION_PARAMS 数据库中的部分内容

JOB_EXECUTION_ID	TYPE_CD	KEY_NAME	STRING_VAL	DATE_VAL	IDENTIFYING
1	DATE	executionDate		2017-11-28 00:00:00	Y

到目前为止，我们已经反复说明了识别性参数是影响作业标识的因素。这也在暗示我们，还有一些非识别性参数。确实如此。从 Spring Batch 2.2 开始，用户可以指定哪些作业参数会影响作业的标识，这是十分有用的。例如，给定一个作业，它使用与作业实例一致的执行日期(因此应该是识别性参数)，但你期望能够基于运行时的多种条件，在不同作业的执行中修改其他的参数(例如，输入文件名)。为了表明作业参数是非识别性参数，可以将-作为前缀，参见代码清单 4-7。

代码清单 4-7 将作业参数标识为非识别性参数

```
java -jar demo.jar executionDate(date)=2017/11/28 -name=Michael
```

在代码清单 4-7 中，executionDate 是识别性参数，所以它将被用于确定当前运行是现有作业实例的一部分，还是会触发创建新的作业实例。然而，name 并不是识别性参数。如果作业在第一次执行时使用的 executionDate 为 2017/11/28 并且失败，那么可以使用同样的 executionDate，并修改 name 为 John 来重新启动这个作业，Spring Batch 会在现有作业实例的下方创建一次新的作业执行。

1. 访问作业参数

现在你已经知道了如何给批处理作业传递参数，那么如何访问它们？如果快速地查看一下 ItemReader、ItemProcessor、ItemWriter 和 Tasklet 等接口，你就能注意到，所有方法都不接收 JobParameters

作为参数。根据试图访问参数的位置，这里有一些不同的选择。

- **ChunkContext**：如果查看 HelloWorld Tasklet，就会看到 execute 方法接收两个参数。第一个参数是 org.springframework.batch.core.StepContribution，它包含当前还没有提交的事务的相关信息(写入次数、读取次数等)。第二个参数是 ChunkContext 实例，它提供了作业在执行时的状态。如果处在 Tasklet 中，那么其中将包含你当前正在处理的块(chunk)的所有信息，比如步骤和作业的相关信息。ChunkContext 有一个指向 org.springframework.batch.core.scope.context.StepContext 的引用，其中包含了 JobParameters。
- **延迟绑定**：对于 Spring Batch 框架中任何既不是步骤也个不是作业的部分，获得参数句柄的最简单方法是通过 Spring 配置进行注入。假定 JobParameters 是不可变的，那么在引导期间进行绑定则非常有意义。

代码清单 4-8 展示了更新过的 HelloWorld 作业，我们在输出时使用了 name 参数。作为示例，代码清单 4-8 展示了从 ChunkContext 中访问参数的方法。

代码清单 4-8　访问 Spring 配置中的 JobParameters

```
...
@EnableBatchProcessing
@SpringBootApplication
public class HelloWorldJob {

    @Autowired
    private JobBuilderFactory jobBuilderFactory;

    @Autowired
    private StepBuilderFactory stepBuilderFactory;

    @Bean
    public Job job() {
        return this.jobBuilderFactory.get("basicJob")
                .start(step1())
                .build();
    }

    @Bean
    public Step step1() {
        return this.stepBuilderFactory.get("step1")
                .tasklet(helloWorldTasklet())
                .build();
    }

    @Bean
    public Tasklet helloWorldTasklet() {

        return (contribution, chunkContext) -> {
            String name = (String) chunkContext.getStepContext()
                .getJobParameters()
                .get("name");

            System.out.println(String.format("Hello, %s!", name));
            return RepeatStatus.FINISHED;
        };
```

```
    }
    public static void main(String[] args) {
        SpringApplication.run(HelloWorldJob.class, args);
    }
}
```

尽管 Spring Batch 把作业参数保存在了 JobParameters 类的实例中，但是当通过这种方式获取参数时，getJobParameters()返回的类型是 Map<String, Object>。因此，在前面的示例中必须进行类型转换。

代码清单 4-9 展示了不引用任何 JobParameters 代码，而是使用 Spring 的延迟绑定功能把作业参数注入组件中的方法。除了使用 Spring 的 EL(Expression Language，表达式语言)来传值以外，任何通过延迟绑定功能来配置的 Bean 的作用域都必须设置为步骤或作业。

代码清单 4-9　通过延迟绑定功能获取作业参数

```
...
@EnableBatchProcessing
@SpringBootApplication
public class HelloWorldJob {

    @Autowired
    private JobBuilderFactory jobBuilderFactory;

    @Autowired
    private StepBuilderFactory stepBuilderFactory;

    @Bean
    public Job job() {
        return this.jobBuilderFactory.get("basicJob")
                .start(step1())
                .build();
    }

    @Bean
    public Step step1() {
        return this.stepBuilderFactory.get("step1")
                .tasklet(helloWorldTasklet(null))
                .build();
    }

    @StepScope
    @Bean
    public Tasklet helloWorldTasklet(
                @Value("#{jobParameters['name']}") String name) {
            return (contribution, chunkContext) -> {
                System.out.println(String.format("Hello, %s!", name));
                return RepeatStatus.FINISHED;
            };
    }

    public static void main(String[] args) {
        SpringApplication.run(HelloWorldJob.class, args);
    }
}
```

代码清单 4-10 展示了作用域为步骤的 Bean 的配置(允许进行延迟绑定)。

代码清单 4-10　作用域为步骤的 Bean 的配置

```
...
@StepScope
@Bean
public Tasklet helloWorldTasklet(
        @Value("#{jobParameters['name']}") String name) {

    return (contribution, chunkContext) -> {
        System.out.println(String.format("Hello, %s!", name));
        return RepeatStatus.FINISHED;
    };
}
...
```

Spring Batch 包含的自定义步骤和作业作用域，让延迟绑定变得相对容易了一些。它们要做的事情，就是把 Bean 的创建延迟到执行步骤或作业的作用域中，从而允许从命令行或其他来源获取作业参数，并在创建 Bean 时将它们注入。

有了给作业传递参数和使用参数的能力后，接下来讨论 Spring Batch 框架中的两个特定于参数的功能：校验参数以及在运行时递增给定的参数。下面首先讨论如何校验参数，因为在之前的示例中已提到过。

2. 校验参数

当软件从外部获取输入时，要确保输入符合期望。在 Web 世界里，人们会使用客户端 JavaScript 或多种服务器端框架来校验用户的输入，而批处理参数的校验与它们并无二致。幸运的是，Spring 将参数校验变得异常简单。只需要实现 org.springframework.batch.core.JobParametersValidator 接口并在作业中进行配置即可。代码清单 4-11 展示了 Spring Batch 中的一个参数校验器。

代码清单 4-11　一个参数校验器

```
...
public class ParameterValidator implements JobParametersValidator {

    @Override
    public void validate(JobParameters parameters) throws JobParametersInvalidException {
        String fileName = parameters.getString("fileName");

        if(!StringUtils.hasText(fileName)) {
            throw new JobParametersInvalidException("fileName parameter is missing");
        }
        else if(!StringUtils.endsWithIgnoreCase(fileName, "csv")) {
            throw new JobParametersInvalidException("fileName parameter does " +
                                    "not use the csv file extension");
        }
    }
}
```

如你所见，这里最重要的方法是 validation 方法。由于这个方法返回 void，因此只要没有抛出 JobParametersInvalidException 异常，就认为校验已通过。在这个示例中，如果缺少 fileName 参数，或者要校验的文件不以.csv 结尾，就会抛出上述异常，并且不会执行作业。

除了实现自定义的参数校验器(就像前面那样)以外，Spring Batch 还提供了如下参数校验器，用来确定传入了所有必需的参数：org.springframework.batch.core.job.DefaultJobParametersValidator。为了使用这个参数校验器，可以像配置自定义的参数校验器一样配置它。DefaultJobParametersValidator 有两个可选的依赖项：requiredKeys 和 optionalKeys。它们两个都是 String 数组，里面保存了参数名称的列表，分别对应必需或可选的参数。代码清单 4-12 展示了 DefaultJobParametersValidator 的配置样例。

代码清单 4-12　BatchConfiguration.java 中的 DefaultJobParametersValidator 配置

```
...
@Bean
public JobParametersValidator validator() {
    DefaultJobParametersValidator validator = new DefaultJobParametersValidator();

    validator.setRequiredKeys(new String[] {"fileName"});
    validator.setOptionalKeys(new String[] {"name"});

    return validator;
}
...
```

在代码清单 4-12 中，DefaultJobParametersValidator 将 fileName 配置为必需的参数。作业在试图运行时，如果缺少作业参数 fileName，那么校验就会失败。这里还配置了可选键 name。如此一来，能够传递给作业的参数就只能是 fileName 和 name 了。如果传入任何其他的参数，校验就会失败。如果没有配置可选键(仅仅配置了必需键)，那么除了必需键之外，还可以传入任何键的组合并通过校验。需要特别注意的是，除了参数的存在性之外，DefaultJobParametersValidator 不会执行任何其他校验。任何更加健壮的逻辑都必须通过自定义 JobParametersValidator 的实现来完成。

为了真正地实现这两个校验器，我们需要配置作业以使用它们。回到本章一直在使用的 HelloWorld 作业，我们可以把 JobParametersValidator 添加到这个作业中，Spring Batch 将在这个作业开始时执行它们。然而，这里有个小小的问题。我们想使用两个校验器，但是 JobBuilder 中用来配置校验器的方法只能接收一个 JobParametersValidator 实例。幸运的是，Spring Batch 为这种情况提供了 CompositeJobParametersValidator。代码清单 4-13 展示了更新过的作业配置，其中使用了校验器。

代码清单 4-13　使用 JobParameters 校验器配置作业

```
...
@EnableBatchProcessing
@SpringBootApplication
public class HelloWorldJob {

    @Autowired
    private JobBuilderFactory jobBuilderFactory;

    @Autowired
    private StepBuilderFactory stepBuilderFactory;

    @Bean
    public CompositeJobParametersValidator validator() {
        CompositeJobParametersValidator validator =
```

```java
                    new CompositeJobParametersValidator();

        DefaultJobParametersValidator defaultJobParametersValidator =
                    new DefaultJobParametersValidator(
                            new String[] {"fileName"},
                            new String[] {"name"});

        defaultJobParametersValidator.afterPropertiesSet();

        validator.setValidators(
                Arrays.asList(new ParameterValidator(),
                        defaultJobParametersValidator));

        return validator;
}

@Bean
public Job job() {
    return this.jobBuilderFactory.get("basicJob")
                .start(step1())
                .validator(validator())
                .build();
}

@Bean
public Step step1() {
    return this.stepBuilderFactory.get("step1")
                .tasklet(helloWorldTasklet(null, null))
                .build();
}

@StepScope
@Bean
public Tasklet helloWorldTasklet(
            @Value("#{jobParameters['name']}") String name,
            @Value("#{jobParameters['fileName']}") String fileName) {

    return (contribution, chunkContext) -> {

            System.out.println(
                    String.format("Hello, %s!", name));
            System.out.println(
                    String.format("fileName = %s", fileName));
            return RepeatStatus.FINISHED;
        };
}

public static void main(String[] args) {
    SpringApplication.run(HelloWorldJob.class, args);
}
}
```

在构建了这个应用之后,如果在执行时不传入必需的参数 fileName,或者 fileName 的格式不正确(文件名的末尾不是.csv),就会抛出异常,并且作业也不会运行。执行命令 java -jar target/Chapter04-0.0.1-SNAPSHOT.jar 时的输出如代码清单 4-14 所示。

代码清单 4-14　JobParameters 校验失败时的输出

```
...
2019-01-16 15:48:20.638 INFO 4023 --- [      main] o.s.b.a.b.
JobLauncherCommandLineRunner    : Running default command line with: []
2019-01-16 15:48:20.689 INFO 4023 --- [      main]
ConditionEvaluationReportLoggingListener :

Error starting ApplicationContext. To display the conditions report re-run your application
with 'debug' enabled.
2019-01-16 15:48:20.696 ERROR 4023 --- [      main] o.s.boot.SpringApplication :
Application run failed

java.lang.IllegalStateException: Failed to execute CommandLineRunner
    at org.springframework.boot.SpringApplication.callRunner(SpringApplication.
java:816)
    [spring-boot-2.1.2.RELEASE.jar!/:2.1.2.RELEASE]
    at
...
Caused by: org.springframework.batch.core.JobParametersInvalidException: fileName
parameter is missing
    at com.example.Chapter04.batch.ParameterValidator.validate(ParameterValidator.
java:33)~[classes!/:0.0.1-SNAPSHOT]
    at org.springframework.batch.core.job.CompositeJobParametersValidator.
validate(CompositeJobParametersValidator.java:49)
    ~[spring-batch-core-4.1.1.RELEASE.jar!/:4.1.1.RELEASE]
    at
...
```

如果使用相同的代码，并且在执行时仅仅传入必需的参数，那么作业将会运行，但在 Tasklet 中打印 Hello 的 System.out 将输出 Hello 和 null。必须运行命令 java -jar target/Chapter04-0.0.1-SNAPSHOT.jarfileName=foo.csv name=Michael 并提供两个参数，才能按预期运行作业。代码清单 4-15 展示了在提供了所有参数时最终的输出。

代码清单 4-15　提供了所有参数时的输出

```
...
2019-01-16 15:48:41.124 INFO 4044 --- [      main] o.s.b.a.b.
JobLauncherCommandLineRunner    : Running default command line with:
[fileName=foo.csv, name=bar]
2019-01-16 15:48:41.216 INFO 4044 --- [      main] o.s.b.c.l.support.
SimpleJobLauncher    : Job: [SimpleJob: [name=basicJob]] launched with the
following parameters : [{name=bar, fileName=foo.csv}]
2019-01-16 15:48:41.249 INFO 4044 --- [      main] o.s.batch.core.job.
SimpleStepHandler    : Executing step: [step1]
Hello, bar!
fileName = foo.csv!
2019-01-16 15:48:41.320 INFO 4044 --- [      main] o.s.b.c.l.support.
SimpleJobLauncher    : Job: [SimpleJob: [name=basicJob]] completed with the following
parameters: [{name=bar, fileName=foo.csv}] and the following status: [COMPLETED]
...
```

3. 递增给定的参数

到目前为止，运行作业时会有如下限制：对于给定的识别性参数，只能运行作业一次。如果一直在按照示例操作，那么当第二次尝试使用与代码清单 4-4 相同的参数运行相同的作业时，可能就会遇到这种情况。然而，这里使用了 JobParametersIncrementer，这将成为漏洞。

Spring Batch 提供了 org.springframework.batch.core.JobParametersIncrementer 接口，以允许为给定的作业生成唯一的参数。可以为每一次运行添加一个时间戳。你很可能有一些其他的业务逻辑，并且在每次运行作业时需要递增参数。Spring Batch 框架为 JobParametersIncrementer 接口提供了唯一的实现，可使用默认的名称 run.id 增加类型为 long 的参数。

代码清单 4-16 展示了 JobParametersIncrementer 的配置方式：把对参数的引用添加到本章使用的作业中。

代码清单 4-16　在作业中使用 JobParametersIncrementer

```
...
@Bean
public CompositeJobParametersValidator validator() {
    CompositeJobParametersValidator validator =
            new CompositeJobParametersValidator();

    DefaultJobParametersValidator defaultJobParametersValidator =
            new DefaultJobParametersValidator(
                        new String[] {"fileName"},
                        new String[] {"name", "run.id"});
    defaultJobParametersValidator.afterPropertiesSet();

    validator.setValidators(
            Arrays.asList(new ParameterValidator(),
                    defaultJobParametersValidator));
    return validator;
}
@Bean
public Job job() {
    return this.jobBuilderFactory.get("basicJob")
            .start(step1())
            .validator(validator())
            .incrementer(new RunIdIncrementer())
            .build();
}
...
```

你将会注意到，在示例作业的配置中，不仅要让作业接收 RunIdIncrementer，还必须更新 JobParametersValidator，使之允许引入新参数。

一旦配置了 JobParametersIncrementer(在本例中是 Spring Batch 框架提供的 org.springframework.batch.core.launch.support.RunIdIncrementer)，通过使用代码清单 4-17 所示的参数，就可以按照自己的意愿多次运行作业。

代码清单 4-17　用于运行作业和递增参数的命令

```
java -jar target/Chapter04-0.0.1-SNAPSHOT.jar fileName=foo.csv name=Michael
```

在运行这个作业三四次之后，查看数据表 BATCH_JOB_EXECUTION_PARAMS，你就会看到 Spring Batch 在执行作业时使用了三个参数：字符串类型的参数 name 和值 Michael、字符串类型的参数 fileName 和值 foo.csv，以及 long 类型的参数 run.id。run.id 的值每次都在变化：每次执行时都增加

1,如代码清单 4-18 所示。

代码清单 4-18　在执行三次后 RunIdIncrementer 的结果

```
mysql> select job_execution_id as id, type_cd as type, key_name as name, string_val,
long_val, identifying from SPRING_BATCH.BATCH_JOB_EXECUTION_PARAMS;
+----+--------+----------+------------+----------+-------------+
| id | type   | name     | string_val | long_val | identifying |
+----+--------+----------+------------+----------+-------------+
|  1 | STRING | name     | Michael    |        0 | Y           |
|  1 | LONG   | run.id   |            |        1 | Y           |
|  1 | STRING | filename | foo.csv    |        0 | Y           |
|  2 | STRING | name     | Michael    |        0 | Y           |
|  2 | STRING | filename | foo.csv    |        0 | Y           |
|  2 | LONG   | run.id   |            |        2 | Y           |
|  3 | STRING | name     | Michael    |        0 | Y           |
|  3 | STRING | filename | foo.csv    |        0 | Y           |
|  3 | LONG   | run.id   |            |        3 | Y           |
+----+--------+----------+------------+----------+-------------+
9 rows in set (0.00 sec)
```

前面提到过，你可能希望在每次运行作业时都有一个时间戳。在那些每天运行一次的作业中，这很常见。为此，你需要自行创建 JobParametersIncrementer 的实现。配置和执行与前面介绍的相同。不过，此时将使用 DailyJobTimestamper 而不是 RunIdIncrementer，如代码清单 4-19 所示。

代码清单 4-19　DailyJobTimestamper.java

```java
...
public class DailyJobTimestamper implements JobParametersIncrementer {
    @Override
    public JobParameters getNext(JobParameters parameters) {

        return new JobParametersBuilder(parameters)
            .addDate("currentDate", new Date())
            .toJobParameters();
    }
}
```

创建递增器后，你需要把它添加到作业中。你还需要更新参数校验，移除 RunIdIncrementer，以及添加递增器将会引入的 currentDate 参数。代码清单 4-20 展示了更新过的作业配置。

代码清单 4-20　更新作业以使用 DailyJobTimestamper

```java
@EnableBatchProcessing
@SpringBootApplication
public class HelloWorldJob {

    @Autowired
    private JobBuilderFactory jobBuilderFactory;

    @Autowired
    private StepBuilderFactory stepBuilderFactory;

    @Bean
    public CompositeJobParametersValidator validator() {
```

```java
            CompositeJobParametersValidator validator = 
                    new CompositeJobParametersValidator();

            DefaultJobParametersValidator defaultJobParametersValidator = 
                    new DefaultJobParametersValidator(
                            new String[] {"fileName"},
                            new String[] {"name", "currentDate"});

            defaultJobParametersValidator.afterPropertiesSet();

            validator.setValidators(
                    Arrays.asList(new ParameterValidator(),
                            defaultJobParametersValidator));

            return validator;
        }

        @Bean
        public Job job() {
            return this.jobBuilderFactory.get("basicJob")
                    .start(step1())
                    .validator(validator())
                    .incrementer(new DailyJobTimestamper())
                    .build();
        }

        @Bean
        public Step step1() {
            return this.stepBuilderFactory.get("step1")
                    .tasklet(helloWorldTasklet(null, null))
                    .build();
        }

        @StepScope
        @Bean
        public Tasklet helloWorldTasklet(
                @Value("#{jobParameters['name']}") String name,
                @Value("#{jobParameters['fileName']}") String fileName) {

            return (contribution, chunkContext) -> {

                System.out.println(
                        String.format("Hello, %s!", name));
                System.out.println(
                        String.format("fileName = %s", fileName));
                return RepeatStatus.FINISHED;
            };
        }

        public static void main(String[] args) {
            SpringApplication.run(HelloWorldJob.class, args);
        }
}
```

构建之后，就可以使用如下命令执行作业：java -jar target/Chapter04-0.0.1-SNAPSHOT.jar fileName=foo.csv name=Michael，并且在空的数据库中你会看到代码清单 4-12 所示的结果。

代码清单 4-21　使用 DailyJobTimestamper 之后的 BATCH_JOB_EXECUTION_PARAMS 表

```
mysql> select job_execution_id as id, type_cd as type, key_name as name, string_val
as s_val, date_val as d_val, identifying from SPRING_BATCH.BATCH_JOB_EXECUTION_PARAMS;
+----+--------+-------------+---------+---------------------+-------------+
| id | type   | name        | s_val   | d_val               | identifying |
+----+--------+-------------+---------+---------------------+-------------+
|  1 | STRING | name        | Michael | 1969-12-31 18:00:00 | Y           |
|  1 | DATE   | currentDate |         | 2019-01-16 16:40:55 | Y           |
|  1 | STRING | fileName    | foo.csv | 1969-12-31 18:00:00 | Y           |
+----+--------+-------------+---------+---------------------+-------------+
3 rows in set (0.00 sec)
```

显然，作业参数是 Spring Batch 框架的重要组成部分。它们允许在运行时为作业指定一些值。它们还用来唯一地标识作业的一次运行。在本书中，我们将更多地使用它们，例如配置运行作业的日期和重新处理错误文件。现在我们来看看作业级别的另一个强大特性：作业监听器。

4.2.3　使用作业监听器

每个作业都有生命周期。事实上，Spring Batch 的每个方面都有定义良好的生命周期，这允许我们在生命周期的不同阶段通过提供钩子(Hook)来添加额外的逻辑。在执行作业时，JobExecutionListener 接口提供了两个回调方法：beforeJob(JobExecution jobExecution)和 afterJob(JobExecution jobExecution)。这两个回调方法在作业的生命周期中应分别尽可能早或晚地执行。于是，我们可以在许多不同的用例中使用这两个回调方法。

- 通知：Spring Cloud Task 提供的 JobExecutionListener 将通过消息队列发出消息，以通知其他系统某个作业已经启动或结束。
- 初始化：如果某些准备工作需要在作业运行之前发生，那么 beforeJob 回调方法适合执行这些逻辑。
- 清理：许多作业都涉及清理工作，它们需要在作业运行之后发生(例如，删除或归档文件)。这些清理工作不影响作业的成功与失败，但仍需要执行。afterJob 回调方法是执行这些任务的完美场所。

创建作业监听器的方法有两个。第一个是实现 org.springframework.batch.core.JobExecutionListener 接口。这个接口有两个重要方法：beforeJob 和 afterJob，其中的每个方法都把 JobExecution 作为参数。你已经猜到，这两个方法分别会在作业运行之前和之后执行。关于 afterJob 方法，需要注意的一件重要事情是：不管作业完成的状态如何，它都会被调用。正因如此，你可能需要评估作业结束时的状态，以决定接下来做什么。代码清单 4-22 展示了一个简单的作业监听器，这个作业监听器打印了作业运行前后的一些信息以及作业完成时的状态。

代码清单 4-22　JobLoggerListener.java

```
...
public class JobLoggerListener implements JobExecutionListener {

    private static String START_MESSAGE = "%s is beginning execution";
    private static String END_MESSAGE =
            "%s has completed with the status %s";
```

```
    @Override
    public void beforeJob(JobExecution jobExecution) {
        System.out.println(String.format(START_MESSAGE,
                    jobExecution.getJobInstance().getJobName()));
    }
    @Override
    public void afterJob(JobExecution jobExecution) {
        System.out.println(String.format(END_MESSAGE,
                    jobExecution.getJobInstance().getJobName(),
                    jobExecution.getStatus()));
    }
}
```

为了在作业中使用这个新的作业监听器,你只需要在 JobBuilder 上简单地调用 .listener 方法,如代码清单 4-23 所示。

代码清单 4-23　使用了 JobLoggerListener 的作业

```
...
@Bean
public Job job() {

    return this.jobBuilderFactory.get("basicJob")
            .start(step1())
            .validator(validator())
            .incrementer(new DailyJobTimestamper())
            .listener(new JobLoggerListener())
            .build();
}
...
```

执行更新过的代码时,Spring Batch 会在作业中执行任何额外的处理之前,自动调用 beforeJob 方法,并且在所有其他的处理完成后调用 afterJob 方法。代码清单 4-24 展示了更新后的输出。

代码清单 4-24　JobExecutionListener 的输出

```
...
019-01-16 21:22:25.094 INFO 9006 --- [    main] o.s.b.a.b.
JobLauncherCommandLineRunner    : Running default command line with: [fileName=foo.csv,
name=Michael]
2019-01-16 21:22:25.186 INFO 9006 --- [    main] o.s.b.c.l.support.
SimpleJobLauncher    : Job: [SimpleJob: [name=basicJob]] launched with the following
parameters:[{name=Michael, currentDate=1547695345140, fileName=foo.csv}]
basicJob is beginning execution
2019-01-16 21:22:25.217 INFO 9006 --- [    main] o.s.batch.core.job.
SimpleStepHandler    : Executing step: [step1]
Hello, Michael!
fileName = foo.csv
basicJob has completed with the status COMPLETED
2019-01-16 21:22:25.281 INFO 9006 --- [    main] o.s.b.c.l.support.
SimpleJobLauncher    : Job: [SimpleJob: [name=basicJob]] completed with the following
parameters:[{name=Michael, currentDate=1547695345140, fileName=foo.csv}] and the
following status:
[COMPLETED]
...
```

第 4 章　理解作业和步骤

就像 Spring 中的所有对象一样，如果能够实现某个接口，那么很可能会有注解让事情变得更容易。创建作业监听器也不例外。为此，Spring Batch 提供了@BeforeJob 和@AfterJob 注解。当使用这两个注解时，唯一的区别是不再需要实现 JobExecutionListener 接口，如代码清单 4-25 所示。

代码清单 4-25　JobLoggerListener.java

```java
...
public class JobLoggerListener {

    private static String START_MESSAGE = "%s is beginning execution";

    private static String END_MESSAGE = "%s has completed with the status %s";

    @BeforeJob
    public void beforeJob(JobExecution jobExecution) {
        System.out.println(String.format(START_MESSAGE,
                    jobExecution.getJobInstance().getJobName()));
    }

    @AfterJob
    public void afterJob(JobExecution jobExecution) {
        System.out.println(String.format(END_MESSAGE,
                    jobExecution.getJobInstance().getJobName(),
                    jobExecution.getStatus()));
    }
}
```

使用注解的代码与前面的接口略有不同。Spring Batch 需要对作业监听器进行封装才能将它们注入作业中。我们可通过使用 JobListenerFactoryBean 做到这一点，参见代码清单 4-26，这将产生与前面相同的输出。

代码清单 4-26　在 BatchConfiguration.java 中配置作业监听器

```java
...
@Bean
public Job job() {

    return this.jobBuilderFactory.get("basicJob")
                .start(step1())
                .validator(validator())
                .incrementer(new DailyJobTimestamper())
                .listener(JobListenerFactoryBean.getListener(
                        new JobLoggerListener()))
                .build();
}
...
```

作业监听器十分有用，它们可以在作业的某些地方执行逻辑。作业监听器也可用于许多其他的批处理难题，例如步骤、读取器、写入器等。本书后续章节中将介绍它们各自的组件。现在，还有关于作业的最后一部分没有介绍：执行上下文。

4.2.4 执行上下文

批处理过程在本质上是有状态的。它们需要知道当前所处的步骤，并且需要知道截至目前那个步骤已经处理了多少记录。这些以及其他的状态元素不仅对于当前处理至关重要，而且对于重启之前遇到处理失败的批处理也很重要。例如，假设有一个批处理过程，它每天晚上要处理 100 万笔交易，但有一天晚上降到 90 万笔交易。即使是定期提交，那么在重启时，如何知道从哪里继续呢？重新建立执行状态的想法可能令人生畏，这就是 Spring Batch 为何要进行处理的原因。

你在前面已经了解到 JobExecution 表示的是作业的一次尝试执行。随着 JobExecution 在作业或步骤中执行，状态会发生变化。作业的状态可在作业执行的 ExecutionContext (执行上下文)中维护[1]。

思考一下 Web 应用是如何存储状态的，通常是使用 HttpSession。ExecutionContext 在本质上是批处理作业的会话(session)。仅仅通过持有简单的键-值对，ExecutionContext 便提供了在作业中安全地存储状态的一种方式。Web 应用的会话与 ExecutionContext 的区别是：在执行作业的过程中有多个执行上下文。每个 JobExecution 有一个 ExecutionContext，每个 StepExecution 也有一个 ExecutionContext(你在本章的后面可以看到)。这可以让数据处于合适级别的作用域内(要么是特定于步骤的数据，要么是针对作业的全局数据)。图 4-3 展示了这些元素之间的关系。

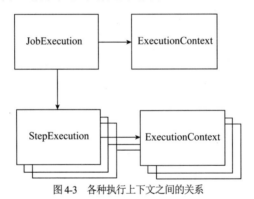

图 4-3　各种执行上下文之间的关系

ExecutionContext 提供了一种"安全"的方式来存储数据。存储是安全的，这是因为进入 ExecutionContext 的任何内容都会被持久化到作业存储库中。让我们看看如何把数据添加到 ExecutionContext 中，如何从 ExecutionContext 中检索数据，以及 ExecutionContext 在数据库中是什么样子。

4.2.5 操作 ExecutionContext

前面提到过，ExecutionContext 是 JobExecution 或 StepExecution 的一部分。因此，为了得到 ExecutionContext 句柄，需要访问 JobExecution 或 StepExecution。代码清单 4-27 展示了在 HelloWorld Tasklet 中获取 ExecutionContext 句柄的方式。

---

[1] 此处忽略了那些以客户端形式(Cookie、胖客户端等)存储状态的 Web 框架。

第 4 章　理解作业和步骤

代码清单 4-27　把用户名添加到作业的执行上下文中(一)

```
...
public class HelloWorld implements Tasklet {
    private static final String HELLO_WORLD = "Hello, %s";

    public RepeatStatus execute( StepContribution step,
                                 ChunkContext context ) throws Exception {
        String name = (String) context.getStepContext()
                            .getJobParameters()
                            .get("name");

        ExecutionContext jobContext = context.getStepContext()
                                        .getStepExecution()
                                        .getJobExecution()
                                        .getExecutionContext();
        jobContext.put("user.name", name);

        System.out.println( String.format(HELLO_WORLD, name) );
        return RepeatStatus.FINISHED;
    }
}
```

注意，必须执行一些遍历才能得到作业的 ExecutionContext：在本例中，方法是从块到步骤再到作业，在作用域树中向上移动。如果查看 StepContext 的 API，你就会看到 getJobExecutionContext() 方法，这个方法返回的 Map<String, Object>代表了作业的执行上下文的当前状态。尽管这是访问当前值的一种便捷方式，但在使用时存在以下限制：你对 StepContext.getJobExecutionContext()返回的 Map 所做的更新，并不会持久化到真正的执行上下文中。因此，在发生错误时，你对这个 Map 所做的任何修改(并不会更改真正的 ExecutionContext)都会丢失。

代码清单 4-27 展示了如何使用作业的执行上下文，但步骤的执行上下文的获取和操作方式与之前相同。这种情况下是从 StepExecution 中获取 ExecutionContext，而不是从 JobExecution 中获取。代码清单 4-28 展示了更新过的代码，使用的是步骤的 ExecutionContext 而不是作业的 ExecutionContext。

代码清单 4-28　把用户名添加到作业的执行上下文中(二)

```
...
public class HelloWorld implements Tasklet {
    private static final String HELLO_WORLD = "Hello, %s";
    public RepeatStatus execute( StepContribution step,
                                 ChunkContext context ) throws Exception {
        String name = (String) context.getStepContext()
                            .getJobParameters()
                            .get("name");

        ExecutionContext jobContext = context.getStepContext()
                                        .getStepExecution()
                                        .getExecutionContext();
        jobContext.put("user.name", name);

        System.out.println( String.format(HELLO_WORLD, name) );
        return RepeatStatus.FINISHED;
    }
}
```

操作作业执行的 ExecutionContext 的另一种方式，是把步骤执行的 ExecutionContext 中的键提升到作业执行的 ExecutionContext 中。如果想在步骤之间共享数据，并且除非第一步就取得成功，否则不希望共享这些数据，那么这种方式十分有用。用来执行这种提升的机制是通过 ExecutionContextPromotionListener 监听器完成的。代码清单 4-29 展示了这个监听器在批处理作业中是如何配置的，这里提升了 name 键(假定把 name 键放在了步骤的 ExecutionContext 中)。

代码清单 4-29　把用户名添加到作业的执行上下文中(三)

```
...
public class BatchConfiguration {

    @Autowired
    public JobBuilderFactory jobBuilderFactory;

    @Autowired
    public StepBuilderFactory stepBuilderFactory;

    @Bean
    public Job job() {
        return this.jobBuilderFactory.get("job")
                    .start(step1())
                    .next(step2())
                    .build();
    }

    @Bean
    public Step step1() {
          this.stepBuilderFactory.get("step1")
                      .tasklet(new HelloTasklet())
                      .listener(promotionListener())
                      .build();
    }

    @Bean
    public Step step2() {
          this.stepBuilderFactory.get("step2")
                      .tasklet(new GoodByeTasklet())
                      .build();
    }

    @Bean
    public StepExecutionListener promotionListener() {
        ExecutionContextPromotionListener listener = new
                    ExecutionContextPromotionListener();
        listener.setKeys(new String[] {"name"});
        return listener;
    }
}
```

在代码清单 4-29 中，配置的 promotionListner 会在步骤的 ExecutionContext 中查找 name 键。在步骤成功完成后，如果能够找到 name 键，那么 name 键会被复制到作业执行的 ExecutionContext 中。默认情况下，如果没有找到，那么不会发生任何事情。但也可配置监听器，使得在没有找到 name 键时抛出异常。

第 4 章　理解作业和步骤

访问 ExecutionContext 的最后一种方式是通过 ItemStream 实例，这将在本书后续章节中讨论。

ExecutionContext 的持久化

在处理作业的过程中，Spring Batch 会在提交每个块时持久化状态，其中一部分是保存作业和当前步骤的 ExecutionContext。第 2 章简单提到了数据表的布局。下面查看并执行代码清单 4-30 中的作业，看看在数据表中会持久化什么样的值。

代码清单 4-30　给作业的 ExecutionContext 添加用户名

```
...
@EnableBatchProcessing
@Configuration
public class BatchConfiguration {

    @Autowired
    private JobBuilderFactory jobBuilderFactory;

    @Autowired
    private StepBuilderFactory stepBuilderFactory;

    @Bean
    public Job job() {
        return this.jobBuilderFactory.get("job")
                    .start(step1())
                    .build();
    }

    @Bean
    public Step step1() {
        return this.stepBuilderFactory.get("step1")
                    .tasklet(helloWorldTasklet())
                    .build();
    }

    @Bean
    public Tasklet helloWorldTasklet() {
        return new HelloWorld();
    }

    public static class HelloWorld implements Tasklet {
        private static final String HELLO_WORLD = "Hello, %s";

        public RepeatStatus execute( StepContribution step,
                    ChunkContext context ) throws Exception {
            String name = (String) context.getStepContext()
                                    .getJobParameters()
                                    .get("name");

            ExecutionContext jobContext = context.getStepContext()
                            .getStepExecution()
                            .getExecutionContext();
            jobContext.put("name", name);

            System.out.println( String.format(HELLO_WORLD, name) );
```

61

```
            return RepeatStatus.FINISHED;
      }
   }
}
```

在使用 name 参数和参数值 Michael 运行后，BATCH_JOB_EXECUTION_CONTEXT 表中的内容如表 4-2 所示。

表 4-2 BATCH_JOB_EXECUTION_CONTEXT 表中的内容

JOB_EXECUTION_ID	SHORT_CONTEXT	SERIALIZED_CONTEXT
1	{"batch.taskletType":"io.spring.batch.demo.configuration.BatchConfiguration$HelloWorld","name":"Michael","batch.stepType":"org.springframework.batch.core.step.tasklet.TaskletStep"}	NULL

表 4-2 包含三列。第一列是引用，指向 ExecutionContext 相关的 JobExecution。第二列是作业的 ExecutionContext 的 JSON 表示，这个字段会在处理的时候进行更新。最后一列是 SERIALIZED_CONTEXT，用于包含序列化的 Java 对象。SERIALIZED_CONTEXT 只会在作业运行或失败时填充。

请注意，SHORT_CONTEXT 列包含字符串"name" : "Michael"和一些其他的字段。这些字段(比如 batch.taskletType 和 batch.stepType)都是 Spring Cloud Data Flow 将要使用的值。

至此，我们讨论了 Spring Batch 中作业的不同部分。然而，为创建有效的作业，要求作业中至少有一个步骤。接下来我们讨论 Spring Batch 框架中的下一个主要部分：步骤。

4.3 使用步骤

如果作业定义了完整的处理过程，那么步骤就是作业的构建块。步骤是独立、有序的批处理器。将步骤称为批处理器是有原因的。步骤包含一个工作单元需要的所有东西。步骤处理自己的输入，有自己的处理器，并且处理自己的输出。步骤中有自包含的事务。步骤还被设计为相互独立。这允许开发人员按照自己的需要自由地构建作业。

与深入研究作业时一样，本节将以相同的方式深入研究步骤。你将了解 Spring Batch 如何在步骤中按块进行分解处理，以及如何在这种执行风格中处理事务。你还会看到在作业中配置步骤的大量示例，包括如何控制步骤之间的流以及执行步骤时的条件控制。最后，你会为银行对账单作业配置所需的步骤。了解所有这些内容后，我们首先看看步骤如何处理数据。

4.3.1 Tasklet 和基于块的处理

一般来说，批处理过程都是关于数据处理的。批处理作业中的一些工作仅仅需要执行单个命令，可能是用于清理目录的 shell 脚本，也可能是用于删除临时表内容的 SQL 语句。其他的工作则需要在一个大的数据集中进行迭代：一次读取一条记录，执行某种类型的逻辑，然后将它们写入某种类型的存储介质。Spring Batch 同时支持两种处理模型。

到目前为止，本书使用的都是第一种处理模型：Tasklet 模型。我们使用过的 Tasklet 接口允许开发人员创建一个代码块，这个代码块可在事务的作用域内重复执行，直到 Tasklet.execute 方法返回 RepeatStatus.FINISHED 为止[1]。

第二种处理模型是基于块的处理(chunk-based processing)。基于块的步骤包含两个或三个主要组件：一个 ItemReader、一个可选的 ItemProcessor 以及一个 ItemWriter。使用了这些组件后，Spring Batch 就会按块或按组处理记录。每个块都会在自己的事务中执行，这允许 Spring Batch 在遇到失败情形时，能够从最后一次执行成功的事务之后重新启动。

使用这三个主要组件，框架会执行三个循环。第一个循环是 ItemReader，用于把块中所有想要处理的数据读入内存。第二个循环是可选的 ItemProcessor，如果配置了 ItemProcessor，那么将会循环处理读入内存的项，并把每一项传入 ItemProcessor。最后一个循环是 ItemWriter，所有的项会在一次调用中传给 ItemWriter，它们可以被一次性写出。对 ItemWriter 的这一次调用允许通过批量的物理写入进行 I/O 优化。图 4-4 展示了基于块的处理是如何工作的。

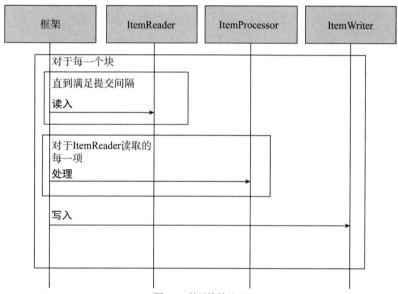

图 4-4　基于块的处理

在你从本书中学习关于步骤、读取器、写入器和可伸缩性的更多内容时，始终要记住 Spring Batch 采用的是基于块的处理模型。下面继续深入研究如何配置作业的构建块：步骤。

4.3.2　步骤的配置

你现在已经认识到，作业实际上只不过是步骤的容器。通过一定的配置，可以从一个步骤转换到另一个步骤。你对这种方式是不是很熟悉？这就是状态机。从根本上讲，Spring Batch 就是状态机，步骤代表了状态，通过一组转换的集合，就能够从一个状态转到下一个状态。下面我们从最常用的步骤类型 Tasklet 开始查看步骤。

---

1 到目前为止，使用了 Tasklet 的每个示例都在第一次执行后就返回了 RepeatStatus.FINISHED，所以我们并没有演示在 Tasklet 中进行迭代的可能性。

Tasklet 步骤

Spring Batch 有两种风格的步骤，Tasklet 是其中之一。你可能对 Tasklet 最熟悉，因为你几乎在每个作业中都用到了它。不同之处在于，在本例中，你正在编写自己的代码作为 Tasklet 来执行。使用 MethodInvokingTaskletAdapter 是定义 Tasklet 步骤的方式之一。在本例中，我们允许 Spring 把处理转发给你的代码，这使你可以开发普通的 POJO，并且将其作为步骤。

创建 Tasklet 步骤的另一种方式是实现 Tasklet 接口，就像在第 2 章创建 HelloWorld Tasklet 时使用的做法一样，其中实现了 Tasklet 接口要求的 execute 方法，返回了 RepeatStatus 对象，并且通知 Spring Batch 在完成处理之后应该做什么。从技术上讲，由于 Tasklet 接口是函数式接口，因此还可以使用 lambda 来实现 Tasklet。代码清单 4-31 展示了 HelloWorld Tasklet 的 lambda 配置。

代码清单 4-31　HelloWorld Tasklet

```
@EnableBatchProcessing
@Configuration
public class BatchConfiguration {

    @Autowired
    private JobBuilderFactory jobBuilderFactory;

    @Autowired
    private StepBuilderFactory stepBuilderFactory;

    @Bean
    public Job job() {
        return this.jobBuilderFactory.get("job")
                            .start(step1())
                            .build();
    }

    @Bean
    public Step step1() {
        return this.stepBuilderFactory.get("step1")
                        .tasklet((stepContribution, chunkContext) -> {
                            System.out.println("Hello, World!");
                            return RepeatStatus.FINISHED;
                        })
                        .build();
    }
}
```

当 Tasklet 实现中的处理完成时，会返回一个 org.springframework.batch.repeat.RepeatStatus 对象。这里有两个选项：RepeatStatus.CONTINUABLE 和 RepeatStatus.FINISHED。乍一看，这两个选项容易让人混淆。返回 RepeatStatus.CONTINUABLE 并不意味着作业可以继续，而是在告诉 Spring Batch 再次运行 Tasklet。例如，你可能想在一个循环中执行特定的 Tasklet，直到满足给定的条件为止，同时你还想使用 Spring Batch 来跟踪 Tasklet 的执行次数、转换次数等。Tasklet 可以返回 RepeatStatus.CONTINUABLE，直到满足某个条件为止。如果返回了 RepeatStatus.FINISHED，那么意味着这个 Tasklet 的处理已经完成(无论是否成功)，并将继续下一段处理。

4.3.3 理解其他类型的 Tasklet

到目前为止，尽管每个示例都使用了 Tasklet 接口的自定义实现，但这并不是使用 Tasklet 接口的唯一方式。Spring Batch 还提供了 Tasklet 的其他实现，比如：CallableTaskletAdapter、MethodInvokingTaskletAdapter 和 SystemCommandTasklet。下面首先看看 CallableTaskletAdapter。

1. CallableTaskletAdapter

org.springframework.batch.core.step.tasklet.CallableTaskletAdapter 适配器用来配置 java.util.concurrent.Callable<RepeatStatus>接口的实现。如果你对这个接口还不熟悉，那么可以认为 Callable<V>接口与 java.lang.Runnable 接口很相似，它们都在新的线程中执行。然而，与 java.lang.Runnable 接口既不返回值也不抛出检查型异常(Checked Exception)不同，Callable<V>接口可以返回一个值(在本例中是 RepeatStatus)，并且能够抛出检查型异常。

CallableTaskletAdapter 适配器的实现其实极其简单：调用 Callable 对象的 call()方法，并且返回 call()方法的返回值，仅此而已。显然，如果想在执行步骤的线程之外的其他线程中执行逻辑，就可以使用这个适配器。查看代码清单 4-32，其中将 CallableTaskletAdapter 配置为普通的 Spring Bean，然后在步骤中将其注册为 Tasklet。CallableTaskletAdapter 确实有一个依赖项，即这个可调用对象本身。

代码清单 4-32　使用 CallableTaskletAdapter

```
...
@EnableBatchProcessing
@SpringBootApplication
public class CallableTaskletConfiguration {

    @Autowired
    private JobBuilderFactory jobBuilderFactory;

    @Autowired
    private StepBuilderFactory stepBuilderFactory;

    @Bean
    public Job callableJob() {
        return this.jobBuilderFactory.get("callableJob")
                    .start(callableStep())
                    .build();
    }

    @Bean
    public Step callableStep() {
        return this.stepBuilderFactory.get("callableStep")
                    .tasklet(tasklet())
                    .build();
    }

    @Bean
    public Callable<RepeatStatus> callableObject() {

        return () -> {
            System.out.println("This was executed in another thread");
```

```
            return RepeatStatus.FINISHED;
        };
    }

    @Bean
    public CallableTaskletAdapter tasklet() {
        CallableTaskletAdapter callableTaskletAdapter =
                    new CallableTaskletAdapter();

        callableTaskletAdapter.setCallable(callableObject());

        return callableTaskletAdapter;
    }

    public static void main(String[] args) {
        SpringApplication.run(CallableTaskletConfiguration.class, args);
    }
}
```

使用 CallableTaskletAdapter 时,需要注意一件事情:尽管 Tasklet 在与步骤不同的线程中执行,但这并不会将步骤的执行并行化。只有在 Callable 对象返回有效的 RepeatStatus 对象后,步骤的执行才被认定为已完成。在步骤被认定为已完成之前,你在这个流程中配置的其他步骤都不会执行。在本书后续章节中,你会看到多种并行化处理方式,包括并行执行步骤。

2. MethodInvokingTaskletAdapter

Tasklet 的下一种实现是 org.springframework.batch.core.step.tasklet.MethodInvokingTaskletAdapter 适配器。这个适配器与 Spring 框架中的许多实用工具十分类似。它允许你在作业中以 Tasklet 的形式在另一个类中执行早已存在的方法。例如,你已经有了一个服务,它用来执行一段逻辑,你希望在批处理作业中只运行这个服务一次。你可以使用 MethodInvokingTaskletAdapter 来调用这个服务,而不是实现一个仅封装了服务调用的 Tasklet 接口。代码清单 4-33 展示了 MethodInvokingTaskletAdapter 适配器的一个配置示例。

代码清单 4-33 使用 MethodInvokingTaskletAdapter

```
...
@EnableBatchProcessing
@SpringBootApplication
public class MethodInvokingTaskletConfiguration {

    @Autowired
    private JobBuilderFactory jobBuilderFactory;
    @Autowired
    private StepBuilderFactory stepBuilderFactory;

    @Bean
    public Job methodInvokingJob() {
        return this.jobBuilderFactory.get("methodInvokingJob")
                    .start(methodInvokingStep())
                    .build();
    }

    @Bean
```

```
    public Step methodInvokingStep() {
        return this.stepBuilderFactory.get("methodInvokingStep")
                    .tasklet(methodInvokingTasklet())
                    .build();
    }

    @Bean
    public MethodInvokingTaskletAdapter methodInvokingTasklet() {
        MethodInvokingTaskletAdapter methodInvokingTaskletAdapter =
                    new MethodInvokingTaskletAdapter();

        methodInvokingTaskletAdapter.setTargetObject(service());
        methodInvokingTaskletAdapter.setTargetMethod("serviceMethod");

        return methodInvokingTaskletAdapter;
    }

    @Bean
    public CustomService service() {
        return new CustomService();
    }

    public static void main(String[] args) {
        SpringApplication.run(MethodInvokingTaskletConfiguration.class, args);
    }
}
```

代码清单 4-33 中引用的 CustomService 只是一个简单的 POJO。这个 POJO 调用了 System.out.println() 方法并打印了一段文本以表明自己被调用，参见代码清单 4-34。

代码清单 4-34　CustomService

```
...
public class CustomService {

    public void serviceMethod() {
        System.out.println("Service method was called");
    }
}
```

代码清单 4-33 展示的适配器配置示例指定了一个对象和一个方法。使用这种配置，适配器就会调用不带参数的方法并返回结果 ExitStatus.COMPLETED，除非指定的这个方法也返回 org.springframework.batch.core.ExitStatus 类型。如果确实返回了 ExitStatus 类型，那么这个方法的返回值会从 Tasklet 中返回。如果想配置一组静态参数，那么可以使用本章前面提到的用于传递参数的延迟绑定方法，如代码清单 4-35 所示。

代码清单 4-35　使用带参数的 MethodInvokingTaskletAdapter

```
...
@EnableBatchProcessing
@SpringBootApplication
public class MethodInvokingTaskletConfiguration {

    @Autowired
```

```java
    private JobBuilderFactory jobBuilderFactory;

    @Autowired
    private StepBuilderFactory stepBuilderFactory;

    @Bean
    public Job methodInvokingJob() {
        return this.jobBuilderFactory.get("methodInvokingJob")
                .start(methodInvokingStep())
                .build();
    }

    @Bean
    public Step methodInvokingStep() {
        return this.stepBuilderFactory.get("methodInvokingStep")
                .tasklet(methodInvokingTasklet(null))
                .build();
    }

    @StepScope
    @Bean
    public MethodInvokingTaskletAdapter methodInvokingTasklet(
            @Value("#{jobParameters['message']}") String message) {

        MethodInvokingTaskletAdapter methodInvokingTaskletAdapter =
                new MethodInvokingTaskletAdapter();

        methodInvokingTaskletAdapter.setTargetObject(service());
        methodInvokingTaskletAdapter.setTargetMethod("serviceMethod");
        methodInvokingTaskletAdapter.setArguments(new String[] {message});

        return methodInvokingTaskletAdapter;
    }

    @Bean
    public CustomService service() {
        return new CustomService();
    }

    public static void main(String[] args) {
        SpringApplication.run(MethodInvokingTaskletConfiguration.class, args);
    }
}
```

为了让代码清单 4-35 中的代码能够工作，你还需要更新 CustomService，让其接收消息并打印。代码清单 4-36 展示了这部分更新。

代码清单 4-36　使用带参数的 CustomService

```java
...
public class CustomService {

    public void serviceMethod(String message) {
        System.out.println(message);
    }
}
```

3. SystemCommandTasklet

Spring 提供的最后一种 Tasklet 实现是 org.springframework.batch.core.step.tasklet.SystemCommandTasklet。你可能已经猜到，它用来执行系统命令！指定的系统命令是异步执行的。正因为如此，代码清单 4-37 中展示的超时时间(单位为毫秒)非常重要。代码清单 4-37 中的 setInterruptOnCancel 属性是可选的，它用来向 Spring Batch 表明，当作业异常退出时是否要结束与系统进程关联的线程。

代码清单 4-37　使用 SystemCommandTasklet

```
...
@EnableBatchProcessing
@SpringBootApplication
public class SystemCommandJob {

    @Autowired
    private JobBuilderFactory jobBuilderFactory;

    @Autowired
    private StepBuilderFactory stepBuilderFactory;

    @Bean
    public Job job() {
        return this.jobBuilderFactory.get("systemCommandJob")
                    .start(systemCommandStep())
                    .build();
    }

    @Bean
    public Step systemCommandStep() {
        return this.stepBuilderFactory.get("systemCommandStep")
                    .tasklet(systemCommandTasklet())
                    .build();
    }

    @Bean
    public SystemCommandTasklet systemCommandTasklet() {
        SystemCommandTasklet systemCommandTasklet = new SystemCommandTasklet();

        systemCommandTasklet.setCommand("rm -rf /tmp.txt");
        systemCommandTasklet.setTimeout(5000);
        systemCommandTasklet.setInterruptOnCancel(true);

        return systemCommandTasklet;
    }

    public static void main(String[] args) {
        SpringApplication.run(SystemCommandJob.class, args);
    }
}
```

SystemCommandTasklet 有多个参数，它们会影响系统命令的执行效果。代码清单 4-38 展示了一个更健壮的示例。

代码清单4-38 使用带有完整环境配置的SystemCommandTasklet

```java
...
@EnableBatchProcessing
@SpringBootApplication
public class AdvancedSystemCommandJob {

    @Autowired
    private JobBuilderFactory jobBuilderFactory;

    @Autowired
    private StepBuilderFactory stepBuilderFactory;

    @Bean
    public Job job() {
        return this.jobBuilderFactory.get("systemCommandJob")
                .start(systemCommandStep())
                .build();
    }

    @Bean
    public Step systemCommandStep() {
        return this.stepBuilderFactory.get("systemCommandStep")
                .tasklet(systemCommandTasklet())
                .build();
    }

    @Bean
    public SystemCommandTasklet systemCommandTasklet() {
        SystemCommandTasklet tasklet = new SystemCommandTasklet();
        tasklet.setCommand("touch tmp.txt");
        tasklet.setTimeout(5000);
        tasklet.setInterruptOnCancel(true);

        // Change this directory to something appropriate for your environment
        tasklet.setWorkingDirectory("/Users/mminella/spring-batch");

        tasklet.setSystemProcessExitCodeMapper(touchCodeMapper());
        tasklet.setTerminationCheckInterval(5000);
        tasklet.setTaskExecutor(new SimpleAsyncTaskExecutor());
        tasklet.setEnvironmentParams(new String[] {
                "JAVA_HOME=/java",
                "BATCH_HOME=/Users/batch"});

        return tasklet;
    }

    @Bean
    public SimpleSystemProcessExitCodeMapper touchCodeMapper() {
        return new SimpleSystemProcessExitCodeMapper();
    }

    public static void main(String[] args) {
        SpringApplication.run(AdvancedSystemCommandJob.class, args) ;
    }
}
```

代码清单 4-38 所示的配置中多了 5 个参数。

- setWorkingDirectory：这是执行命令时所在的目录。在本例中，相当于在执行真正的命令之前，执行 cd ~/spring-batch。
- setSystemProcessExitCodeMapper：取决于执行的命令，系统码的意义可能不同。这个属性允许你使用 org.springframework.batch.core.step.tasklet.SystemProcessExitCodeMapper 接口的实现，将系统码映射为 Spring Batch 的状态值。Spring 默认提供了两个实现：org.springframework.batch.core.step.tasklet.ConfigurableSystemProcessExitCodeMapper 允许自行配置映射；org.springframework.batch.core.step.tasklet.SimpleSystemProcessExitCodeMapper 会在系统码为 0 时返回 ExitStatus.FINISHED，否则返回 ExitStatus.FAILED。
- setTerminationCheckInterval：由于系统命令在默认情况下以异步的方式执行，因此 Tasklet 会定期检查它们是否已完成。这个属性的值默认是 1 秒，但你也可以将其配置为其他值(单位为毫秒)。
- setTaskExecutor：这个属性允许你自行配置 TaskExecutor 以执行系统命令。强烈建议不要配置同步的任务执行器，因为如果系统命令引起了问题，就可能会锁定作业。
- setEnvironmentParams：这个属性指定了可以在命令执行之前进行设置的环境参数列表。

你在前面已经看到了 Spring Batch 中有许多不同的 Tasklet 类型。现在我们来看看另一种最常用的步骤类型：基于块的步骤。

4. 基于块的步骤

如前所述，块是以它们的提交间隔定义的。如果提交间隔是 50 个条目，那么作业会读取 50 个条目，处理 50 个条目，然后一次性写入 50 个条目。代码清单 4-39 展示了如何为面向块的处理配置基本步骤。

代码清单 4-39　BatchConfiguration.java

```
...
@EnableBatchProcessing
@Configuration
public class BatchConfiguration {

    @Autowired
    private JobBuilderFactory jobBuilderFactory;

    @Autowired
    private StepBuilderFactory stepBuilderFactory;

    @Bean
    public Job job() {
        return this.jobBuilderFactory.get("job")
                .start(step1())
                .build();
    }

    @Bean
    public Step step1() {
        return this.stepBuilderFactory.get("step1")
                .<String, String>chunk(10)
```

```
                    .reader(itemReader(null))
                    .writer(itemWriter(null))
                    .build();
    }

    @Bean
    @StepScope
    public FlatFileItemReader<String> itemReader(
            @Value("#{jobParameters['inputFile']}") Resource inputFile) {

        return new FlatFileItemReaderBuilder<String>()
                    .name("itemReader")
                    .resource(inputFile)
                    .lineMapper(new PassThroughLineMapper())
                    .build();
    }

    @Bean
    @StepScope
    public FlatFileItemWriter<String> itemWriter(
            @Value("#{jobParameters['outputFile']}") Resource outputFile) {

        return new FlatFileItemWriterBuilder<String>()
                    .name("itemWtiter")
                    .resource(outputFile)
                    .lineAggregator(new PassThroughLineAggregator<>())
                    .build();
    }
}
```

代码清单 4-39 可能有些让人望而生畏，但我们还是将重点放在作业和最开始的步骤配置上。BatchConfiguration.java 文件的其余部分是基本的 ItemReader 和 ItemWriter，第 7 和 9 章将分别介绍它们。如果查看代码清单 4-39 中的作业，就会发现步骤是从 StepBuilderFactory 中的 StepBuilder 开始的。然后你可以断定这是一个基于块的步骤，因为调用了 chunk 方法。在本例中，传入的参数 10 是提交间隔。作业会在处理 10 条记录后提交一次。这个基于块的步骤在调用 build 方法之前设置了一个读取器(ItemReader 接口的实现)和一个写入器(ItemWriter 接口的实现)。

注意，提交间隔(Commit Interval)很重要。提交间隔在本例中设置为 10，这意味着只有在读取和处理 10 条记录后，才会写入这些记录。如果在处理 9 条记录之后发生了错误，那么 Spring Batch 就会回滚当前的块(事务)，并且将作业标记为失败。如果提交间隔设置为 1，那么作业就会读取一个条目，处理一个条目，然后写这个条目。从本质上讲，这又回到了基于单个条目的处理。这里存在的问题是：在提交间隔中不止有一个条目被持久化。作业的状态也会在作业存储库中更新。你会在本书的后面试验提交间隔，但你现在需要知道，在写入数据时，合理并尽可能设置好提交间隔是很重要的。

接下来，我们将详细讨论基于块的步骤的组件。

5. 配置块的大小

由于基于块的处理是 Spring Batch 的基础，因此理解如何配置块的大小以便充分利用这个重要特性是很重要的。下面介绍用来配置块的大小的两个选项：静态的提交次数和 CompletionPolicy 接口的实现。其他的与错误处理相关的块配置选项会在相应的内容中进行讨论。

代码清单 4-40 展示了一个基本的只配置了读取器、写入器和提交间隔的块。读取器是 ItemReader 接口的实现，写入器是 ItemWriter 接口的实现。在本书的后面，每个接口都有专门的章节用来进行介绍，所以此处不会深入细节。你只需要知道，它们分别提供了步骤的输入和输出功能。提交间隔定义了每个块中条目的数量(在本例中是 10 个条目)。

代码清单 4-40　配置一个基本的块

```
...
@EnableBatchProcessing
@SpringBootApplication
public class ChunkJob {

    @Autowired
    private JobBuilderFactory jobBuilderFactory;

    @Autowired
    private StepBuilderFactory stepBuilderFactory;

    @Bean
    public Job chunkBasedJob() {
        return this.jobBuilderFactory.get("chunkBasedJob")
                    .start(chunkStep())
                    .build();
    }

    @Bean
    public Step chunkStep() {
        return this.stepBuilderFactory.get("chunkStep")
                    .<String, String>chunk(1000)
                    .reader(itemReader())
                    .writer(itemWriter())
                    .build();
    }

    @Bean
    public ListItemReader<String> itemReader() {
        List<String> items = new ArrayList<>(100000);

        for (int i = 0; i < 100000; i++) {
            items.add(UUID.randomUUID().toString());
        }

        return new ListItemReader<>(items);
    }

    @Bean
    public ItemWriter<String> itemWriter() {
        return items -> {
            for (String item : items) {
                System.out.println(">> current item = " + item);
            }
        };
    }
```

```
    public static void main(String[] args) {
        SpringApplication.run(ChunkJob.class, args);
    }
}
```

尽管我们通常通过设定固定的提交间隔来定义块的大小，如代码清单 4-40 所示，但这并不总是最佳选择。例如，作业处理的块并不总是具有相同的大小(需要在单个事务中处理某个账户的所有事务)。Spring Batch 为你提供了以编程的方式定义块被处理完毕的能力，只需要实现 org.springframework.batch.repeat.CompletionPolicy 接口即可。

CompletionPolicy 接口允许通过实现决策逻辑来决定块是否已处理完毕。Spring Batch 提供了这个接口的多种实现。默认的 org.springframework.batch.repeat.policy.SimpleCompletionPolicy 会统计处理过的条目的数目，当达到配置的阈值时，就将这个块标记为已完成。另一种开箱即用的实现是 org.springframework.batch.repeat.policy.TimeoutTerminationPolicy，它允许你为块配置超时时间，这样它就可以在一段时间后优雅地退出。所谓"优雅地退出"，是指当块被确认为已完成时，所有的事务处理还会正常继续进行。

可以推断，有时候，超时时间本身就足以决定块何时处理完成。TimeoutTerminationPolicy 更可能作为 org.springframework.batch.repeat.policy.CompositeCompletionPolicy 的一部分被使用。这种策略允许配置多个条件以决定一个块是否完成。当使用 CompositeCompletionPolicy 时，只要有策略认为一个块已完成，那么这个块就会被标记为已完成。代码清单 4-41 展示了如何使用超时时间(3 毫秒)和提交次数(200)来确认块已完成。

代码清单 4-41　结合使用超时时间和提交次数来确认块已完成

```
...
@EnableBatchProcessing
@SpringBootApplication
public class ChunkJob {

    @Autowired
    private JobBuilderFactory jobBuilderFactory;

    @Autowired
    private StepBuilderFactory stepBuilderFactory;

    @Bean
    public Job chunkBasedJob() {
        return this.jobBuilderFactory.get("chunkBasedJob")
                    .start(chunkStep())
                    .build();
    }

    @Bean
    public Step chunkStep() {
        return this.stepBuilderFactory.get("chunkStep")
                    .<String, String>chunk(completionPolicy())
                    .reader(itemReader())
                    .writer(itemWriter())
                    .build();
    }
```

```java
@Bean
public ListItemReader<String> itemReader() {
    List<String> items = new ArrayList<>(100000);

    for (int i = 0; i < 100000; i++) {
        items.add(UUID.randomUUID().toString());
    }
    return new ListItemReader<>(items);
}

@Bean
public ItemWriter<String> itemWriter() {
    return items -> {
        for (String item : items) {
            System.out.println(">> current item = " + item);
        }
    };
}

@Bean
public CompletionPolicy completionPolicy() {
    CompositeCompletionPolicy policy = new CompositeCompletionPolicy();

    policy.setPolicies(
            new CompletionPolicy[] {
                            new TimeoutTerminationPolicy(3),
                            new SimpleCompletionPolicy(1000) });

    return policy;
}

public static void main(String[] args) {
    SpringApplication.run(ChunkJob.class, args);
}
```

注意，如果执行前面两个示例，那么从表面上看不会有任何不同。但事实并非如此。第一个示例(参见代码清单 4-40)中有 101 次提交，但第二个示例中(参见代码清单 4-41)有大约 191 次提交[1]，这是添加 TimeoutTerminationPolicy 后带来的影响。

使用现有的 CompletionPolicy 接口的实现并不是决定块大小的唯一选择，你还可以自定义 CompletionPolicy 接口的实现。

CompletionPolicy 接口需要 4 个方法：两个版本的 isComplete 方法以及 start 和 update 方法。如果查看这个接口的实现类的生命周期，那么第一个被调用的方法就是 start。这个方法用于初始化策略，以便通知开始处理块。需要重点注意的是，CompletionPolicy 接口的实现类是有状态的，并且应该能够通过内部状态来决定块是否已处理完。start 方法位于块处理代码的开头，可用来把内部状态重置为实现类所需的状态。以 SimpleCompletionPolicy 为例，start 方法会首先把内部的计数器重置为 0。回到前面的那个 SimpleCompletionPolicy 示例，update 方法会在处理每一个条目之后，递增内部计数器。块处理代

---

1 由于 TimeoutTerminationPolicy 的基于时间的本质，此处的提交次数会随着作业运行环境的不同而变化。

码的最后是两个 isComplete 方法。isComplete 方法的第一个方法签名接收 RepeatContext 参数，这个实现是为了使用内部状态来决定块是否已处理完；第二个方法签名接收 RepeatContext 和 RepeatStatus 参数，这个实现基于传入的状态(RepeatStatus)来决定块是否已处理完。代码清单 4-42 展示了 CompletionPolicy 的实现，一旦处理的条目的随机数目小于 20，就认为块已处理完；代码清单 4-43 展示了相应的配置。

代码清单 4-42　块大小随机的 CompletionPolicy 实现

```
...
public class RandomChunkSizePolicy implements CompletionPolicy {

    private int chunksize;
    private int totalProcessed;
    private Random random = new Random();

    @Override
    public boolean isComplete(RepeatContext context,
                RepeatStatus result) {
        if(RepeatStatus.FINISHED == result) {
            return true;
        }
        else {
            return isComplete(context);
        }
    }

    @Override
    public boolean isComplete(RepeatContext context) {
        return this.totalProcessed >= chunksize;
    }

    @Override
    public RepeatContext start(RepeatContext parent) {
        this.chunksize = random.nextInt(20);
        this.totalProcessed = 0;

        System.out.println("The chunk size has been set to " +
                        this.chunksize);
        return parent;
    }

    @Override
    public void update(RepeatContext context) {
        this.totalProcessed++;
    }
}
```

代码清单 4-43　RandomChunkSizePolicy 的配置

```
...
@EnableBatchProcessing
@SpringBootApplication
public class ChunkJob {

    @Autowired
    private JobBuilderFactory jobBuilderFactory;
```

```java
@Autowired
private StepBuilderFactory stepBuilderFactory;

@Bean
public Job chunkBasedJob() {
    return this.jobBuilderFactory.get("chunkBasedJob")
                .start(chunkStep())
                .build();
}

@Bean
public Step chunkStep() {
    return this.stepBuilderFactory.get("chunkStep")
                .<String, String>chunk(randomCompletionPolicy())
                .reader(itemReader())
                .writer(itemWriter())
                .build();
}

@Bean
public ListItemReader<String> itemReader() {
    List<String> items = new ArrayList<>(100000);

    for (int i = 0; i < 100000; i++) {
            items.add(UUID.randomUUID().toString());
    }

    return new ListItemReader<>(items);
}

@Bean
public ItemWriter<String> itemWriter() {
    return items -> {
            for (String item : items) {
                    System.out.println(">> current item = " + item);
            }
    };
}

@Bean
public CompletionPolicy randomCompletionPolicy() {
    return new RandomChunkSizePolicy();
}

public static void main(String[] args) {
    SpringApplication.run(ChunkJob.class, args);
}
}
```

在执行代码清单 4-43 中的作业后，你可以在整个输出中看到每个新块在启动时打印出来的块大小，并且可以看到输出行之间的条目数量，以查看 CompletionPolicy 对块大小的影响。

在了解错误处理时，你将探索块配置的剩余部分，其中将涵盖有关重试和跳过逻辑的内容，剩下的大多数内容都将围绕这两部分展开。本章接下来将要介绍的关于步骤的内容也存在于作业中，比如步骤监听器。

6. 步骤监听器

本章在前面讨论作业监听器时，介绍过两个可触发的事件：作业的开始和结束。步骤监听器包含相似的事件类型(步骤的开始和结束)，但它们是针对独立的步骤而不是整个作业而言。本小节将介绍 org.springframework.batch.core.StepExecutionListener 和 org.springframework.batch.core.ChunkListener 接口，它们两者都允许在步骤和块的开始及结束部分进行逻辑处理。请注意，步骤监听器有两个名称：StepExecutionListener 和 StepListener。StepListener 只是一个标记接口，所有与步骤相关的接口都会扩展它。

StepExecutionListener 和 ChunkListener 提供了类似于 JobExecutionListener 接口中的方法。正如你所期待的那样，StepExecutionListener 提供了 beforeStep 和 afterStep 方法，ChunkListener 提供了 beforeChunk 和 afterChunk 方法。除了 afterStep 方法以外，其他所有方法的返回值都为空。afterStep 方法会返回 ExitStatus，因为步骤监听器可以在返回作业之前，自行修改步骤返回的 ExitStatus。当作业不只根据某个操作是否成功来确定处理是否成功时，这个特性非常有用，例如在导入文件后进行一些基本的完整性检查(判断写入数据库的记录数量是否正确等)。通过注解配置，步骤监听器能够保持一致性，可使用 Spring Batch 提供的@BeforeStep、@AfterStep、@BeforeChunk 和@AfterChunk 注解来简化实现。代码清单 4-44 展示的 StepExecutionListener 示例就使用注解来标识方法。

代码清单 4-44　记录步骤的开始和结束

```
...
public class LoggingStepStartStopListener {

    @BeforeStep
    public void beforeStep(StepExecution stepExecution) {
        System.out.println(stepExecution.getStepName() + " has begun!");
    }

    @AfterStep
    public ExitStatus afterStep(StepExecution stepExecution) {
        System.out.println(stepExecution.getStepName() + " has ended!");

        return stepExecution.getExitStatus();
    }
}
```

所有步骤监听器的配置都已在步骤的配置中合并为单个列表。代码清单 4-45 配置了你在前面编写的 LoggingStepStartStopListener。

代码清单 4-45　配置 LoggingStepStartStopListener

```
...
@EnableBatchProcessing
@SpringBootApplication
public class ChunkJob {

    @Autowired
    private JobBuilderFactory jobBuilderFactory;
```

```java
@Autowired
private StepBuilderFactory stepBuilderFactory;

@Bean
public Job chunkBasedJob() {
    return this.jobBuilderFactory.get("chunkBasedJob")
                .start(chunkStep())
                .build();
}

@Bean
public Step chunkStep() {
    return this.stepBuilderFactory.get("chunkStep")
                .<String, String>chunk(1000)
                .reader(itemReader())
                .writer(itemWriter())
                .listener(new LoggingStepStartStopListener())
                .build();
}

@Bean
public ListItemReader<String> itemReader() {
    List<String> items = new ArrayList<>(100000);

    for (int i = 0; i < 100000; i++) {
        items.add(UUID.randomUUID().toString());
    }

    return new ListItemReader<>(items);
}

@Bean
public ItemWriter<String> itemWriter() {
    return items -> {
        for (String item : items) {
            System.out.println(">> current item = " + item);
        }
    };
}

@Bean
public CompletionPolicy randomCompletionPolicy() {
    return new RandomChunkSizePolicy();
}

public static void main(String[] args) {
    SpringApplication.run(ChunkJob.class, args);
}
}
```

如你所见，在 Spring Batch 框架的几乎每个级别都可以使用监听器，以允许挂起批处理作业。它们通常不仅可用于执行组件的预处理或者评估某个组件的结果，还可用于错误处理。

到目前为止，尽管所有的步骤都是按顺序处理的，但这并不是 Spring Batch 的需求。接下来你将了解如何通过简单的逻辑来决定执行哪个步骤，以及如何将步骤流外部化(Externalize)以进行重用。

4.3.4 步骤流

到目前为止，我们对作业的处理都是针对单个文件进行的。你已经把步骤排好，并且一个接一个地执行。如果这是执行步骤的唯一方法，那么 Spring Batch 的功能将非常有限。事实是，Spring Batch 框架为你提供了一组健壮的选项来自定义作业的处理流程。

首先，我们介绍如何决定将要执行的下一个步骤，或者如何决定是否执行给定的步骤。这可以通过 Spring Batch 的条件逻辑来实现。

1. 条件逻辑

在 Spring Batch 的作业中，可通过使用 StepBuilder 的 next 方法来指定步骤的执行顺序。如果想修改执行顺序，那也很容易：使用转变(transition)即可。如代码清单 4-46 所示，可以通过构建器在正常情况下将作业从 firstStep 导向 successStep；或者在步骤返回的 ExitStatus 为 FAILED 时，将作业导向 failureStep。

代码清单 4-46　步骤执行中的 If/Else 逻辑

```
...
@EnableBatchProcessing
@SpringBootApplication
public class ConditionalJob {

    @Autowired
    private JobBuilderFactory jobBuilderFactory;

    @Autowired
    private StepBuilderFactory stepBuilderFactory;

    @Bean
    public Tasklet passTasklet() {
        return (contribution, chunkContext) -> {
            return RepeatStatus.FINISHED;
            //            throw new RuntimeException("This is a failure");
        };
    }

    @Bean
    public Tasklet successTasklet() {
        return (contribution, context) -> {
            System.out.println("Success!");
            return RepeatStatus.FINISHED;
        };
    }

    @Bean
    public Tasklet failTasklet() {
        return (contribution, context) -> {
            System.out.println("Failure!");
            return RepeatStatus.FINISHED;
        };
    }
```

第 4 章　理解作业和步骤

```java
@Bean
public Job job() {
    return this.jobBuilderFactory.get("conditionalJob")
                .start(firstStep())
                .on("FAILED").to(failureStep())
                .from(firstStep()).on("*").to(successStep())
                .end()
                .build();
}

@Bean
public Step firstStep() {
    return this.stepBuilderFactory.get("firstStep")
                .tasklet(passTasklet())
                .build();
}

@Bean
public Step successStep() {
    return this.stepBuilderFactory.get("successStep")
                .tasklet(successTasklet())
                .build();
}

@Bean
public Step failureStep() {
    return this.stepBuilderFactory.get("failureStep")
                .tasklet(failTasklet())
                .build();
}

public static void main(String[] args) {
    SpringApplication.run(ConditionalJob.class, args);
}
```

on 方法能让 Spring Batch 根据步骤的 ExitStatus(返回状态)来决定接下来该做什么。需要重点注意的是，本章前面出现过 org.springframework.batch.core.ExitStatus 和 org.springframework.batch.core.BatchStatus。其中，BatchStatus 是 JobExecution 或 StepExecution 的属性，用于标识作业或步骤的当前状态。ExitStatus 是当作业或步骤结束时返回给 Spring Batch 的值。Spring Batch 根据返回的 ExitStatus 来决定接下来怎么转变。所以，代码清单 4-46 中的示例相当于："如果 firstStep 的退出码不是 FAILED，那么执行 successStep，否则执行 failureStep。"

由于 ExitStatus 的值仅仅是字符串，因此可以通过使用通配符让事情变得更有意思。Spring Batch 允许在 on 条件中使用如下两个通配符。

- *：用于匹配一个或多个字符。例如，C*能够匹配 C、COMPLETE 以及 CORRECT。
- ?：用于匹配单个字符。在本例中，?AT 能够匹配 CAT 或 KAT，但不能匹配 THAT。

尽管根据 ExitStatus 可以决定接下来做什么，但这种方法并不总是奏效。例如，如果想在当前步骤中跳过所有记录，但又不想执行某个步骤的话，该怎么办呢？仅仅从 ExitStatus 是无法获知这种情况的。

81

> **注意** 在配置转变时,Spring Batch 能够提供帮助。Spring Batch 会自动地按照从最严格到最不严格的顺序对转变进行排序,并按照这种顺序应用它们。

Spring Batch 支持以编程的方式决定下一步做什么,方法是实现 org.springframework.batch.core.job.flow.JobExecutionDecider 接口。这个接口只有 decide 方法,该方法接收 JobExecution 和 StepExecution 参数并返回 FlowExecutionStatus(用于封装 BatchStatus 和 ExitStatus)。有了用于计算的 JobExecution 和 StepExecution,就有了用来决定作业下一步该做什么的所有信息。代码清单 4-47 展示的 JobExecutionDecider 实现可用来随机地决定下一个步骤应该是什么。

代码清单 4-47　RandomDecider

```
...
public class RandomDecider implements JobExecutionDecider {

    private Random random = new Random();

    public FlowExecutionStatus decide(JobExecution jobExecution,
            StepExecution stepExecution) {

        if (random.nextBoolean()) {
            return new
                FlowExecutionStatus(FlowExecutionStatus.COMPLETED.getName());
        } else {
            return new
                FlowExecutionStatus(FlowExecutionStatus.FAILED.getName());
        }
    }
}
```

为了使用 RandomDecider,需要为步骤配置名为 decider 的特性。decider 用来引用实现了 JobExecutionDecider 接口的 Spring Bean。代码清单 4-48 展示了配置好的 RandomDecider。可以看到,这里已将 decider 的返回值映射为可执行的步骤。

代码清单 4-48　步骤执行中的 If/Else 逻辑

```
...
@EnableBatchProcessing
@SpringBootApplication
public class ConditionalJob {

    @Autowired
    private JobBuilderFactory jobBuilderFactory;

    @Autowired
    private StepBuilderFactory stepBuilderFactory;

    @Bean
    public Tasklet passTasklet() {
        return (contribution, chunkContext) -> RepeatStatus.FINISHED;
    }

    @Bean
    public Tasklet successTasklet() {
```

```java
        return (contribution, context) -> {
            System.out.println("Success!");
            return RepeatStatus.FINISHED;
        };
    }

    @Bean
    public Tasklet failTasklet() {
        return (contribution, context) -> {
            System.out.println("Failure!");
            return RepeatStatus.FINISHED;
        };
    }

    @Bean
    public Job job() {
        return this.jobBuilderFactory.get("conditionalJob")
                    .start(firstStep())
                    .next(decider())
                    .from(decider())
                    .on("FAILED").to(failureStep())
                    .from(decider())
                    .on("*").to(successStep())
                    .end()
                    .build();
    }

    @Bean
    public Step firstStep() {
        return this.stepBuilderFactory.get("firstStep")
                    .tasklet(passTasklet())
                    .build();
    }

    @Bean
    public Step successStep() {
        return this.stepBuilderFactory.get("successStep")
                    .tasklet(successTasklet())
                    .build();
    }

    @Bean
    public Step failureStep() {
        return this.stepBuilderFactory.get("failureStep")
                    .tasklet(failTasklet())
                    .build();
    }

    @Bean
    public JobExecutionDecider decider() {
        return new RandomDecider();
    }

    public static void main(String[] args) {
        SpringApplication.run(ConditionalJob.class, args);
    }
}
```

现在，你已经知道了可以通过顺序或使用条件逻辑，把处理过程从一个步骤导向另一个步骤。但是，你并不总是想进入另一个步骤。你可能想要结束或暂停作业。

2. 结束作业

之前介绍过，JobInstance 只能成功一次，并且 JobInstance 是通过作业名称和传入的参数来标识的。正因为如此，如果想要通过编程的方式结束作业，那么你需要知道作业结束时的状态。事实上，在 Spring Batch 中以编程的方式结束作业时，作业会有以下三种状态。

- Completed(完成)：这个状态用来告诉 Spring Batch 作业处理已成功结束。一个 JobInstance 在完成后，就不能再以同样的参数重新运行。
- Failed(失败)：在这种情况下，作业并没有成功完成。Spring Batch 允许使用同样的参数重新执行处于失败状态的作业。
- Stopped(停止)：可以重启处于停止状态的作业。对于停止的作业，有意思的地方是，尽管没有发生过错误，但作业可以从停止的地方重启。对于在步骤之间有人工干预或者存在其他检查及处理的情况，停止状态特别有用。

务必注意，Spring Batch 是根据步骤的 ExitStatus 来决定持久化到 JobRepository 中的哪个 BatchStatus 的。ExitStatus 可以从步骤、块或作业中返回。BatchStatus 在 StepExecution 或 JobExecution 中维护，并且在 JobRepository 中持久化。下面介绍如何使用"完成"状态来结束作业。

为了根据步骤的退出状态将作业的结束状态配置为"完成"，需要使用构建器的 end 方法。在这种状态下，不能以同样的参数再次执行同一个作业。在代码清单 4-49 中，end 标签有一个特性，该特性声明了 ExitStatus 的值，从而将作业触发为结束状态。

代码清单 4-49　以"完成"状态结束作业

```
...
@EnableBatchProcessing
@SpringBootApplication
public class ConditionalJob {

    @Autowired
    private JobBuilderFactory jobBuilderFactory;

    @Autowired
    private StepBuilderFactory stepBuilderFactory;

    @Bean
    public Tasklet passTasklet() {
        return (contribution, chunkContext) -> {
                return RepeatStatus.FINISHED;
//              throw new RuntimeException("Causing a failure");
        };
    }

    @Bean
    public Tasklet successTasklet() {
        return (contribution, context) -> {
                System.out.println("Success!");
                return RepeatStatus.FINISHED;
        };
```

```
    }

    @Bean
    public Tasklet failTasklet() {
        return (contribution, context) -> {
                System.out.println("Failure!");
                return RepeatStatus.FINISHED;
        };
    }

    @Bean
    public Job job() {
        return this.jobBuilderFactory.get("conditionalJob")
                .start(firstStep())
                .on("FAILED").end()
                .from(firstStep()).on("*").to(successStep())
                .end()
                .build();
    }

    @Bean
    public Step firstStep() {
        return this.stepBuilderFactory.get("firstStep")
                .tasklet(passTasklet())
                .build();
    }

    @Bean
    public Step successStep() {
        return this.stepBuilderFactory.get("successStep")
                .tasklet(successTasklet())
                .build();
    }

    @Bean
    public Step failureStep() {
        return this.stepBuilderFactory.get("failureStep")
                .tasklet(failTasklet())
                .build();
    }

    @Bean
    public JobExecutionDecider decider() {
        return new RandomDecider();
    }

    public static void main(String[] args) {
        SpringApplication.run(ConditionalJob.class, args);
    }
}
```

一旦运行了 conditionalJob，就将如你期望的那样，BATCH_STEP_EXECUTION 表将包含步骤返回的 ExitStatus，而 BATCH_JOB_EXECUTION 表将包含 COMPLETED，而无论执行路径是什么。

对于"失败"状态，Spring Batch 允许使用相同的参数重新运行作业，配置与上面类似，但不使用 end 方法，而是使用 fail 方法，如代码清单 4-50 所示。

代码清单 4-50　以"失败"状态结束作业

```
...
@EnableBatchProcessing
@SpringBootApplication
public class ConditionalJob {

    @Autowired
    private JobBuilderFactory jobBuilderFactory;

    @Autowired
    private StepBuilderFactory stepBuilderFactory;

    @Bean
    public Tasklet passTasklet() {
        return (contribution, chunkContext) -> {
//            return RepeatStatus.FINISHED;
            throw new RuntimeException("Causing a failure");
        };
    }

    @Bean
    public Tasklet successTasklet() {
        return (contribution, context) -> {
            System.out.println("Success!");
            return RepeatStatus.FINISHED;
        };
    }

    @Bean
    public Tasklet failTasklet() {
        return (contribution, context) -> {
            System.out.println("Failure!");
            return RepeatStatus.FINISHED;
        };
    }

    @Bean
    public Job job() {
        return this.jobBuilderFactory.get("conditionalJob")
                .start(firstStep())
                .on("FAILED").fail()
                .from(firstStep()).on("*").to(successStep())
                .end()
                .build();
    }

    @Bean
    public Step firstStep() {
        return this.stepBuilderFactory.get("firstStep")
                .tasklet(passTasklet())
                .build();
    }
```

```
@Bean
public Step successStep() {
    return this.stepBuilderFactory.get("successStep")
                .tasklet(successTasklet())
                .build();
}

@Bean
public Step failureStep() {
    return this.stepBuilderFactory.get("failureStep")
                .tasklet(failTasklet())
                .build();
}

@Bean
public JobExecutionDecider decider() {
    return new RandomDecider();
}

public static void main(String[] args) {
    SpringApplication.run(ConditionalJob.class, args);
}
}
```

当使用代码清单 4-50 中的配置重新执行 conditionalJob 作业时，结果会略有不同。此时，如果 firstStep 以 FAILURE ExitStatus 结束，那么这个作业在 jobRepository 中会被标识为"失败"，这意味着可以使用相同的参数重新执行这个作业。

当以编程方式结束作业时，作业的最后一种状态是"停止"。在这种情况下，可以重启作业，作业将从配置的步骤那里重启，如代码清单 4-51 所示。

代码清单 40-51　以"停止"状态结束作业

```
...
@EnableBatchProcessing
@SpringBootApplication
public class ConditionalJob {
private JobBuilderFactory jobBuilderFactory;

    @Autowired
    private StepBuilderFactory stepBuilderFactory;

    @Bean
    public Tasklet passTasklet() {
        return (contribution, chunkContext) -> {
//            return RepeatStatus.FINISHED;
            throw new RuntimeException("Causing a failure");
        };
    }

    @Bean
    public Tasklet successTasklet() {
        return (contribution, context) -> {
            System.out.println("Success!");
            return RepeatStatus.FINISHED;
```

87

```java
            };
        }

        @Bean
        public Tasklet failTasklet() {
            return (contribution, context) -> {
                System.out.println("Failure!");
                return RepeatStatus.FINISHED;
            };
        }

        @Bean
        public Job job() {
            return this.jobBuilderFactory.get("conditionalJob")
                    .start(firstStep())
                    .on("FAILED").stopAndRestart(successStep())
                    .from(firstStep())
                    .on("*").to(successStep())
                    .end()
                    .build();
        }

        @Bean
        public Step firstStep() {
            return this.stepBuilderFactory.get("firstStep")
                    .tasklet(passTasklet())
                    .build();
        }

        @Bean
        public Step successStep() {
            return this.stepBuilderFactory.get("successStep")
                    .tasklet(successTasklet())
                    .build();
        }

        @Bean
        public Step failureStep() {
            return this.stepBuilderFactory.get("failureStep")
                    .tasklet(failTasklet())
                    .build();
        }

        @Bean
        public JobExecutionDecider decider() {
            return new RandomDecider();
        }

        public static void main(String[] args) {
            SpringApplication.run(ConditionalJob.class, args);
        }
    }
```

当使用以上最终配置执行 conditionalJob 时，可以使用相同的参数重新执行这个作业。但是这一次，如果进入 FAILED 执行路径，那么在重启作业时，会从 successStep 开始执行。

作业的配置可能非常复杂，从步骤到步骤的执行流程并不仅仅是作业配置的又一层配置，执行流程也可在可重用的组件中进行配置。接下来将讨论如何把步骤流封装到可重用的组件中。

3. 外部化流

你已经知道，步骤可以定义为 Bean，这样就可以从给定的作业中抽取步骤的定义，使其成为可重用的组件。步骤的顺序也是如此。在 Spring Batch 中，外部化(Externalize)步骤顺序的方式有三种。第一种方式是创建步骤流，步骤流是独立的步骤序列。第二种方式是使用 FlowStep，尽管配置类似，但是持久化到 JobRepository 中的状态略有不同。最后一种方式实际上是在一个作业中调用另一个作业。

下面首先介绍第一种方式。流(Flow)与作业类似，并且使用与作业类似的配置方式。代码清单 4-52 展示了使用流构建器定义流的方式：为流构建器传递 id，并且在作业中引用。

代码清单 4-52　定义流

```
...
@EnableBatchProcessing
@SpringBootApplication
public class FlowJob {

    @Autowired
    private JobBuilderFactory jobBuilderFactory;

    @Autowired
    private StepBuilderFactory stepBuilderFactory;

    @Bean
    public Tasklet loadStockFile() {
        return (contribution, chunkContext) -> {
            System.out.println("The stock file has been loaded");
            return RepeatStatus.FINISHED;
        };
    }

    @Bean
    public Tasklet loadCustomerFile() {
        return (contribution, chunkContext) -> {
            System.out.println("The customer file has been loaded");
            return RepeatStatus.FINISHED;
        };
    }

    @Bean
    public Tasklet updateStart() {
        return (contribution, chunkContext) -> {
            System.out.println("The start has been updated");
            return RepeatStatus.FINISHED;
        };
    }

    @Bean
    public Tasklet runBatchTasklet() {
        return (contribution, chunkContext) -> {
            System.out.println("The batch has been run");
```

```java
                return RepeatStatus.FINISHED;
            };
    }

    @Bean
    public Flow preProcessingFlow() {
        return new FlowBuilder<Flow>("preProcessingFlow").start(loadFileStep())
                    .next(loadCustomerStep())
                    .next(updateStartStep())
                    .build();
    }

    @Bean
    public Job conditionalStepLogicJob() {
        return this.jobBuilderFactory.get("conditionalStepLogicJob")
                    .start(preProcessingFlow())
                    .next(runBatch())
                    .end()
                    .build();
    }

    @Bean
    public Step loadFileStep() {
        return this.stepBuilderFactory.get("loadFileStep")
                    .tasklet(loadStockFile())
                    .build();
    }

    @Bean
    public Step loadCustomerStep() {
        return this.stepBuilderFactory.get("loadCustomerStep")
                    .tasklet(loadCustomerFile())
                    .build();
    }

    @Bean
    public Step updateStartStep() {
        return this.stepBuilderFactory.get("updateStartStep")
                    .tasklet(updateStart())
                    .build();
    }

    @Bean
    public Step runBatch() {
        return this.stepBuilderFactory.get("runBatch")
                    .tasklet(runBatchTasklet())
                    .build();
    }

    public static void main(String[] args) {
        SpringApplication.run(HelloWorldJob.class, args);
    }
}
```

当执行包含流的作业并查看 JobRepository 时,你就会看到流中的步骤是作为作业的一部分被记录的,就像它们之前在步骤中配置的那样。最终,无论是使用流配置步骤,还是在作业中配置步骤,在 JobRepository 看来,它们并无二致。

外部化步骤顺序的第二种方式是使用流步骤。在这种方式下,流的配置是相同的,但并不是通过把流传给 JobBuilder 来配置流的执行,而是将流封装到一个步骤中,并把这个步骤传给 JobBuilder。代码清单 4-53 展示了如何使用 FlowStep 来配置代码清单 4-52 中的示例。

代码清单 4-53　使用 FlowStep

```
...
@EnableBatchProcessing
@SpringBootApplication
public class FlowJob {

    @Autowired
    private JobBuilderFactory jobBuilderFactory;

    @Autowired
    private StepBuilderFactory stepBuilderFactory;

    @Bean
    public Tasklet loadStockFile() {
        return (contribution, chunkContext) -> {
                System.out.println("The stock file has been loaded");
                return RepeatStatus.FINISHED;
        };
    }

    @Bean
    public Tasklet loadCustomerFile() {
        return (contribution, chunkContext) -> {
                System.out.println("The customer file has been loaded");
                return RepeatStatus.FINISHED;
        };
    }

    @Bean
    public Tasklet updateStart() {
        return (contribution, chunkContext) -> {
                System.out.println("The start has been updated");
                return RepeatStatus.FINISHED;
        };
    }

    @Bean
    public Tasklet runBatchTasklet() {
        return (contribution, chunkContext) -> {
                System.out.println("The batch has been run");
                return RepeatStatus.FINISHED;
        };
    }

    @Bean
    public Flow preProcessingFlow() {
        return new FlowBuilder<Flow>("preProcessingFlow").start(loadFileStep())
```

```
                .next(loadCustomerStep())
                .next(updateStartStep())
                .build();
    }

    @Bean
    public Job conditionalStepLogicJob() {
        return this.jobBuilderFactory.get("conditionalStepLogicJob")
                .start(intializeBatch())
                .next(runBatch())
                .build();
    }
    @Bean
    public Step intializeBatch() {
        return this.stepBuilderFactory.get("initalizeBatch")
                .flow(preProcessingFlow())
                .build();
    }

    @Bean
    public Step loadFileStep() {
        return this.stepBuilderFactory.get("loadFileStep")
                .tasklet(loadStockFile())
                .build();
    }

    @Bean
    public Step loadCustomerStep() {
        return this.stepBuilderFactory.get("loadCustomerStep")
                .tasklet(loadCustomerFile())
                .build();
    }

    @Bean
    public Step updateStartStep() {
        return this.stepBuilderFactory.get("updateStartStep")
                .tasklet(updateStart())
                .build();
    }

    @Bean
    public Step runBatch() {
        return this.stepBuilderFactory.get("runBatch")
                .tasklet(runBatchTasklet())
                .build();
    }

    public static void main(String[] args) {
        SpringApplication.run(HelloWorldJob.class, args);
    }
}
```

那么，把流传给 JobBuilder 和 FlowStep 有何区别？这要看在 JobRepository 中发生了什么。为 JobBuilder 使用 flow 方法得到的结果，与通过在作业中配置步骤得到的结果相同。使用 FlowStep 会添加一条附加记录。当使用 FlowStep 时，Spring Batch 会把包含流的步骤作为单独的步骤来记录。这有什么好处？主要的好处在于可以实现监控和报告的目的。使用 FlowStep 就可以查看将流作为整体产生的影响，否则就必须聚合各个独立的步骤。

外部化步骤顺序的最后一种方式是完全不外部化它们。在这种情况下，可以在一个作业中调用另一个作业，而不是创建流。FlowStep 会创建 StepExecutionContext 来执行流以及其中的步骤，JobStep 与之类似，JobStep 会为步骤创建 JobExecutionContext 并调用外部作业。代码清单 4-54 展示了 JobStep 的配置。

代码清单 4-54　使用 JobStep

```
...
@EnableBatchProcessing
@SpringBootApplication
public class JobJob {

    @Autowired
    private JobBuilderFactory jobBuilderFactory;

    @Autowired
    private StepBuilderFactory stepBuilderFactory;

    @Bean
    public Tasklet loadStockFile() {
        return (contribution, chunkContext) -> {
                System.out.println("The stock file has been loaded");
                return RepeatStatus.FINISHED;
        };
    }

    @Bean
    public Tasklet loadCustomerFile() {
        return (contribution, chunkContext) -> {
                System.out.println("The customer file has been loaded");
                return RepeatStatus.FINISHED;
        };
    }

    @Bean
    public Tasklet updateStart() {
        return (contribution, chunkContext) -> {
                System.out.println("The start has been updated");
                return RepeatStatus.FINISHED;
        };
    }

    @Bean
    public Tasklet runBatchTasklet() {
        return (contribution, chunkContext) -> {
                System.out.println("The batch has been run");
                return RepeatStatus.FINISHED;
        };
    }

    @Bean
    public Job preProcessingJob() {
```

```java
        return this.jobBuilderFactory.get("preProcessingJob")
                    .start(loadFileStep())
                    .next(loadCustomerStep())
                    .next(updateStartStep())
                    .build();
    }

    @Bean
    public Job conditionalStepLogicJob() {
        return this.jobBuilderFactory.get("conditionalStepLogicJob")
                    .start(intializeBatch())
                    .next(runBatch())
                    .build();
    }

    @Bean
    public Step intializeBatch() {
        return this.stepBuilderFactory.get("initalizeBatch")
                    .job(preProcessingJob())
                    .parametersExtractor(new DefaultJobParametersExtractor())
                    .build();
    }

    @Bean
    public Step loadFileStep() {
        return this.stepBuilderFactory.get("loadFileStep")
                    .tasklet(loadStockFile())
                    .build();
    }

    @Bean
    public Step loadCustomerStep() {
        return this.stepBuilderFactory.get("loadCustomerStep")
                    .tasklet(loadCustomerFile())
                    .build();
    }

    @Bean
    public Step updateStartStep() {
        return this.stepBuilderFactory.get("updateStartStep")
                    .tasklet(updateStart())
                    .build();
    }

    @Bean
    public Step runBatch() {
        return this.stepBuilderFactory.get("runBatch")
                    .tasklet(runBatchTasklet())
                    .build();
    }

    public static void main(String[] args) {
        SpringApplication.run(HelloWorldJob.class, args);
    }
}
```

第 4 章　理解作业和步骤

你可能想知道代码清单 4-54 中的 jobParametersExtractor 是什么。当启动作业时，作业是通过作业的名称和参数来识别的。在这里，我们并没有手动给子作业(preProcessingJob)传递参数；而是定义了一个类，用于从父作业的 JobParameters 或 ExecutionContext(默认的 DefaultJobParameterExtractor 会检查这两个地方)中抽取参数，并把这些参数传给子作业。抽取器(Extractor)可以从作业参数 job.stockFile 和 job.customerFile 中获取值，将它们作为参数传给 preProcessingJob。

当 preProcessingJob 执行时，这个子作业会使用与其他作业一样的方式在 JobRepository 中得到标识。preProcessingJob 有自己的作业实例、执行上下文以及相关的数据库记录。

■ 注意　为了运行代码清单 4-54 中的示例，你必须在 application.properties 中配置属性 spring.batch.job.names 为 conditionalStepLogicJob，以阻止 Spring Boot 在启动时自动执行 preProcessingJob。

关于使用 JobStep 方式的警告是：这看起来似乎是处理作业的一种不错的方式，在创建独立的作业后，再使用主作业将它们串起来，但这可能会严重限制对处理过程的控制。在真实世界中，暂停批处理或者基于外部因素跳过作业的情况并不罕见(例如，其他部门不能按时提供文件，因而无法在要求的时间窗口内完成处理)。管理作业的功能只存在于单作业级别，当管理使用此功能创建的整个作业树时会存在问题，所以应该避免使用。以这种方式将作业连接在一起，并将它们作为主作业执行，会严重限制你处理此类情况的能力，所以也应该避免使用。

4.4　本章小结

本章涵盖大量的内容。你首先学习了作业是什么，并了解了作业的生命周期，你还掌握了如何配置作业，以及如何通过作业参数与之交互。接下来，你学习了如何编写和配置作业监听器，在作业开始和结束时执行逻辑，以及如何使用作业和步骤的执行上下文。

步骤是作业的构建块。你在学习步骤的同时，探索了 Spring Batch 中最重要的概念之一：基于块的处理。你学习了如何配置块以及用来控制它们的一些高级方式(通过策略等)。你还了解了步骤监听器，以及如何在步骤开始和结束时使用它们执行逻辑。最后，你了解了如何对步骤进行排序：使用基本排序或逻辑定义来决定下一步应该执行什么步骤。

作业和步骤是 Spring Batch 框架中的结构化组件，它们用于设计流程。本书后续大部分内容将涵盖构成这些结构化组件的所有不同部分。

第 5 章

作业存储库和元数据

当编写有限的处理过程时，以没有 UI 的独立方式(stand-alone)执行处理过程并非难事。Spring Boot 的 CommandLineRunner 允许开发人员编写执行单个函数的 Spring Boot 应用，由于这个函数可以包含你所能想到的任何业务逻辑并执行它们，因此我们并不需要 Spring Batch 做这些事情。

然而，事情变得有趣的地方，恰恰就是出错的地方。如果批处理作业在运行时发生了错误，那么如何恢复？作业如何能知道在错误发生时进行到了哪一步，在作业重启时又会发生什么？状态管理是处理大量数据时必然经历的阶段，并且也是 Spring Batch 提供的重要特性之一。本书在前面提到过，Spring Batch 在作业存储库中维护作业的运行状态。当重启作业或重试某个条目时，可使用这些信息来决定如何继续。

作业存储库有助于批处理的另一个方面是监控。在企业环境中，能够查看作业进行到哪一步以及发现一些趋势因素(例如运维时长、因为发生错误而重试的条目数量等)是至关重要的。事实上，Spring Batch 收集了这些信息，以便能够更容易地完成这种趋势统计。

本章将详细地介绍作业存储库。你将首先学习如何通过使用关系数据库或内存存储库来为大多数环境配置作业存储库。在配置了作业存储库之后，你将学习如何利用 JobExplorer 和 JobOperator 来使用作业存储库中存储的作业信息。

5.1 作业存储库是什么

在 Spring Batch 中提到作业存储库时，可能是在引用 JobRepository 接口或由这个接口使用的用于持久化数据的数据存储。除了可能配置 JobRepository 的实例之外，用户几乎不会与这个接口发生交互，所以本章将通过 JobRepository 的一个实现，重点介绍 Spring Batch 提供的数据存储。Spring Batch 提供了两个开箱即用的用于批处理作业的数据存储，它们分别是内存数据库和关系数据库。下面首先介绍关系数据库。

5.1.1 使用关系数据库

在 Spring Bath 中，作业存储库默认是关系数据库，这样就可以利用 Spring Batch 提供的一组数据表来持久化批处理中的元数据。下面看看图 5-1 所示的作业存储库模式。

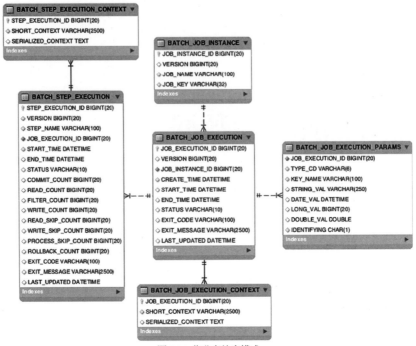

图 5-1　作业存储库模式

如图 5-1 所示,作业存储库中包含如下 6 个表:

- BATCH_JOB_INSTANCE
- BATCH_JOB_EXECUTION
- BATCH_JOB_EXECUTION_PARAMS
- BATCH_JOB_EXECUTION_CONTEXT
- BATCH_STEP_EXECUTION
- BATCH_STEP_EXECUTION_CONTEXT

图 5-1 所示的作业存储库模式实际上是从 BATCH_JOB_INSTANCE 表开始的。如前所述,当第一次执行作业时,会使用一组唯一的识别性参数创建一个作业实例。这个作业实例代表作业的一次逻辑运行。表 5-1 展示了 BATCH_JOB_INSTANCE 表的组成部分。

表 5-1　BATCH_JOB_INSTANCE 表

字　　段	描　　述
JOB_EXECUTION_ID	表的主键
VERSION	用于乐观锁[1]的数据记录的版本
JOB_NAME	作业的名称
JOB_KEY	作业名和识别性参数的哈希值,用于唯一地表示一个作业实例

---

[1] 如果想了解有关领域驱动设计中的版本和实体的更多信息,请阅读 Eric Evans 的 *Domain Driven Design* 一书(Addison-Wesley 已于 2003 年出版)。

BATCH_JOB_EXECUTION 表中的记录代表批处理作业的每一次物理运行。每次启动作业时，都会创建一条新的记录，并且在执行作业的过程中会定期地更新这条记录。表 5-2 展示了 BATCH_JOB_EXECUTION 表的组成部分。

表 5-2　BATCH_JOB_EXECUTION 表

字　　段	描　　述
JOB_EXECUTION_ID	表的主键
VERSION	用于乐观锁的数据记录的版本
JOB_INSTANCE_ID	指向 BATCH_JOB_INSTANCE 表的外键
CREATE_TIME	记录创建的时间
START_TIME	作业执行的开始时间
END_TIME	作业执行的完成时间
STATUS	作业执行的批处理状态
EXIT_CODE	作业执行的退出码
EXIT_MESSAGE	与 EXIT_CODE 关联的消息(可能是栈跟踪信息)
LAST_UPDATED	最后一次更新记录的时间

与 BATCH_JOB_EXECUTION 表相关联的表有 3 个。首先是 BATCH_JOB_EXECUTION_CONTEXT 表。第 4 章已经介绍过 ExecutionContext 及其如何用来保存状态。但为了能在执行中有用(例如在启动时)，Spring Batch 需要将 ExecutionContex 持久化，而 BATCH_JOB_EXECUTION_CONTEXT 表就是持久化 JobExecution 的 ExecutionContext 的地方。表 5-3 展示了 BATCH_JOB_EXECUTION_CONTEXT 表的组成部分。

表 5-3　BATCH_JOB_EXECUTION_CONTEXT 表

字　　段	描　　述
JOB_EXECUTION_ID	表的主键
SHORT_CONTEXT	SERIALIZED_CONTEXT 的缩减版本
SERIALIZED_CONTEXT	真正的序列化之后的 ExecutionContext

注意这里有几种用于序列化 ExecutionContext 的方法。在 Spring Batch 4 之前，XStream 的 JSON 工具是首选的方法，也是 Spring Batch 框架默认使用的方法。然而，在 Spring Batch 4 出现后，XStream 的 JSON 支持并不能完全满足 Spring Batch 的需要，因此默认使用的方法被修改为 Jackson 2。本章后面将介绍如何自定义 ExecutionContext 的序列化配置。

与 BATCH_JOB_EXECUTION 表相关联的第二个表是 BATCH_JOB_EXECUTION_PARAMS 表。这个表中保存了每次执行作业时的参数。前面提到过，识别性的作业参数用于决定一次运行是否需要新的作业实例。然而，BATCH_JOB_EXECUTION_PARAMS 表实际上保存了传给作业的所有参数。在重启时，只有识别性参数会被自动传入。表 5-4 展示了 BATCH_JOB_EXECUTION_PARAMS 表的组成部分。

表 5-4 BATCH_JOB_EXECUTION_PARAMS 表

字段	描述
JOB_EXECUTION_ID	表的主键
TYPE_CODE	一个字符串，用来表明参数值的类型
KEY_NAME	参数的名称
STRING_VAL	当类型是 String 时的参数值
DATE_VAL	当类型是 Date 时的参数值
LONG_VAL	当类型是 long 时的参数值
DOUBLE_VAL	当类型是 double 时的参数值
IDENTIFYING	一个标志，用来表明某个参数是否为识别性参数

除了用于定义作业的元数据的表之外，作业存储库还使用了另外两个表。这两个表用来保存步骤的元数据。第一个表是 BATCH_STEP_EXECUTION，它不仅负责维护步骤的开始、结束和状态等元数据(与用于作业的 BATCH_JOB_EXECUTION 表相同)，还用于持久化各种计数，以便对步骤进行分析，比如持久化读取的数量、处理的数量、写入的数量、跳过的数量等数据点。表 5-5 展示了 BATCH_STEP_EXECUTION 表的组成部分。

表 5-5 BATCH_STEP_EXECUTION 表

字段	描述
STEP_EXECUTION_ID	表的主键
VERSION	用于乐观锁的数据记录的版本
STEP_NAME	步骤的名称
JOB_EXECUTION_ID	指向 BATCH_JOB_EXECUTION 表的外键
START_TIME	步骤执行的开始时间
END_TIME	步骤执行的完成时间
STATUS	步骤的批处理状态
COMMIT_COUNT	在步骤执行中提交的事务数量
READ_COUNT	读入的条目数量
FILTER_COUNT	由于 ItemProcessor 返回为 null 而过滤掉的条目数量
WRITE_COUNT	写入的条目数量
READ_SKIP_COUNT	由于在 ItemReader 中抛出异常而被跳过的条目数量
PROCESS_SKIP_COUNT	由于在 ItemProcessor 中抛出异常而被跳过的条目数量
WRITE_SKIP_COUNT	由于在 ItemWriter 中抛出异常而被跳过的条目数量
ROLLBACK_COUNT	在步骤执行中被回滚的事务数量
EXIT_CODE	步骤的退出码
EXIT_MESSAGE	步骤执行返回的消息或栈跟踪信息
LAST_UPDATED	最后一次更新记录的时间

作业存储库使用的最后一个表是 BATCH_STEP_EXECUTION_CONTEXT 表。就像 JobExecution

有用于存储组件状态的 ExecutionContext 一样,StepExecution 也有用于以上目的的 ExecutionContext。StepExecution 的 ExecutionContext 用来存储步骤级别的组件状态。在深入研究诸如 ItemReader 和 ItemWriter 的组件时,我们将详细地介绍具体用法。表 5-6 展示了 BATCH_STEP_EXECUTION_CONTEXT 表的组成部分。

表 5-6 BATCH_STEP_EXECUTION_CONTEXT 表

字 段	描 述
STEP_EXECUTION_ID	表的主键
SHORT_CONTEXT	SERIALIZED_CONTEXT 的缩减版本
SERIALIZED_CONTEXT	真正序列化的 ExecutionContext

5.1.2 使用内存存储库

在开发 Spring Batch 作业或运行单元测试时,因配置额外的数据库带来的麻烦可能大于得到的好处。正因为如此,Spring Batch 提供了使用 java.util.Map 实例作为数据存储的 JobRepository 实现。稍后我们将介绍如何配置这种类型的 JobRepository 以及如何自定义数据库配置。

■ 注意 基于 Map 的 JobRepository 并不适用于生产环境。如果期望在运行批处理作业时不使用外部数据库,那么应该使用有较好的多线程和事务支持能力的内存数据库(例如 H2 或 HSQLDB)。

5.2 配置批处理基础设施

在使用了 @EnableBatchProcessing 注解后,Spring Batch 提供了一种开箱即用的、不需要额外配置的 JobRepository。然而,很多时候我们需要自定义 JobRepository。本节将介绍 Spring Batch 基础设施的所有自定义事宜,包括使用 BatchConfigurer 接口配置 JobRepository。

5.2.1 BatchConfigurer 接口

BatchConfigurer 作为策略接口提供了自定义 Spring Batch 基础设施组件的能力。在使用了 @EnableBatchProcessing 注解后,即可获取每个 Spring Batch 基础设施组件的实例。从本质上讲,除了提供一些 Bean,这个注解还做了以下工作:首先通过 BatchConfigurer 接口的实现来创建这些 Bean,然后通过 SimpleBatchConfiguration 将它们添加到 Spring ApplicationContext 中。在绝大多数情况下,并不需要直接使用 SimpleBatchConfiguration。然而,通过 BatchConfigurer 可以自定义暴露出来的组件。下面首先介绍 BatchConfigurer 接口本身,参见代码清单 5-1。

代码清单 5-1 BatchConfigurer 接口

```
public interface BatchConfigurer {

    JobRepository getJobRepository() throws Exception;

    PlatformTransactionManager getTransactionManager() throws Exception;
```

```
JobLauncher getJobLauncher() throws Exception;

JobExplorer getJobExplorer() throws Exception;
}
```

BatchConfigurer 接口的每个方法都为 Spring Batch 基础设施提供了一个主要组件。前面已经讨论过 JobRepository 和 JobLauncher 是什么。在 Spring Batch 框架提供的所有事务管理中，都会使用 BatchConfigurer 接口提供的 PlatformTransactionManager。最后，JobExplorer 提供了针对作业存储库中所保存数据的只读视图。

大多数时候，你并不需要自行实现整个接口。Spring Batch 使用 DefaultBatchConfigurer 提供了这些组件的所有默认选项。在典型情况下，只需要重写一两个组件的配置即可。因此通常情况下，扩展 DefaultBatchConfigurer 并重写合适的方法会更容易一些。下面看看自定义 BatchConfigurer 所提供组件的一些常见方法，以及自定义这些组件的原因。

5.2.2 自定义 JobRepository

JobRepository 可通过 FactoryBean 来创建，毫不意外，名称是 JobRepositoryFactoryBean，可以使用的设置器如表 5-7 所示。

表 5-7 配置 JobRepositoryFactoryBean

设置器的名称	描述
setClobType(int type)	使用 java.sql.Type 类型的值来指明用于 CLOB 列的类型
setDatabaseType(String dbType)	配置数据库的类型。通常不需要设置，因为 Spring Batch 能够自动地尝试识别数据库的类型
setDataSource(DataSource dataSource)	与 JobRepository 一起使用的数据源
setIncrementerFactory(DataFieldMaxValueIncrementerFactory incrementerFactory)	用于递增大多数表的主键
setIsolationLevelForCreate(String isoltationLevelForCreate)	当创建 JobExecution 实例时使用的事务序列化级别，默认是 ISOLATION_SERIALIZABLE
setJdbcOperations(JdbcOperations jdbcTemplate)	用于设置 JdbcOperations 实例。如果没有提供，就使用相关的 setter 方法
setLobHandler(LobHandler lobHandler)	仅在需要特殊处理 LOB 的旧版 Oracle 中使用
setMaxVarCharLength(int maxLength)	用于裁剪退出消息(包括步骤和作业)和缩减版执行上下文列的长度。除非修改 Spring Batch 提供的模式，否则不用设置
setSerializer(ExecutionContextSerializer serializer)	配置要使用的 ExecutionContextSerializer 接口的实现，用于序列化 JobExecution 和 StepExecution 的 ExecutionContext
setTablePrefix(String tablePrefix)	除了默认的 BATCH_ 之外，允许用户为所有的表配置新的前缀
setTransactionManager(PlatformTransactionManager transactionManager)	当使用多个数据库时，为了保持数据库同步，需要使用支持两阶段提交的事务管理器(Transaction Manager)
setValidateTransactionState(boolean validateTransactionState)	一个布尔标志，用于指明在创建 JobExecution 时是否检查现有事务。默认为 true，因为通常会出错

扩展 DefaultBatchConfigurer 并重写 createJobRepository()方法的最常见场景发生在 ApplicationContext

有多个 DataSource 时。例如，如果有两个 DataSource，它们分别用于业务数据和 JobRepository，那么需要显式地配置 JobRepository 所使用的 DataSource。代码清单 5-2 展示了一个自定义的 JobRepository，它是通过扩展 DefaultBatchConfigurer 并重写 createJobRepository() 方法来实现的。

代码清单 5-2　一个自定义的 JobRepository

```
...
public class CustomBatchConfigurer extends DefaultBatchConfigurer {

    @Autowired
    @Qualifier("repositoryDataSource")
    private DataSource dataSource;

    @Override
    protected JobRepository createJobRepository() throws Exception {
        JobRepositoryFactoryBean factoryBean = new JobRepositoryFactoryBean();

        factoryBean.setDatabaseType(DatabaseType.MYSQL.getProductName());
        factoryBean.setTablePrefix("FOO_");
        factoryBean.setIsolationLevelForCreate("ISOLATION_REPEATABLE_READ");
        factoryBean.setDataSource(this.dataSource);

        factoryBean.afterPropertiesSet();

        return factoryBean.getObject();
    }
}
```

浏览代码清单 5-2，你会看到 CustomBatchConfigurer 扩展了 DefaultBatchConfigurer，因此无须重新实现接口的所有方法。我们装入(autowire)一个 DataSource，将其命名为 repositoryDataSource。在这里，代码清单 5-2 假定在 ApplicationContext 的其他地方也有一个类型为 DataSource 且名称为 repositoryDataSource 的可装入 Bean。然后，你可以看到 DefaultBatchConfigurer#createJobRepository() 被重写了，DefaultBatchConfigurer 实际上用它来创建真正的 JobRepository。在这里的实现中，我们还创建了一个 JobRepository，不过也自定义了一些默认设置，特别是：指定了数据库类型，配置了数据表的前缀为 FOO_ 而不是默认的 BATCH_，设置了用于创建实体的事务隔离级别为 ISOLATION_REPEATABLE_READ 而不是默认的 ISOLATION_SERIALIZED，最后还设置了刚才装入的 DataSource。

需要特别注意的是，Spring 容器并不会直接调用 DefaultBatchConfigurer 的任何 create*方法用于定义 Bean。正因为如此，我们需要调用 InitializingBean#afterPropertiesSet()和 FactoryBean#getObject() (Spring 容器通常会为我们做这两件事)。

JobRepository 并不是能够通过 BatchConfigurer 进行自定义的唯一组件。通常，用于批处理应用的事务管理器 TransactionManager 是由 Spring Boot 提供的，因此并不需要我们做很多事情。然而，如果想自定义事务管理器或者应用包含了多个事务管理器，就要使用 BatchConfigurer 来指定具体使用哪一个。5.2.3 节将介绍事务管理器是如何工作的。

5.2.3　自定义 TransactionManager

Spring Batch 需要重度地使用事务。Spring Batch 把事务作为核心组件，因为 TransactionManager 是 Spring Batch 框架中的重要组件之一。查看 TransactionManager 的所有配置选项已超出本书的讨论范畴。代码清单 5-3 展示了如何通过扩展 DefaultBatchConfigurer 来指定使用哪一个 TransactionManager，这是十分有用的。

代码清单 5-3　自定义 TransactionManager

```
...
public class CustomBatchConfigurer extends DefaultBatchConfigurer {

    @Autowired
    @Qualifier("batchTransactionManager")
    private PlatformTransactionManager transactionManager;

    @Override
    public PlatformTransactionManager getTransactionManager() {
        return this.transactionManager;
    }
}
```

在代码清单 5-3 中，对 BatchConfigurer#getTransactionManager()方法的调用显式地返回了一个 PlatformTransactionManager，这个事务管理器已在其他地方定义，并且可由批处理过程使用。你将注意到，这里并没有重写 DefaultBatchConfigurer 中的被保护的方法。这是因为，如果没有创建过 DataSourceTransactionManager，那么默认情况下 DefaultBatchConfigurer 会在 DataSource 的 setter 方法中创建 DataSourceTransactionManager。这是 BatchConfigurer 提供的以这种方式处理的唯一组件。

5.2.4　自定义 JobExplorer

JobRepository 提供了用于持久化以及从保存作业状态的底层数据库中获取数据的 API。但我们有时只想对外暴露这些数据的只读视图。JobExplorer 正好提供了批处理中元数据的只读视图。

由于只读视图 JobExplorer 中的数据与 JobRepository 操作的数据相同，因此底层的数据访问层实际上是相同的。JobRepository 和 JobExplorer 共享了很多常见的数据访问对象(DAO)。因此，在用于自定义 JobRepository 和 JobExplorer 的选项中，所有涉及读取数据库的属性都是相同的。表 5-8 列出了 JobExplorerFactoryBean 可以使用的设置器。

表 5-8　配置 JobExplorerFactoryBean

设置器的名称	描述
setDataSource(DataSource dataSource)	与 JobRepository 一起使用的数据源
setJdbcOperations(JdbcOperations jdbcTemplate)	用于设置 JdbcOperations 实例。如果没有提供，就使用与数据源相关的 setter 方法
setLobHandler(LobHandler lobHandler)	仅在需要特殊处理 LOB 的旧版 Oracle 中使用
setSerializer(ExecutionContextSerializer serializer)	配置要使用的 ExecutionContextSerializer 接口的实现，用于序列化 JobExecution 和 StepExecution 的 ExecutionContext
setTablePrefix(String tablePrefix)	为数据表配置新的前缀(默认是 BATCH_)

代码清单 5-4 展示了一个自定义的 JobExplorer，它与代码清单 5-2 中配置的 JobRepository 对应。

代码清单 5-4　自定义 JobExplorer

```
...
public class CustomBatchConfigurer extends DefaultBatchConfigurer {

    @Autowired
    @Qualifier("batchTransactionManager")
    private DataSource dataSource;

    @Override
    protected JobExplorer createJobExplorer() throws Exception {
        JobExplorerFactoryBean factoryBean = new JobExplorerFactoryBean();

        factoryBean.setDataSource(this.dataSource);
        factoryBean.setTablePrefix("FOO_");

        factoryBean.afterPropertiesSet();

        return factoryBean.getObject();
    }
}
```

与代码清单 5-2 中自定义 JobRepository 的行为类似，代码清单 5-4 使用了相同的配置：DataSource、序列化器和数据表前缀。同样，由于 BatchConfigurer 接口的方法不会直接暴露给 Spring 容器，因此需要调用 InitializerBean#afterPropertiesSet()和 FactoryBean#getObject()方法。

> **注意**　由于 JobRepository 和 JobExplorer 使用相同的底层数据存储，因此最佳实践是同时配置它们以保持一致。

5.2.5　自定义 JobLauncher

JobLauncher 是启动 Spring Batch 作业的入口。大部分情况下，当使用 Spring Batch 提供的 JobLauncher，通过 Spring Boot 的默认机制运行作业时，并不需要自定义 JobLauncher。不过，如果想要展示以其他方式启动作业的能力(例如，通过 Spring MVC 应用中的控制器)，那么可能需要调整 JobLauncher 的工作方式。表 5-9 列出了 JobLauncher 可以使用的设置器。

表 5-9　配置 JobLauncher

设置器的名称	描述
setJobRepository(JobRepository jobRepository)	配置待使用的 JobRepository
setTaskExecutor(TaskExecutor taskExecutor)	配置用于 JobLauncher 的 TaskExecutor，默认是 SyncTaskExecutor

有了这些组件后，就可以对 BatchConfigurer 进行自定义了(包括自定义 JobRepository、PlatformTransactionManager、JobLauncher 和 JobExplorer)，另外每一个接口也都可以自行实现。本章已经介绍了 Spring Batch 提供的用于自定义这些实现的开箱即用方式。然而，我们还需要了解一下如何配置 Spring Batch 基础设施中的另一部分——数据库。配置数据库连接信息以及用来初始化 Spring Batch 数据库的模式十分有必要。5.2.6 节将介绍如何配置 Spring Batch 以完成这个目标。

5.2.6 配置数据库

Spring Boot 能够让简单的事情变得更简单，比如配置数据库。为了配置数据库以使用 Spring Batch，只需要将数据库驱动添加到类路径(classpath)中，并配置合适的属性即可。在本书中，我们将使用 HSQLDB 作为内存数据库，而在需要外部数据库的用例中使用 MySQL。

在将数据库驱动添加到项目中之后，需要使用 Spring Boot 支持的机制之一来配置一组属性。可使用的机制包括 application.properties、application.yml、环境变量以及命令行参数。对于本书而言，大部分情况下将使用 application.yml。代码清单 5-5 展示了如何通过 Spring Boot 属性来配置 MySQL 数据库。

代码清单 5-5　配置数据库

```
spring:
  datasource:
    driverClassName: com.mysql.cj.jdbc.Driver
    url: jdbc:mysql://localhost:3306/spring_batch
    username: 'root'
    password: 'myPassword'
  batch:
    initialize-schema: always
```

在代码清单 5-5 中，有 4 个选项是自解释的：数据库驱动的类名、URL、用户名和密码。对于任何数据库，这都是标准配置。属性 spring.batch.initialize-schema 被设置为 always，从而让 Spring Boot 执行 Spring Batch 的模式脚本。这个属性有如下三个可能的取值。

- always：在每次运行应用时，都会运行模式脚本。在这种情况下，由于会忽略可能发生的错误，并且 Spring Batch 的 SQL 文件并不包含删除表的语句，因此这是开发环境中最容易的情况。
- never：永远不执行模式脚本。
- embedded：只在使用嵌入式数据库时执行模式脚本。在每次启动时都使用干净的数据库实例。

本节介绍了如何配置 Spring Batch 的基础设施。大部分基础设施都是用来管理和查询元数据的。但只有使用它们，这些元数据才有用。

5.3　使用元数据

尽管 Spring Batch 通过一组 DAO 来访问作业存储库中的表，但是它们提供了更加实用的 API。本节将介绍 Spring Batch 提供的用来暴露作业存储库中数据的方法。我们已经知道如何配置，但 Spring Batch 提供的用于访问元数据的主要方法是 JobExplorer。

JobExplorer

为了访问作业存储库中的历史数据和活跃数据，你首先需要了解 org.springframework.batch.core.explore.JobExplorer。尽管在访问已保存的作业执行信息时，大部分会通过 JobRepository 来进行，但 JobExplorer 可以直接访问数据库，如图 5-2 所示。

第 5 章 作业存储库和元数据

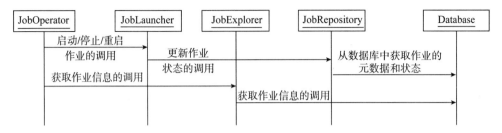

图 5-2 作业管理组件之间的关系

JobExplorer 的基本用途是提供对作业存储库中数据的只读访问，JobExplorer 提供了 7 个接口，可以用来获取作业的实例信息和执行信息。表 5-10 列出了 JobExplorer 提供的方法。

表 5-10 JobExplorer 提供的方法

方法	描述
java.util.Set<JobExecution>findRunningJob Executions (java.lang.String jobName)	返回不包含结束时间的所有 JobExecution
List<JobInstance>findJobInstancesByName (java.lang.String name, int start, int count)	根据提供的名称返回一页的 JobInstance
JobExecution getJobExecution(java.lang. Long executionId)	返回由 id 标识的 JobExecution；如果找不到，就返回 null
java.util.List<JobExecution>getJobExecutions (JobInstance instance)	返回与提供的 JobInstance 相关的所有 JobExecution 的列表
JobInstance getJobInstance(java.lang.Long instanceId)	返回由 id 标识的 JobInstance；如果找不到，就返回 null
java.util.List<JobInstance>getJobInstances (java.lang.String jobName, int start, int count)	返回从指定索引开始的 JobInstance 区间
int getJobInstanceCount(String jobName)	对于给定的作业，返回已经创建的 JobInstance 数量
java.util.List<java.lang.String>getJobNames()	以字母顺序返回作业存储库中具有唯一性的所有作业
StepExecution getStepExecution(java. lang.Long jobExecutionId, java.lang.Long stepExecutionId)	基于 StepExecution 的 id 以及 JobExecution 的 id，返回指定的 StepExecution

如你所见，整个作业存储库都能够通过 JobExplorer 接口暴露的方法来访问。为了查看 JobExplorer 如何工作，可以将其注入 Tasklet 中并进行探索。之后，你就可以看到 JobExplorer 的用途。代码清单 5-6 配置了新的 Tasklet，并且还使用了注入的 JobExplorer。

代码清单 5-6 配置 ExploringTasklet 和 JobExplorer

```
...
@EnableBatchProcessing
@SpringBootApplication
public class DemoApplication {

    @Autowired
    private JobBuilderFactory jobBuilderFactory;
```

107

```
@Autowired
private StepBuilderFactory stepBuilderFactory;

@Autowired
private JobExplorer jobExplorer;

@Bean
public Tasklet explorerTasklet() {
    return new ExploringTasklet(this.jobExplorer);
}

@Bean
public Step explorerStep() {
    return this.stepBuilderFactory.get("explorerStep")
            .tasklet(explorerTasklet())
            .build();
}

@Bean
public Job explorerJob() {
    return this.jobBuilderFactory.get("explorerJob")
            .start(explorerStep())
            .build();
}

public static void main(String[] args) {
    SpringApplication.run(DemoApplication.class, args);
}
```

JobExplorer 配置完毕后，就可以用来做很多事情。在 Spring Batch 框架中，可以使用 JobExplorer 和第 4 章介绍的 RunIdIncrementer 来查看 run.id 参数的值。JobExplorer 在 Spring Cloud Data Flow 服务器中的另一个用途是在启动新的实例前，确定当前是否有正在运行的作业。在本例中，我们将生成一份简单的报告，用来度量这个作业运行了多少个作业实例，执行了多少次以及执行结果。

在代码清单 5-7 中，我们首先得到了当前作业的名称，然后查询了所有运行过的 JobInstance。需要注意的是，这个调用返回的结果中包含当前的 JobInstance。于是，我们打印了这个作业包含的 JobExecution 的数量。之后对于每个 JobInstance，使用 JobExplorer 获取与之关联的 JobExecution 并显示结果。

代码清单 5-7　ExploringTasklet

```
...
public class ExploringTasklet implements Tasklet {

    private JobExplorer explorer;

    public ExploringTasklet(JobExplorer explorer) {
        this.explorer = explorer;
    }

    public RepeatStatus execute(StepContribution stepContribution,
            ChunkContext chunkContext) {
```

```
        String jobName = chunkContext.getStepContext().getJobName();

        List<JobInstance> instances =
                explorer.getJobInstances(jobName,
                        0,
                        Integer.MAX_VALUE);

        System.out.println(
                String.format("There are %d job instances for the job %s",
                instances.size(),
                jobName));

        System.out.println("They have had the following results");
        System.out.println("***********************************");

        for (JobInstance instance : instances) {
            List<JobExecution> jobExecutions =
                    this.explorer.getJobExecutions(instance);

        System.out.println(
                String.format("Instance %d had %d executions",
                        instance.getInstanceId(),
                        jobExecutions.size()));

        for (JobExecution jobExecution : jobExecutions) {
                System.out.println(
                        String.format("\tExecution %d resulted in Exit
                                Status %s",
                                jobExecution.getId(),
                                jobExecution.getExitStatus()));
            }
        }

        return RepeatStatus.FINISHED;
    }
}
```

有了这些代码和配置后，多次运行这个作业，就能够看到与代码清单 5-8 类似的输出。

代码清单 5-8　ExplorerJob 的输出

```
2019-01-18 00:01:27.392 INFO 35356 --- [    main] o.s.b.c.l.support.
SimpleJobLauncher       : Job: [SimpleJob: [name=explorerJob]] launched with the
following parameters: [{1=1}]
2019-01-18 00:01:27.423 INFO 35356 --- [    main] o.s.batch.core.job.
SimpleStepHandler       : Executing step: [explorerStep]
There are 2 job instances for the job explorerJob
They have had the following results
***********************************
Instance 2 had 1 executions
    Execution 2 resulted in Exit Status exitCode=UNKNOWN;exitDescription=
Instance 1 had 1 executions
    Execution 1 resulted in Exit Status exitCode=COMPLETED;exitDescription=
2019-01-18 00:01:27.517 INFO 35356 --- [    main] o.s.b.c.l.support.
SimpleJobLauncher       : Job: [SimpleJob: [name=explorerJob]] completed with the
following parameters: [{1=1}] and the following status: [COMPLETED]
2
```

本节介绍了如何通过 JobExplorer 来访问作业存储库中的数据,并使用 API(例如 JobExplorer)以安全的方式访问数据。

5.4 本章小结

Spring Batch 提供的管理作业元数据以及在错误处理中维护作业状态的能力,是在企业级批处理中使用 Spring Batch 的主要原因或主要原因之一。Spring Batch 不仅提供了强大的错误处理能力,并且允许批处理过程能够基于作业中发生的其他事件来确定该做什么。在第 6 章,随着深入研究如何在多种环境中启动、停止和重启作业,我们将进一步使用这些元数据。

第6章
运行作业

Spring Boot 改变了我们对运行 Java 应用的看法,这非常令人惊喜。在 Spring Boot 出现之前,你是在 servlet 容器中还是在应用服务器上运行应用?你可能选择使用可执行的 Jar 文件来部署应用。如果是这样,就必须通过脚本或其他诸如 Shade Maven 的插件工具来构建类路径。有了 Spring Boot 之后,基本可以忽略这种方式了。现在,大部分应用是通过启动 Spring Boot 生成的可执行的 Jar 文件来引导的。

Spring Boot 还提供了运行 Spring Batch 作业的设施。在本章,我们将介绍 Spring Boot 如何让 Spring Batch 作业变得容易。不过,默认的设施并不是启动 Spring Batch 作业的唯一方式,所以我们将探索启动 Spring Batch 作业的各个组件,这样就可以自行开发启动作业的机制。

既然 Spring Batch 已经能够启动作业,那么为什么还要自行编写代码来启动作业呢?原因在于,Spring Boot 只能处理简单的用例,也就是在启动时执行批处理作业。然而,这并不适用于所有情况。例如,一种很常见的情况是,批处理作业需要根据时间表,在给定的时间执行。在与不同的调度器集成时,就需要编写自定义的代码(本章在后面会介绍原因)。此外,你还可能想在某种形式的事件发生后启动作业。

启动 Spring Batch 作业并不意味着能够运行作业。停止正在运行的作业,是作业执行中的另一重要环节。如果作业运行失败,那么重新启动作业也是执行生产级批处理过程的关键部分。当一个花费数小时并处理百万条记录的作业发生故障时,你并不想从头开始处理这个作业。Spring Batch 提供了重启功能,允许作业从停止的地方重新启动。本章将介绍批处理作业的停止和重启。

我们首先看看 Spring Boot 如何处理批处理作业的启动。

6.1 使用 Spring Boot 启动作业

到目前为止,我们都是通过 Spring Boot 的原生功能来启动作业的。然而,我们还没有真正了解工作原理。本节将介绍如何使用 Spring Boot 启动作业。

Spring Boot 有两个用于在启动时执行逻辑的接口:CommandLineRunner 和 ApplicationRunner。这两个接口都只有一个方法,作用是允许在应用上下文(ApplicationContext)刷新和就绪后调用并执行一段代码。当同时使用 Spring Boot 和 Spring Batch 时,也将使用如下特殊的 CommandLineRunner:

JobLauncherCommandLineRunner。

JobLauncherCommandLineRunner 使用 Spring Batch 的 JobLauncher 执行作业。本章在后面会深入介绍 JobLauncher 接口。目前你只需要知道,这个接口了解如何启动 Spring Batch 作业。当 Spring Boot 执行配置在 ApplicationContext 中的所有 CommandLineRunner 时,如果在类路径中存在 spring-boot-starter-batch,那么 JobLauncherCommandLineRunner 就会运行上下文中的所有作业。到目前为止,我们的所有示例都使用这种机制来运行批处理作业。然而,Spring Boot 确实提供了一些值得一试的配置选项。

第一个配置选项用来定义 Spring Boot 启动哪些作业。Spring Boot 的超级 Jar(Uber Jar)可能包含不止一个作业。例如,如果想在某个 REST 调用或某种类型的事件后执行批处理作业,那么可能不需要在应用启动时执行作业。为此,在 application.yml 中将 Spring Boot 暴露的属性 spring.batch.job.enable 设置为 false 即可(这个属性默认为 true)。在代码清单 6-1 中,如果在设置 spring.batch.job.enabled 为 false 后执行代码,就会看到作业并没有运行,上下文在创建后将立即关闭。

代码清单 6-1 不会运行的作业

```
@EnableBatchProcessing
@SpringBootApplication
public class NoRunJob {

    @Autowired
    private JobBuilderFactory jobBuilderFactory;

    @Autowired
    private StepBuilderFactory stepBuilderFactory;

    @Bean
    public Job job() {
        return this.jobBuilderFactory.get("job")
                    .start(step1())
                    .build();
    }

    @Bean
    public Step step1() {
        return this.stepBuilderFactory.get("step1")
                    .tasklet((stepContribution, chunkContext) -> {
                        System.out.println("step1 ran!");
                        return RepeatStatus.FINISHED;
                    }).build();
    }

    public static void main(String[] args) {
        SpringApplication application = new SpringApplication(NoRunJob.class);

        Properties properties = new Properties();
        properties.put("spring.batch.job.enabled", false);
        application.setDefaultProperties(properties);

        application.run(args);
    }
}
```

尽管代码清单 6-1 配置了一个作业，但是 Spring Boot 并不会运行这个作业，因为我们已将这个作业配置为不在启动时运行。

你在使用 Spring Boot 时可能遇到的另一种情况是：虽然在上下文中定义了多个作业，但是只想在启动时运行其中若干指定的作业。例如，使用一个父作业启动其他作业。在这种情况下，由于父作业会编排其他作业，因此你期望 Spring Boot 仅仅启动父作业或主作业(Master Job)。Spring Boot 支持这种用法。为了在 Spring Boot 中配置哪些作业在启动时运行，可使用 spring.batch.job.names 来标识在启动时执行的作业。Spring Boot 会使用这个以逗号分隔的作业列表，然后按顺序执行。

Spring Boot 并非执行作业的唯一方式。你还可以自行编写触发作业执行的机制。6.2 节将介绍如何使用 REST API 启动作业。

6.2 使用 REST API 启动作业

现在，REST API 是最流行的暴露功能的方式。同样，我们可以很容易地使用它们来启动作业。这里没有开箱即用的启动作业的 REST API，这意味着我们必须自行编写。但是，如何以编程的方式启动 Spring Batch 作业呢？JobLauncher 提供了这种功能。

JobLauncher 是用来启动作业的接口，它只有一个方法 run(Job job, JobParameters jobParameters)，该方法接收两个参数：一个将要执行的作业以及传给这个作业的作业参数。代码清单 6-2 展示了 JobLauncher 接口。

代码清单 6-2　JobLauncher 接口

```
public interface JobLauncher {

    public JobExecution run(Job job, JobParameters jobParameters) throws
                              JobExecutionAlreadyRunningException,
                              JobRestartException,
                              JobInstanceAlreadyCompleteException,
                              JobParametersInvalidException;
}
```

Spring Batch 已经提供了如下开箱即用的 JobLauncher 实现：SimpleJobLauncher。在绝大多数情况下，JobLauncher 能解决所有的启动需求。JobLauncher 能判断作业的运行是现有 JobInstance 的一部分，还是一次全新的运行，然后执行相应的操作。

> **注意**　SimpleJobLauncher 并不会处理传入的作业参数，因此，如果作业使用了 JobParametersIncrementer，那么需要在将作业参数传入 SimpleJobLauncher 之前应用。

JobLauncher 既没有规定作业是同步运行还是异步运行，也没有什么强烈的倾向。可以通过配置作业使用的 TaskExecutor 来决定作业运行的方式。默认情况下，由于使用了同步的 TaskExecutor，因此 SimpleJobLauncher 会以同步的方式执行作业(与调用者在相同的线程中)。然而，如果想释放当前线程(例如，想在作业开始后返回 REST 调用)，那么使用异步的 TaskExecutor 实现可能是更好的选择。

为了使用 REST API 启动应用，需要创建一个新项目。这个新项目将来自 Spring Initializr，并且需要下面的依赖项：

- Batch
- MySQL
- JDBC
- Web

在创建这个新项目后,需要配置 application.yml,使批处理作业不在应用启动时运行(因为我们只想让它们在调用 REST API 时才运行),并且配置数据库。代码清单 6-3 展示了这个新项目的 application.yml。

代码清单 6-3　application.yml

```
spring:
  batch:
    job:
      enabled: false
    initialize-schema: always
  datasource:
    driverClassName: com.mysql.cj.jdbc.Driver
    url: jdbc:mysql://localhost:3306/spring_batch
    username: 'root'
    password: 'p@ssw0rd'
    platform: mysql
```

我们将在创建 REST API 时使用 SimpleJobLauncher 来启动作业。在使用了 @EnableBatchProcessing 注解后,这会很方便。Spring Batch 提供了开箱即用的 SimpleJobLauncher,因此我们不需要做什么。假定这就是启动作业所需的一切,下面我们看看控制器部分,控制器将把作业的名称和参数作为请求的参数,然后启动合适的作业。代码清单 6-4 展示了完整的应用。

代码清单 6-4　JobLaunchingController 应用

```
...
@EnableBatchProcessing
@SpringBootApplication
public class RestApplication {

    @Autowired
    private JobBuilderFactory jobBuilderFactory;

    @Autowired
    private StepBuilderFactory stepBuilderFactory;

    @Bean
    public Job job() {
        return this.jobBuilderFactory.get("job")
                .incrementer(new RunIdIncrementer())
                .start(step1())
                .build();
    }

    @Bean
    public Step step1() {
        return this.stepBuilderFactory.get("step1")
                .tasklet((stepContribution, chunkContext) -> {
                    System.out.println("step1 ran today!");
```

```java
                        return RepeatStatus.FINISHED;
                    }).build();
        }
        @RestController
        public static class JobLaunchingController {

            @Autowired
            private JobLauncher jobLauncher;

            @Autowired
            private ApplicationContext context;

            @PostMapping(path = "/run")
            public ExitStatus runJob(@RequestBody JobLaunchRequest request) throws
            Exception {
                    Job job = this.context.getBean(request.getName(), Job.class);

                    return this.jobLauncher.run(job, request.getJobParameters()).
                        getExitStatus();
            }
        }

        public static class JobLaunchRequest {
            private String name;

            private Properties jobParameters;

            public String getName() {
                return name;
            }

            public void setName(String name) {
                this.name = name;
            }

            public Properties getJobParamsProperties() {
                return jobParameters;
            }
            public void setJobParamsProperties(Properties jobParameters) {
                    this.jobParameters = jobParameters;
            }
            public JobParameters getJobParameters() {
                    Properties properties = new Properties();
                    properties.putAll(this.jobParameters);
                    return new JobParametersBuilder(properties).toJobParameters();
            }
        }
        public static void main(String[] args) {
            new SpringApplication(RestApplication.class).run(args);
        }
    }
```

代码清单 6-4 虽然包含很多内容，但是其中大部分你都已十分熟悉。首先是用于运行 Spring Batch 的 Spring Boot 应用的注解(@SpringBootApplication 和@EnableBatchProcessing)。接下来构建了一个批处理作业，这是一个简单的单步骤作业，用于输出 step1 ran today！。从 Spring Batch 的角度看，这个目标并不复杂，但却证明了作业可以运行。之后是两个内部类。第一个内部类是控制器本身，也就是 JobLaunchingController。在这个内部类中，我们装入了 @EnableBatchProcessing 注解提供的 JobLauncher 以及当前的应用上下文 ApplicationContext，以便能够获取作业的 Bean 并在请求中执行。

@PostMapping 用于配置 HTTP POST 方法的 URL(在本例中是/run)与方法的映射。请求体有两个主要组件：要执行的作业以及将要传给作业的可包含任何参数的映射。这种结构将在下一个内部类 (JobLaunchRequest) 中得以建模。我们将调用自己的 API，并传入 JSON 格式的请求体，Spring 则将其映射为 JobLaunchRequest 实例。

控制器中的 runJob 方法做了两件事情。首先是获取将要执行的作业，这是通过 ApplicationContext 得到的。有了作业和作业参数后，就可以将它们传给 JobLauncher 来执行作业。默认情况下，JobLauncher 将以同步的方式执行作业，因此可以把退出状态(ExitStatus)返回给用户。需要注意的是，由于真实场景中的处理量会很大，因此大部分批处理作业都不会很快完成。在这种情况下，选择异步运行会更加合适(此时仅仅返回 JobExecution 的 id)。

最后是用于启动作业的 main 方法。与代码清单 6-1 类似，由于我们不想让批处理作业在 Spring Boot 应用启动时运行，而是通过调用 API 来运行，因此这里把批处理作业配置为不在启动时运行。代码清单 6-5 展示了如何通过 curl 命令启动代码清单 6-4 中的作业。

代码清单 6-5　对代码清单 6-4 执行 curl 命令后的输出

```
$ curl -H "Content-Type: application/json" -X POST -d '{"name":"job", "jobParameters":
{"foo":"bar", "baz":"quix"}}' http://localhost:8080/run
{"exitCode":"COMPLETED","exitDescription":"","running":false}
```

代码清单 6-5 使用 HTTP POST 将 JSON 格式的数据发送给了 http://localhost:8080/run，并设置 Content-Type 为 application/json。如果运行代码清单 6-4 中的应用，并执行代码清单 6-5 中的 curl 命令，就会看到代码清单 6-6 所示的输出。

代码清单 6-6　　REST API 的输出

```
2018-02-08 12:07:56.327 INFO 22104 --- [nio-8080-exec-1] o.s.b.c.l.support.
SimpleJob Launcher     : Job: [SimpleJob: [name=job]] launched with the following
parameters:[{baz=quix, foo=bar}]
2018-02-08 12:07:56.345 INFO 22104 --- [nio-8080-exec-1] o.s.batch.core.job.
SimpleStep Handler    : Executing step: [step1]
step1 ran!
2018-02-08 12:07:56.362 INFO 22104 --- [nio-8080-exec-1] o.s.b.c.l.support.
SimpleJob Launcher     : Job: [SimpleJob: [name=job]] completed with the following
parameters:[{baz=quix, foo=bar}] and the following status: [COMPLETED]
```

注意，REST API 并没有包含任何与重启作业、处理作业参数递增等相关的逻辑。本章在后面会介绍如何重启作业。但是，在结束介绍 REST API 之前，我们先看看如何处理用于后续运行的作业递增参数。

当使用 JobParametersIncrementer 时，JobLauncher 的调用者负责将这些变化应用到参数中。一旦参数被传给作业，它们就是不可变参数了。Spring Batch 在 JobParametersBuilder 上提供了如下用于递增这种参数的便捷方法：getNextJobParameters(Job job)。代码清单 6-7 展示的控制器来自前面的应用，但是做了一些更新，比如调用了 JobParametersBuilder#getNextJobParameters。

代码清单 6-7　在启动作业前递增作业参数

```
...
@Bean
public Job job() {
    return this.jobBuilderFactory.get("job")
                .incrementer(new RunIdIncrementer())
                .start(step1())
                .build();
}
...
@RestController
public static class JobLaunchingController {

    @Autowired
    private JobLauncher jobLauncher;

    @Autowired
    private ApplicationContext context;

    @Autowired
    private JobExplorer jobExplorer;

        @PostMapping(path = "/run")
        public ExitStatus runJob(@RequestBody JobLaunchRequest request) throws Exception
{
            Job job = this.context.getBean(request.getName(), Job.class);
            JobParameters jobParameters =
                        new JobParametersBuilder(request.getJobParameters(),
                                            this.jobExplorer)
                                .getNextJobParameters(job)
                                .toJobParameters();
            return this.jobLauncher.run(job, jobParameters).getExitStatus();
    }
}
...
```

在代码清单 6-7 中，首先是作业的定义，这部分与前面的代码相同。然而，在前一个示例中，事实上并没有激活 RunIdIncrementer。接下来是更新后的控制器 JobLaunchingController。

在控制器 JobLaunchingController 中，这里新添加了一行代码。我们创建了一个新的 JobParameters，并通过方法 JobParametersBuilder#getNextJobParameters(job) 把 run.id 添加到这个 JobParameters 中。这个方法会查看传入的作业，确定是否有 JobParametersIncrementer。如果有，就将其应用到上一次作业执行的 JobParameters。这个方法还会确定这一次的作业执行是否是重新启动，并且以合适的方式处理 JobParameters。如果这两种情况都不存在，那么什么也不会改变。

执行完代码清单 6-7 所示的变更后，如果运行应用并且再次访问，那么输出将略有不同。在代码清单 6-8 中，输出结果中多了 run.id=1。如果再次运行应用，就会看到 run.id=2。这一变化很重要。如果在本例的上一个版本中尝试做同样的事情，就会得到异常，表明使用这些参数的 JobInstance 已经完成。

代码清单 6-8　使用 RunIdIncrementer 运行作业时的输出

```
2018-02-08 16:21:34.658 INFO 22990 --- [nio-8080-exec-1] o.s.b.c.l.support.
```

```
    SimpleJob Launcher        : Job: [SimpleJob: [name=job]] launched with the following
parameters:[{baz=quix, foo=bar, run.id=1}]
    2018-02-08 16:21:34.669 INFO 22990 --- [nio-8080-exec-1] o.s.batch.core.job.
    SimpleStep Handler        : Executing step: [step1]
step1 ran today!
    2018-02-08 16:21:34.679 INFO 22990 --- [nio-8080-exec-1] o.s.b.c.l.support.
    SimpleJob Launcher        : Job: [SimpleJob: [name=job]] completed with the following
parameters:[{baz=quix, foo=bar, run.id=1}] and the following status: [COMPLETED]
```

按需启动作业——无论是通过 java -jar 命令执行超级 Jar，还是调用 REST API——是很有用的。然而，在大多数企业中，批处理是通过调度执行的。接下来将介绍如何把 Spring Batch 作业的执行与第三方库集成起来，比如 Quartz。

6.3 使用 Quartz 进行调度

市面上有许多企业级的调度器，从简单但高效的 crontab，到价值数百万美元的企业级自动化平台。在这里，我们使用名为 Quartz 的开源调度器(www.quartz-scheduler.org)。这个调度器可用于各种规模的 Java 环境，除了强大的功能和可靠的社区支持之外，它还与 Spring Integration 有着深厚的历史渊源，包括有助于执行作业的 Spring Boot 支持。

鉴于 Quartz 调度器涉及的范围很广，本书不会对它面面俱到地进行介绍。相反，本书会简明介绍它是如何工作的，以及如何将它与 Spring 集成起来。图 6-1 展示了 Quartz 调度器的组件以及它们之间的关系。

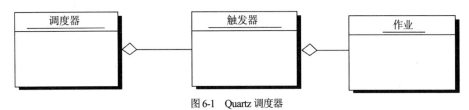

图 6-1 Quartz 调度器

如你所见，Quartz 调度器包括三个主要组件：调度器、触发器和作业。调度器来自 SchedulerFactory，除了充当 JobDetail(指向 Quartz 作业的引用)和触发器的注册中心之外，还负责在触发器被触发时执行作业。这里的作业是可执行的工作单元。触发器定义了作业的运行时间。当触发器被触发时，Quartz 执行作业，此时会创建 JobDetail，用于定义作业的独立执行。

这听起来是不是很熟悉？用于定义 Job 和 JobDetail 对象的这种模型与 Spring Batch 用于定义 Job 和 JobInstance 的方式非常类似。为了将 Quartz 集成到 Spring Batch 处理过程中，你需要完成以下事项：

- 从 Spring Initializr 创建项目，并使用合适的启动器。
- 编写一个 Spring Batch 作业。
- 编写一个 Quartz 作业，并使用 Spring 的 QuartzJobBean 启动这个 Quartz 作业。
- 配置 Spring 提供的 JobDetailBean，用于创建 QuartzJobDetail。
- 配置用于定义作业运行时机的触发器。

为了展示如何使用 Quartz 定期地执行作业，我们首先从 https://start.spring.io 创建一个新项目，然后选择 Batch、MySQL、JDBC 和 Quartz Scheduler 等依赖项。实际上，Quartz 可以把元数据保存在数据库中,但这超出了本书的讨论范畴。

在 IDE 加载了这个新项目后,下一步是创建 Spring Batch 作业。代码清单 6-9 展示的配置可用于我们将要调度的作业。

代码清单 6-9　用于调度作业的配置

```
...
@Configuration
public class BatchConfiguration {

    @Autowired
    private JobBuilderFactory jobBuilderFactory;

    @Autowired
    private StepBuilderFactory stepBuilderFactory;

    @Bean
    public Job job() {
        return this.jobBuilderFactory.get("job")
                    .incrementer(new RunIdIncrementer())
                    .start(step1())
                    .build();
    }

    @Bean
    public Step step1() {
        return this.stepBuilderFactory.get("step1")
                    .tasklet((stepContribution, chunkContext) -> {
                        System.out.println("step1 ran!");
                        return RepeatStatus.FINISHED;
                    }).build();
    }
}
```

代码清单 6-9 中并没有什么新的内容。其中只有一个单步骤的作业,用于向控制台打印 step1 ran!。需要特别注意的是,这里使用了递增器。由于需要多次运行同一个作业,并且无法以其他方式提供具有唯一性的作业参数,因此递增器是一项基本需求。

在定义了作业之后,需要创建 Quartz 作业。这部分代码用来完成在调度事件被触发时启动作业的机制,它们与使用 REST 控制器启动作业时编写的代码相同。代码清单 6-10 展示了 BatchScheduledJob。

代码清单 6-10　BatchScheduledJob

```
...
public class BatchScheduledJob extends QuartzJobBean {

    @Autowired
    private Job job;

    @Autowired
    private JobExplorer jobExplorer;

    @Autowired
    private JobLauncher jobLauncher;
```

```java
    @Override
    protected void executeInternal(JobExecutionContext context) {
        JobParameters jobParameters = new JobParametersBuilder(this.jobExplorer)
                .getNextJobParameters(this.job)
                .toJobParameters();
        try {
            this.jobLauncher.run(this.job, jobParameters);
        }
        catch (Exception e) {
            e.printStackTrace();
        }
    }
}
```

代码清单 6-10 中的 BatchScheduledJob 类扩展了 Spring 的 QuartzJobBean 类。QuartzJobBean 类包含了运行 Quartz 作业所需的大部分样板代码,通过重写 executeInternal(JobExecutionContext context) 方法,我们对其进行了扩展。此处添加的代码,与 REST 控制器中用于启动作业的代码相同。唯一的区别是,此处不需要动态地选择将要运行的作业,因此我们只需要将其装配到 Quart 作业中。每次调度事件被触发时,executeInternal 方法只会被调用一次。

为了配置调度表,你需要完成两件事情。首先是为 Quartz 作业配置一个 Bean。Quartz 提供了 org.quartz.JobBuilder 构建器来帮助你完成这件事。通过使用这个构建器,并提供已经创建好的作业类(在本例中是 BatchScheduledJob),就可以告诉 Quartz:如果作业没有与触发器关联,就不要删除作业的定义,并且让 Quartz 构建 JobDetail。

创建了 JobDetail 之后,就可以创建触发器并进行调度。我们将使用 org.quartz.SimpleScheduleBuilder 创建调度表,每 5 秒启动一次作业并重复 4 次(总共启动 5 次)。JobDetail 是与 Quartz 作业相关的元数据,调度表定义了 JobDetails 的运行频率,触发器用于关联它们二者。我们选择使用 Quartz 的 TriggerBuilder 来创建触发器,并将其传给作业和调度表。代码清单 6-11 展示了这些 Quartz 组件的完整配置。

代码清单 6-11　Quartz 配置

```java
...
@Configuration
public class QuartzConfiguration {

    @Bean
    public JobDetail quartzJobDetail() {
        return JobBuilder.newJob(BatchScheduledJob.class)
                        .storeDurably()
                        .build();
    }

    @Bean
    public Trigger jobTrigger() {
        SimpleScheduleBuilder scheduleBuilder = SimpleScheduleBuilder.simpleSchedule()
                        .withIntervalInSeconds(5).withRepeatCount(4);

        return TriggerBuilder.newTrigger()
                    .forJob(quartzJobDetail())
                    .withSchedule(scheduleBuilder)
                .build();
```

        }
    }

在把这些配置添加到项目后,你就有了用于运行作业的一切。当运行 Spring Boot 应用时,你就会看到代码清单 6-12 所示的输出,并且会重复 4 次。

代码清单 6-12　Quartz 输出

```
...
2018-02-16 12:00:13.723 INFO 78906 --- [main] i.s.b.quartzdemo.QuartzDemoApplication:
Started QuartzDemoApplication in 1.577 seconds (JVM running for 2.05)
2018-02-16 12:00:13.759 INFO 78906 --- [eduler_Worker-1] o.s.b.c.l.support.
SimpleJobLauncher: Job: [SimpleJob: [name=job]] launched with the following parameters:
[{run.id=1}]
2018-02-16 12:00:13.769 INFO 78906 --- [eduler_Worker-1] o.s.batch.core.job.
SimpleStepHandler: Executing step: [step1]
step1 ran!
2018-02-16 12:00:13.779 INFO 78906 --- [eduler_Worker-1] o.s.b.c.l.support.
SimpleJobLauncher: Job: [SimpleJob: [name=job]] completed with the following parameters:
[{run.id=1}] and the following status: [COMPLETED]
...
```

尽管在理想情况下,我们期望作业会自然结束,但是在真实场景中并非如此,我们可能需要停止它们。6.4 节将介绍用于停止正在运行的批处理作业的各种机制。

6.4　停止作业

停止作业的原因有很多,每一个原因都会影响接下来发生的事情。作业有可能自然结束(截至目前,所有作业都是如此)。另外,可能出于某个原因,我们需要以编程的方式在处理过程中停止作业,甚至可能需要从外部停止作业(例如,某些人意识到出现了错误,需要停止作业并修复问题)。当然,错误也可能导致作业停止执行。本节将介绍如何使用 Spring Batch 应对这些场景,以及在发生这些场景时应该做些什么。

6.4.1　自然结束

到目前为止,所有作业都是自然结束的。也就是说,每个作业都运行了自己所有的步骤,直到所有的步骤返回 COMPLETED 状态,作业本身则返回退出码 COMPLETED。那么,对于作业来说,这意味着什么呢?

你已经知道,作业不能以同样的参数运行多次,否则会失败。当作业已经进入 BatchStatus 状态时,不能以同样的 JobParameters 再次创建新的 JobInstance,需要特别注意的是,这决定了如何执行作业。可以使用 JobParametersIncrementer,并基于作业的每次运行递增参数,这是个好主意,尤其对于那些基于调度的作业而言。例如,如果有一个每天都要执行的作业,那么开发能以递增的时间戳为参数的 JobParametersIncrementer 实现就很有用。这样,当作业每次通过调度执行时,作业参数也会相应地增加,正如我们之前看到的那样。

并不是所有的作业都能在每一次执行后自然结束。在有些情况下,你需要根据处理过程中发生的事情来停止作业(例如,在步骤结束时完整性检查失败)。在类似的情况下,你需要以编程的方式停止作业。

6.4.2 以编程方式结束

批处理需要执行一系列的检查和平衡。当处理大量数据时，你需要能够验证正在处理的内容。当用户在 Web 应用中使用错误的地址更新个人信息时，只会影响单个用户。然而，如果作业导入的文件包含一百万条记录，并且在导入步骤结束后发现只有一万条，会怎么样呢？肯定是哪里出错了，你需要在作业继续之前进行修复。本小节将介绍如何以编程的方式结束作业。你首先会看到一个真实的示例，它使用了第 4 章介绍的 stop 转换，你将结合一些新的属性，重启作业。你还会看到如何通过设置标识来结束作业。

1. 使用 stop 转换

我们首先看看如何构造和配置作业，如何使用 stop 转换停止作业，以及如何解决重新启动作业的问题。下面创建一个包含如下三个步骤的作业，看看在实践中该怎么做：

(1) 导入一个简单的交易文件(transaction.csv)，其中的每笔交易都包含了账号、时间戳以及金额大小(正数为贷方，负数为借方)，这个文件以一条包含了记录数量的摘要记录结束。

(2) 将这些交易记录导入数据表之后，再应用于一张单独的账户摘要表，里面包含了账号和当前账户余额。

(3) 生成一个摘要文件(summary.csv)，在里面列出每个账户的账号和余额。

从设计的角度思考这三个步骤时，你会想到在对每个账户应用数据转换之前，需要验证导入的记录数量是否与摘要记录中的记录数量相同。在处理大量的数据时，这种完整性检查会为你节省很多的恢复和重新处理时间。

为了启动作业，下面看看文件格式和数据模型。这个作业的文件格式很简单，它是一个 CSV 文件。因此，无需代码就可以很容易地配置合适的读取器和写入器。代码清单 6-13 展示了这里用到的两个文件(transaction.csv 和 summary.csv)中的示例记录。

代码清单 6-13　示例记录

```
Transaction file:
3985729387,2010-01-08 12:15:26,523.65
3985729387,2010-01-08 1:28:58,-25.93
2

Summary File:
3985729387,497.72
```

这个示例中的数据模型很简单，仅仅包含两个表：TRANSACTION 和 ACCOUNT_SUMMARY，如图 6-2 所示。

为了创建作业，下面从一个崭新的项目开始。从 Spring Initializr 中选择 Batch、JDBC 和 MySQL 作为依赖项以生成项目。有了这些，就可以创建一个配置文件，并在其中包含步骤和作业所需的一切。

和前面的项目一样，在 application.yml 中进行配置以使用 MySQL(参见代码清单 6-3)。对于这个作业，我们需要自定义一些组件。特别是，我们需要一个自定义的 ItemReader、一个自定义的 ItemProcessor、两个领域对象以及一个数据访问对象(DAO)。在把它们(以及其他的组件)组装到批处理作业之前，我们将逐一介绍它们。

第 6 章 运行作业

图 6-2　数据模型

下面从领域对象开始。就像你期望的那样,每个领域对象都与数据库中的表一一对应,并且与接下来使用的文件对应。代码清单 6-14 展示了领域对象 Transaction 和 AccountSummary。

代码清单 6-14　领域对象 Transaction 和 AccountSummary

```
...
public class Transaction {

    private String accountNumber;

    private Date timestamp;

    private double amount;
    public String getAccountNumber() {
        return accountNumber;
    }

    public void setAccountNumber(String accountNumber) {
        this.accountNumber = accountNumber;
    }

    public Date getTimestamp() {
        return timestamp;
    }

    public void setTimestamp(Date timestamp) {
        this.timestamp = timestamp;
    }

    public double getAmount() {
        return amount;
    }

    public void setAmount(double amount) {
        this.amount = amount;
    }
}

...
public class AccountSummary {

    private int id;

    private String accountNumber;
```

```
    private Double currentBalance;

    public int getId() {
        return id;
    }

    public void setId(int id) {
        this.id = id;
    }

    public String getAccountNumber() {
        return accountNumber;
    }

    public void setAccountNumber(String accountNumber) {
        this.accountNumber = accountNumber;
    }

    public Double getCurrentBalance() {
        return currentBalance;
    }
    public void setCurrentBalance(Double currentBalance) {
        this.currentBalance = currentBalance;
    }
}
```

下一个组件是 ItemReader。Spring Batch 包含了健壮的设施用于读取 CSV 文件,那么为什么还要自定义 ItemReader 呢？原因在于步骤的退出状态与读取器的状态关联在了一起。如果读取到的记录数量与文件末尾指定的记录数量不同,就不应该继续处理。所以,我们需要使用自定义的 ItemReader 来封装 FlatFileItemReader。这个自定义的 ItemReader 会对读入的记录进行计数。在遇到文件末尾的记录时,如果期望的记录数量与读入的记录数量匹配,就会记录处理。但是,如果它们不匹配,自定义的 ItemReader 就会提供 AfterStep 方法,用于把退出状态设置为 STOPPED。代码清单 6-15 展示了这是如何完成的。

代码清单 6-15　TransactionReader

```
...
public class TransactionReader implements ItemStreamReader<Transaction> {

    private ItemStreamReader<FieldSet> fieldSetReader;
    private int recordCount = 0;
    private int expectedRecordCount = 0;

    public TransactionReader(ItemStreamReader<FieldSet> fieldSetReader) {
        this.fieldSetReader = fieldSetReader;
    }

    public Transaction read() throws Exception {
        return process(fieldSetReader.read());
    }

    private Transaction process(FieldSet fieldSet) {
        Transaction result = null;
```

```
        if(fieldSet != null) {
            if(fieldSet.getFieldCount() > 1) {
                result = new Transaction();
                result.setAccountNumber(fieldSet.readString(0));
                result.setTimestamp(fieldSet.readDate(1, "yyyy-mm-dd hh:mm:ss"));
                result.setAmount(fieldSet.readDouble(2));

                recordCount++;
            } else {
                expectedRecordCount = fieldSet.readInt(0);
            }
        }

        return result;
    }
    public void setFieldSetReader(ItemStreamReader<FieldSet> fieldSetReader) {
        this.fieldSetReader = fieldSetReader;
    }

    @AfterStep
    public ExitStatus afterStep(StepExecution execution) {
        if(recordCount == expectedRecordCount) {
            return execution.getExitStatus();
        } else {
            return ExitStatus.STOPPED;
        }
    }

    @Override
    public void open(ExecutionContext executionContext) throws ItemStreamException {
        this.fieldSetReader.open(executionContext);
    }

    @Override
    public void update(ExecutionContext executionContext) throws ItemStreamException {
        this.fieldSetReader.update(executionContext);
    }

    @Override
    public void close() throws ItemStreamException {
        this.fieldSetReader.close();
    }
}
```

在代码清单 6-15 的开头，ItemReader#read()方法把数据的读取委托给了注入的一个读取器。在本例中，这个读取器是 FlatFileItemReader。ItemReader 委托的返回类型是 FieldSet，这是因为实际上会有两种格式的记录：一种格式用于导入的数据，另一种格式用于包含记录数量的文件。这里把 ItemReader 委托返回的 FieldSet 传给了 process 方法，并在这个方法中确定记录的格式。如果当前记录中有多个值，那就是一条数据记录。如果当前记录中只有一个值，那就是文件摘要记录。数据记录被转换到 Transaction 实例中并返回。在记录了文件摘要中的值之后，返回 null 以表明文件已经完成处理。

接下来是 StepExecutionListener.afterStep(StepExecution stepExecution)方法的实现，这个方法会在步骤完成时被调用，从而给予我们返回特定的退出状态(ExitStatus)的机会。在这里，这个方法用来查看已经

Spring Batch 权威指南

读入的记录数量,并与保存在文件末尾的记录数量进行对比。如果二者相同,那么返回设置的退出状态,否则返回 ExitStatus.STOPPED。这允许我们在文件无效的情况下,停止作业的执行。

　　TransactionItemReader 的其他方法是 ItemStream 接口的实现。Spring Batch 会自动查看 ItemReader、ItemProcessor 和 ItemWriter,如果是 ItemStream,就自动注册,使得回调方法在合适的时机执行。然而,我们已经有一个实现了 ItemStream 的委托,这个委托的 ItemReader 并不会使用 Spring Batch 显式地进行注册,所以 Spring Batch 不会查看它是否实现了 ItemStream 接口。这给我们留下两个选择:要么显式地将这个委托注册为作业的 ItemStream(这很容易出错,因为要求我们必须记得进行注册);要么在 TransactionItemReader 中实现 ItemStream 接口,并且调用委托中合适的生命周期方法(这也是此处的做法)[1]。

　　下一个自定义组件是 TransactionDao 接口。这个接口只有一个方法:getTransactionsByAccountNumber(String accountNumber),用于返回与传入的账号相关联的交易列表。代码清单 6-16 展示了这个接口的实现。

代码清单 6-16　TransactionDaoSupport

```
public class TransactionDaoSupport extends JdbcTemplate implements TransactionDao {

    public TransactionDaoSupport(DataSource dataSource) {
        super(dataSource);
    }

    @SuppressWarnings("unchecked")
    public List<Transaction> getTransactionsByAccountNumber(String accountNumber) {
        return query(
                "select t.id, t.timestamp, t.amount " +
                "from transaction t inner join account_summary a on " +
                "a.id = t.account_summary_id " +
                "where a.account_number = ?",
                new Object[] { accountNumber },
                (rs, rowNum) -> {
                    Transaction trans = new Transaction();
                    trans.setAmount(rs.getDouble("amount"));
                    trans.setTimestamp(rs.getDate("timestamp"));
                    return trans;
                }
        );
    }
}
```

　　代码清单 6-16 中的 TransactionDAO 接口选取了与传入的 accountNumber 相关联的所有 Transaction (交易)记录。稍后我们会使用一个 ItemProcessor,将所有的交易应用到给定的账户,以确定它们的余额。代码清单 6-17 展示了用于 TransactionDAO 接口的这个 ItemProcessor。

代码清单 6-17　TransactionApplierProcessor

```
...
public class TransactionApplierProcessor implements
            ItemProcessor<AccountSummary, AccountSummary> {
```

---

[1] ItemStream 接口将在第 7 和 9 章中介绍。

第 6 章 运行作业

```
        private TransactionDao transactionDao;

        public TransactionApplierProcessor(TransactionDao transactionDao) {
            this.transactionDao = transactionDao;
        }

        public AccountSummary process(AccountSummary summary) throws Exception {
            List<Transaction> transactions = transactionDao
                    .getTransactionsByAccountNumber(summary.getAccountNumber());

            for (Transaction transaction : transactions) {
                summary.setCurrentBalance(summary.getCurrentBalance()
                    + transaction.getAmount());
            }
            return summary;
        }
    }
```

如代码清单 6-17 所示，对于传给这个 ItemProcessor 的每一条 AccountSummary 记录，都会使用 TransactionDao 接口找出所有的交易，而账户的当前余额会根据交易进行增加或减少。

对于这个作业而言，这就是所有需要自定义的批处理组件。下一步是配置它们。首先配置每个步骤，然后将这些步骤组装进作业。第一个步骤是 importTransactionFileStep，参见代码清单 6-18。

代码清单 6-18　importTransactionFileStep

```
...
    @Bean
    @StepScope
    public TransactionReader transactionReader() {
        return new TransactionReader(fileItemReader(null));
    }

    @Bean
    @StepScope
    public FlatFileItemReader<FieldSet> fileItemReader(
    @Value("#{jobParameters['transactionFile']}") Resource inputFile) {
        return new FlatFileItemReaderBuilder<FieldSet>()
                    .name("fileItemReader")
                    .resource(inputFile)
                    .lineTokenizer(new DelimitedLineTokenizer())
                    .fieldSetMapper(new PassThroughFieldSetMapper())
                    .build();
    }

    @Bean
    public JdbcBatchItemWriter<Transaction> transactionWriter(DataSource dataSource) {
        return new JdbcBatchItemWriterBuilder<Transaction>()
                    .itemSqlParameterSourceProvider(
                        new BeanPropertyItemSqlParameterSourceProvider<>())
                    .sql("INSERT INTO TRANSACTION " +
                        "(ACCOUNT_SUMMARY_ID, TIMESTAMP, AMOUNT) " +
                        "VALUES ((SELECT ID FROM ACCOUNT_SUMMARY " +
                        "         WHERE ACCOUNT_NUMBER = :accountNumber), " +
                        ":timestamp, :amount)")
```

```
                    .dataSource(dataSource)
                    .build();
}

@Bean
public Step importTransactionFileStep() {
    return this.stepBuilderFactory.get("importTransactionFileStep")
                    .<Transaction, Transaction>chunk(100)
                    .reader(transactionReader())
                    .writer(transactionWriter(null))
                    .allowStartIfComplete(true)
                    .listener(transactionReader())
                    .build();
}
...
```

在这个步骤中，我们首先定义了 TransactionReader，也就是代码清单 6-15 中自定义的 ItemReader；接下来配置了 FlatFileItemReader。到了第 7 章，当详细回顾 ItemReader 时我们将会介绍更多的细节。接下来的 JdbcBatchItemWriter 配置了向数据库写入数据的方式，同样，到了第 9 章我们会详细介绍，因此你不必担心细节，目前你只需要知道这是把数据写入数据库的方式即可。代码清单 6-18 中的最后一个 Bean 定义了步骤本身：使用 StepBuilderFactory 得到了一个构建器，配置了一个基于块的步骤，并且使用了刚刚配置的 TransactionReader 和 JDBC 写入器。我们将这个步骤配置为可在作业重启后重新运行。这么做的原因是，如果导入的数据不合法(意味着记录数量与保存在文件末尾的记录数量不匹配)，那么我们希望清除这次导入，并使用合法的文件重新导入。在使步骤能够重新运行后，我们将 TransactionReader 注册为监听器，最后构建了步骤。

第一个步骤用于把文件导入数据库。第二个步骤用于把文件中的交易数据应用于账户，参见代码清单 6-19。

代码清单 6-19　applyTransactionsStep

```
...
@Bean
@StepScope
public JdbcCursorItemReader<AccountSummary> accountSummaryReader(DataSource dataSource) {
    return new JdbcCursorItemReaderBuilder<AccountSummary>()
                    .name("accountSummaryReader")
                    .dataSource(dataSource)
                    .sql("SELECT ACCOUNT_NUMBER, CURRENT_BALANCE " +
                        "FROM ACCOUNT_SUMMARY A " +
                        "WHERE A.ID IN (" +
                        "   SELECT DISTINCT T.ACCOUNT_SUMMARY_ID " +
                        "   FROM TRANSACTION T) " +
                        "ORDER BY A.ACCOUNT_NUMBER")
                    .rowMapper((resultSet, rowNumber) -> {
                        AccountSummary summary = new AccountSummary();

                        summary.setAccountNumber(resultSet.getString("account_number"));
                        summary.setCurrentBalance(resultSet.getDouble("current_balance"));

                        return summary;
                    }).build();
```

```
    }

    @Bean
    public TransactionDao transactionDao(DataSource dataSource) {
        return new TransactionDaoSupport(dataSource);
    }

    @Bean
    public TransactionApplierProcessor transactionApplierProcessor() {
        return new TransactionApplierProcessor(transactionDao(null));
    }

    @Bean
    public JdbcBatchItemWriter<AccountSummary> accountSummaryWriter(DataSource dataSource) {
        return new JdbcBatchItemWriterBuilder<AccountSummary>()
                    .dataSource(dataSource)
                    .itemSqlParameterSourceProvider(
                            new BeanPropertyItemSqlParameterSourceProvider<>())
                    .sql("UPDATE ACCOUNT_SUMMARY " +
                            "SET CURRENT_BALANCE = :currentBalance " +
                            "WHERE ACCOUNT_NUMBER = :accountNumber")
                    .build();
    }
    @Bean
    public Step applyTransactionsStep() {
        return this.stepBuilderFactory.get("applyTransactionsStep")
                    .<AccountSummary, AccountSummary>chunk(100)
                    .reader(accountSummaryReader(null))
                    .processor(transactionApplierProcessor())
                    .writer(accountSummaryWriter(null))
                    .build();
    }
    ...
```

代码清单 6-19 首先定义了 JdbcCursorItemReader，用于从数据库中读取 AccountSummary 记录；然后定义了两个 Bean——TransactionDao 会查找交易，自定义的 ItemProcessor(参见代码清单 6-17)则用于将交易应用于账户；最后使用 JdbcBatchItemWriter 将更新后的账户摘要记录写入数据库。在配置了这些组件后，就可以把它们组装到步骤中。applyTransactionsStep 使用 StepBuilderFactory 获取了一个构建器，配置了一个基于块的步骤(块的大小为 100 条记录)，并且使用了前面配置的 ItemReader、ItemProcessor 和 ItemWriter。

最后一个步骤是 generateAccountSummaryStep，这个步骤实际上复用了 applyTransactionsStep 中的 ItemReader，这是因为它们读取的是相同的数据(只是做了一些不同的事情)。这也是把 accountSummaryReader 设置为步骤作用域(StepScope)的原因，我们需要为每一个步骤创建新的实例。因此，真正需要为 generateAccountSummaryStep 配置的是 ItemWriter 和步骤本身。代码清单 6-20 展示了这些代码。

代码清单 6-20　generateAccountSummaryStep

```
...
    @Bean
    @StepScope
```

```java
public FlatFileItemWriter<AccountSummary> accountSummaryFileWriter(
        @Value("#{jobParameters['summaryFile']}") Resource summaryFile) {
            DelimitedLineAggregator<AccountSummary> lineAggregator =
                new DelimitedLineAggregator<>();
            BeanWrapperFieldExtractor<AccountSummary> fieldExtractor =
                new BeanWrapperFieldExtractor<>();
            fieldExtractor.setNames(new String[] {"accountNumber", "currentBalance"});
            fieldExtractor.afterPropertiesSet();
            lineAggregator.setFieldExtractor(fieldExtractor);
            return new FlatFileItemWriterBuilder<AccountSummary>()
                .name("accountSummaryFileWriter")
                .resource(summaryFile)
                .lineAggregator(lineAggregator)
                .build();
}

@Bean
public Step generateAccountSummaryStep() {
    return this.stepBuilderFactory.get("generateAccountSummaryStep")
            .<AccountSummary, AccountSummary>chunk(100)
            .reader(accountSummaryReader(null))
            .writer(accountSummaryFileWriter(null))
            .build();
}
...
```

代码清单 6-20 首先配置了 FlatFileItemWriter。FlatFileItemWriter 生成了一个 CSV 文件，其中的每一条记录包括了账号和当前余额。接下来组装步骤，方法是使用 StepBuilderFactory 获取构建器，并使用上一个步骤(accountSummaryReader)中的 ItemReader 和刚才设置的 ItemWriter 配置一个基于块的步骤。

最后一部分是配置作业本身。作业需要三个有序的步骤，但是第一个步骤可能返回 STOPPED 退出状态。代码清单 6-21 包含了用于构建可停止作业的代码。

代码清单 6-21　transactionJob

```java
...
@Bean
public Job transactionJob() {
    return this.jobBuilderFactory.get("transactionJob")
            .start(importTransactionFileStep())
            .on("STOPPED").stopAndRestart(importTransactionFileStep())
            .from(importTransactionFileStep()).on("*").to(applyTransactionsStep())
            .from(applyTransactionsStep()).next(generateAccountSummaryStep())
            .end()
            .build();
}
...
```

代码清单 6-21 首先使用 JobBuilderFactory 得到了一个构建器，可将作业配置为从 importTransactionFileStep 开始。接下来，如果退出状态为 STOPPED，那么停止作业，并且在相同的步骤重启作业(从本质上讲，如果作业是以编程方式停止的，就重启作业)。在其他情况下，进入

applyTransactionsStep 步骤，然后从 applyTransactionsStep 步骤转换到 generateAccountSummaryStep 步骤。调用 end()方法是必要的，因为这里使用转换 API 来构建流。最后调用 build()方法以生成作业。

现在，执行作业两次。在第一次执行作业时，使用包含无效的文件摘要记录的 transaction.csv。也就是说，执行作业时，使用的文件包含了 100 条记录和最后一条文件摘要记录。这条文件摘要记录中保存的是除了 100 之外的任何数字(这里可以设置为 20)。当执行作业时，StepListener 会校验出读入的记录数量(100)和期望的记录数量(20)不同，然后返回 ExitStatus.STOPPED。在代码清单 6-22 中，你可以看到作业在控制台中的执行结果。

代码清单 6-22　transactionJob 的第一次执行

```
...
2018-03-01 22:02:35.770 INFO 36810 --- [main] o.s.b.a.b.JobLauncherCommandLineRunner:
Running default command line with: [transactionFile=/data/transactions.csv,
summaryFile=file://Users/mminella/tmp/summary.xml]
2018-03-01 22:02:35.873 INFO 36810 --- [main] o.s.b.c.l.support.SimpleJobLauncher:
Job: [FlowJob: [name=transactionJob]] launched with the following parameters:
[{transactionFile=/data/transactions.csv, summaryFile=file://Users/ mminella/
tmp/summary.xml}]
2018-03-01 22:02:35.918 INFO 36810 --- [main] o.s.batch.core.job.SimpleStepHandler:
Executing step: [importTransactionFileStep]
2018-03-01 22:03:16.435 INFO 36810 --- [main] o.s.b.c.l.support.SimpleJobLauncher:
Job: [FlowJob: [name=transactionJob]] completed with the following parameters:
[{transactionFile=/data/transactions.csv, summaryFile=file://Users/ mminella/
tmp/summary.xml}] and the following status: [STOPPED]
...
```

当作业停止时，删除 TRANSACTION 表的内容，更新交易文件为 100 条记录和文件摘要记录中保存的记录数量为 100。当再次执行作业时，作业就可以成功地执行，参见代码清单 6-23。

代码清单 6-23　transactionJob 的第二次执行

```
...
2018-03-01 22:04:17.102 INFO 36815 --- [main] o.s.b.c.l.support.SimpleJobLauncher:
Job: [FlowJob: [name=transactionJob]] launched with the following parameters:
[{transactionFile=/data/transactions.csv, summaryFile=file://Users/mminella/
tmp/summary.xml}]
2018-03-01 22:04:17.122 INFO 36815 --- [main] o.s.batch.core.job.SimpleStepHandler:
Executing step: [importTransactionFileStep]
2018-03-01 22:05:02.977 INFO 36815 --- [main] o.s.batch.core.job.SimpleStepHandler:
Executing step: [applyTransactionsStep]
2018-03-01 22:05:53.729 INFO 36815 --- [main] o.s.batch.core.job.SimpleStepHandler:
Executing step: [generateAccountSummaryStep]
2018-03-01 22:05:53.822 INFO 36815 --- [main] o.s.b.c.l.support.SimpleJobLauncher:
Job: [FlowJob: [name=transactionJob]] completed with the following parameters:
[{transactionFile=/data/transactions.csv, summaryFile=file://Users/mminella/tmp/
summary.xml}] and the following status: [COMPLETED]
...
```

使用 stop 转换并配置作业中的步骤，使之能够重新执行，从而允许根据作业执行中的检查来修复问题，这是有用的。接下来将重构监听器，使用 StepExecution#setTerminateOnly()方法与 Spring Batch 进行通信，以结束作业。

使用 StepExecution 停止作业

在 transactionJob 示例中，我们以手动方式处理作业的停止，具体是通过使用 StepExecutionListener 的退出状态并且在作业中配置转换来实现的。这种手动方式虽然十分奏效，但是需要配置作业的转换并重写步骤的退出状态。

另一种更简洁的方式是将 afterStep 替换为 beforeStep，并在其中获得 StepExecution 的句柄。一旦能够访问，那么当读取到文件末尾的摘要记录时，就可以调用 StepExecution#setTerminateOnly()方法以设置标识，从而告诉 Spring Batch 在完成当前步骤后结束，参见代码清单 6-24。

代码清单 6-24　使用了 setTerminateOnly()调用的 TransactionReader

```java
...
public class TransactionReader implements ItemStreamReader<Transaction> {

    private ItemStreamReader<FieldSet> fieldSetReader;
    private int recordCount = 0;
    private int expectedRecordCount = 0;

    private StepExecution stepExecution;

    public TransactionReader(ItemStreamReader<FieldSet> fieldSetReader) {
        this.fieldSetReader = fieldSetReader;
    }

    public Transaction read() throws Exception {
        Transaction record = process(fieldSetReader.read());

        return record;
    }

    private Transaction process(FieldSet fieldSet) {
        Transaction result = null;

        if(fieldSet != null) {
            if(fieldSet.getFieldCount() > 1) {
                result = new Transaction();
                result.setAccountNumber(fieldSet.readString(0));
                result.setTimestamp(fieldSet.readDate(1, "yyyy-mm-dd hh:mm:ss"));
                result.setAmount(fieldSet.readDouble(2));

                recordCount++;
            } else {
                expectedRecordCount = fieldSet.readInt(0);

                if(expectedRecordCount != this.recordCount) {
                    this.stepExecution.setTerminateOnly();
                }
            }
        }

        return result;
    }

    @BeforeStep
```

```
public void beforeStep(StepExecution execution) {
    this.stepExecution = execution;
}

@Override
public void open(ExecutionContext executionContext) throws ItemStreamException {
    this.fieldSetReader.open(executionContext);
}

@Override
public void update(ExecutionContext executionContext) throws ItemStreamException {
    this.fieldSetReader.update(executionContext);
}

@Override
public void close() throws ItemStreamException {
    this.fieldSetReader.close();
}
```

尽管代码只是略微简洁了一些(把对记录数量的检查从 afterStep(StepExecution execution) 方法移到了 read() 方法中)，但是在删除了转换配置后，配置也变得更加整洁。代码清单 6-25 展示了更新后的作业配置。

代码清单 6-25　重新配置的 transactionJob

```
...
@Bean
public Job transactionJob() {
    return this.jobBuilderFactory.get("transactionJob")
            .start(importTransactionFileStep())
            .next(applyTransactionsStep())
            .next(generateAccountSummaryStep())
            .build();
}
...
```

现在可以再次执行作业，进行相同的测试(在交易文件中使用错误的记录数量运行一次，然后使用正确的记录数量再运行一次)，你会看到同样的结果。唯一的区别出现在作业的输出中。作业并没有返回 STOPPED 状态，但是 Spring Batch 抛出了 JobInterruptedException 异常，参见代码清单 6-26。

代码清单 6-26　更新后的作业的第一次执行结果

```
2018-03-01 22:25:19.070 INFO 36931 --- [main] o.s.b.c.l.support.SimpleJobLauncher: 
Job: [SimpleJob: [name=transactionJob]] launched with the following parameters: 
[{transactionFile=/data/transactions.csv,summaryFile=file://Users/mminella/tmp
/summary.csv}]
2018-03-01 22:25:19.118 INFO 36931 --- [main] o.s.batch.core.job.SimpleStepHandler: 
Executing step: [importTransactionFileStep]
2018-03-01 22:26:05.265 INFO 36931 --- [main] o.s.b.c.s.ThreadStepInterruptionPolicy: 
Step interrupted through StepExecution
2018-03-01 22:26:05.266 INFO 36931 --- [main] o.s.batch.core.step.AbstractStep:
```

```
Encountered interruption executing step importTransactionFileStep in job
transactionJob : Job interrupted status detected.
2018-03-01 22:26:05.274 INFO 36931 --- [main] o.s.batch.core.job.AbstractJob:
Encountered interruption executing job: Job interrupted by step execution
2018-03-01 22:26:05.277 INFO 36931 --- [main] o.s.b.c.l.support.SimpleJobLauncher:
Job: [SimpleJob: [name=transactionJob]] completed with the following parameters:
[{transactionFile=/data/transactions.csv,summaryFile=file: //Users/mminella/tmp/
summary.csv}] and the following status: [STOPPED]
```

在设计批处理作业时,以编程的方式停止作业是一种重要的手段。遗憾的是,并不是所有的批处理作业都能完美运行,那么 Spring Batch 如何处理错误呢?答案就在 6.3.3 节中。

6.4.3 错误处理

没有完美的作业——就算自己编写的作业也是如此。你可能收到错误的数据,可能忘记执行 null 检查,进而在最坏的情况下导致空指针异常(NullPointerException)。如何使用 Spring Batch 处理错误是很重要的。接下来我们将讨论当批处理中抛出异常时的各种处理方法以及如何实现它们。

作业故障

Spring Boot 的默认行为看起来在情理之中:停止作业,然后回滚当前提交。这是基于块的处理的驱动概念之一,从而允许在作业成功完成后进行提交,并且在重启时能够从上次中断的地方继续。

默认情况下,当抛出任何异常时,Spring Batch 都会认为步骤和作业失败了。使用代码清单 6-27 所示的方式调整 TransactionReader,你就能看到实际发生的情况。此时,在读取了 510 条记录之后,抛出了 org.springframework.batch.item.ParseException 异常,停止了作业,并将作业标记为 FAILED 状态。

代码清单 6-27 抛出异常的 TransactionReader

```
...
public class TransactionReader implements ItemStreamReader<Transaction> {

    private ItemStreamReader<FieldSet> fieldSetReader;
    private int recordCount = 0;
    private int expectedRecordCount = 0;

    private StepExecution stepExecution;

    public TransactionReader(ItemStreamReader<FieldSet> fieldSetReader) {
        this.fieldSetReader = fieldSetReader;
    }

    public Transaction read() throws Exception {
        if(this.recordCount == 25) {
            throw new ParseException("This isn't what I hoped to happen");
        }
        Transaction record = process(fieldSetReader.read());

        return record;
    }

    private Transaction process(FieldSet fieldSet) {
        Transaction result = null;
```

第 6 章　运行作业

```
        if(fieldSet != null) {
           if(fieldSet.getFieldCount() > 1) {
               result = new Transaction();
               result.setAccountNumber(fieldSet.readString(0));
               result.setTimestamp(fieldSet.readDate(1, "yyyy-mm-dd hh:mm:ss"));
               result.setAmount(fieldSet.readDouble(2));

               recordCount++;
           } else {
               expectedRecordCount = fieldSet.readInt(0);

               if(expectedRecordCount != this.recordCount) {
                   this.stepExecution.setTerminateOnly();
               }
           }
        }

        return result;
    }

    @BeforeStep
    public void beforeStep(StepExecution execution) {
       this.stepExecution = execution;
    }

    @Override
    public void open(ExecutionContext executionContext) throws ItemStreamException {
       this.fieldSetReader.open(executionContext);
    }

    @Override
    public void update(ExecutionContext executionContext) throws ItemStreamException {
       this.fieldSetReader.update(executionContext);
    }

    @Override
    public void close() throws ItemStreamException {
       this.fieldSetReader.close();
    }
}
```

　　使用上一次的配置，当执行 transactionJob 时，在读取交易文件的第 25 条记录后，就会抛出 ParseException 异常，然后 Spring Batch 会认为步骤和作业失败了。如果查看控制台，你就会看到抛出了异常，而且作业也停止处理了。

　　通过 StepExecution 停止作业与使用异常停止作业有一个很大的区别，就是作业最后的状态不同。在使用 StepExecution 停止作业时，作业停止之前的步骤是以退出状态 STOPPED 完成的；而在使用异常停止作业时，步骤并没有完成，事实上，异常是在步骤执行过程中抛出的，因此步骤和作业的退出状态被标记为 FAILED。

　　当一个步骤被标记为 FAILED 状态时，Spring Batch 在重新运行这个步骤时并不会从头开始。相反，Spring Batch 足够聪明，它记得异常抛出时正在处理的块。当重启作业时，Spring Batch 将从之前停止的块开始。举个例子：作业处理了 10 个块中的 5 个，每个块都包含 5 个条目，在处理第 2 个块

135

的第 4 个条目时，抛出了异常，当前块的第 1～4 个条目都会回滚，并且当重新启动作业时，Spring Batch 会跳过第一和第二个块。

尽管 Spring Batch 处理异常的默认方法是以失败状态停止作业，但是还有一些其他方法。由于大部分方法都依赖于特定的输入输出场景，因此接下来的几章将结合 I/O(输入/输出)介绍它们。

6.5 控制作业的重启

Spring Batch 提供了许多用于停止和重启作业的设施。然而，作业能不能重启取决于使用者。如果有一个批处理过程，并且在第一个步骤中导入文件，然后作业在第二个步骤中失败，那么你可能并不想重新导入文件。在某些场景下，你可能在重试步骤时只想重试给定的次数。本节将介绍如何将作业配置为可重启，以及如何控制作业的重启。

6.5.1 阻止作业再次执行

到目前为止，当作业失败或停止时都会重新执行。这是 Spring Batch 的默认行为。但是，对于不能再次执行的作业，该怎么办呢？尝试执行一次，如果成功了，那么很好；如果失败了，那就不要再执行了。Spring Batch 为你提供了把作业配置为不可重启的能力，调用 JobBuilder 的 preventRestart() 方法即可。

如果查看 transactionJob 的配置，你就会发现默认情况下可以重启作业。然而，如果调用了 preventRestart()方法(如代码清单 6-28 所示)，那么当作业失败或以某种原因停止时，将不能再次执行作业。

代码清单 6-28　已配置为不可重启的 transactionJob

```
...
@Bean
public Job transactionJob() {
    return this.jobBuilderFactory.get("transactionJob")
                .preventRestart()
                .start(importTransactionFileStep())
                .next(applyTransactionsStep())
                .next(generateAccountSummaryStep())
                .build();
}
...
```

现在，如果在失败后尝试再次执行作业，Spring Batch 就会告诉你 JobInstance 已经存在，并且是不可重启的，如代码清单 6-29 所示。

代码清单 6-29　重新执行一个不可重启的作业的结果

```
2018-03-01 23:08:49.251 INFO 37017 --- [main] ConditionEvaluationReportLoggingListener :
Error starting ApplicationContext. To display the conditions report re-run your
application with 'debug' enabled.
2018-03-01 23:08:49.271 ERROR 37017 --- [main] o.s.boot.SpringApplication :
Application run failed

java.lang.IllegalStateException: Failed to execute CommandLineRunner
```

```
        at org.springframework.boot.SpringApplication.callRunner(SpringApplication.
    java:793)[spring-boot-2.0.0.RELEASE.jar:2.0.0.RELEASE]
        at org.springframework.boot.SpringApplication.callRunners(SpringApplication.
    java:774)[spring-boot-2.0.0.RELEASE.jar:2.0.0.RELEASE]
        at org.springframework.boot.SpringApplication.run(SpringApplication.java:
    335) [springboot-2.0.0.RELEASE.jar:2.0.0.RELEASE]
        at org.springframework.boot.SpringApplication.run(SpringApplication.java:1246)
    [spring-boot-2.0.0.RELEASE.jar:2.0.0.RELEASE]
        at org.springframework.boot.SpringApplication.run(SpringApplication.java:1234)
    [spring-boot-2.0.0.RELEASE.jar:2.0.0.RELEASE]
        at io.spring.batch.transaction_stop.TransactionStopApplication.
    main(TransactionStopApplication.java:20) [classes/:na]
Caused by: org.springframework.batch.core.repository.JobRestartException: JobInstance
already exists and is not restartable
        at org.springframework.batch.core.launch.support.SimpleJobLauncher.
    run(SimpleJobLauncher.java:101) ~[spring-batch-core-4.0.0.RELEASE.jar:4.0.0.
    RELEASE]
        at sun.reflect.NativeMethodAccessorImpl.invoke0(Native Method) ~[na:1.8.0_131]
        at sun.reflect.NativeMethodAccessorImpl.invoke(NativeMethodAccessorImpl.
    java:62)~[na:1.8.0_131]
        at sun.reflect.DelegatingMethodAccessorImpl.invoke(DelegatingMethodAccessorImpl.
    java:43)~[na:1.8.0_131]
        at java.lang.reflect.Method.invoke(Method.java:498) ~[na:1.8.0_131]
```

在某些情况下，只能执行一次作业可能有些极端。Spring Batch 还允许你配置作业能够运行的次数。

6.5.2 配置重启次数

在某些情况下，作业会因为不受控制的因素导致无法成功执行。例如，作业的第一个步骤是从一个网站下载文件，但是这个网站宕机了。如果第一次下载就失败了，那么 10 分钟之后再次尝试可能就会成功。不过，你可能不想无休止地尝试下载。正因为如此，你可能想将作业配置为仅仅执行 5 次。在第 5 次之后，就不再执行。

Spring Batch 在步骤而不是作业层次提供了这种功能。回到前面的 transactionJob 示例，如果在导入输入文件时只想尝试两次，那么可使用代码清单 6-30 所示的方式修改步骤的配置。

代码清单 6-30　在导入输入文件时只允许尝试两次

```
...
@Bean
public Step importTransactionFileStep() {
    return this.stepBuilderFactory.get("importTransactionFileStep")
                    .startLimit(2)
                    .<Transaction, Transaction>chunk(100)
                    .reader(transactionReader())
                    .writer(transactionWriter(null))
                    .allowStartIfComplete(true)
                    .listener(transactionReader())
                    .build();
}
...
```

在这个示例中，由于使用.startLimit(int limit)将执行次数限制为 2，因此如果试图重启作业的次数超过 1 次，就不能再次执行作业。初始运行已经占用了一次机会，因而只允许再尝试一次。如果试图再次执行作业，就会得到 org.springframework.batch.core.StartLimitExceededException 异常，如代码清单 6-31 所示。

代码清单 6-31　重新执行 transactionJob 超过一次时的结果

```
...
2018-03-01 23:12:17.205 ERROR 37027 --- [main] o.s.batch.core.job.AbstractJob:
Encountered fatal error executing job

org.springframework.batch.core.StartLimitExceededException: Maximum start limit
exceeded for step: importTransactionFileStepStartMax: 2
        at org.springframework.batch.core.job.SimpleStepHandler.shouldStart(SimpleStepHandler.
java:229)~[spring-batch-core-4.0.0.RELEASE.jar:4.0.0.RELEASE]
...
```

用于决定批处理作业重新执行时的行为的是 allowStartIfComplete() 方法，具体用法你在前面已经见识过。

6.5.3　重新运行一个完整的步骤

Spring Batch 框架规定作业在使用相同的识别性参数时，只能成功执行一次。这是无法改变的。不过，以上规定并不一定适用于步骤。

你可以重写 Spring Batch 框架的默认配置，并多次执行已经完成的步骤。在前面，我们可通过使用 transactionJob 来达到这个目的。为了在步骤完成后，让 Spring Batch 框架还能够再次执行某个步骤，你需要在 StepBuilder 上使用 allowStartIfComplete 方法。代码清单 6-32 展示了一个示例。

代码清单 6-32　配置一个在完成后可再次执行的步骤

```
...
@Bean
public Step importTransactionFileStep() {
    return this.stepBuilderFactory.get("importTransactionFileStep")
            .allowStartIfComplete(true)
            .<Transaction, Transaction>chunk(100)
            .reader(transactionReader())
            .writer(transactionWriter(null))
            .allowStartIfComplete(true)
            .listener(transactionReader())
            .build();
}
...
```

在这个示例中，当作业在前一次执行中失败或停止后，如果第二次执行步骤，步骤就会重新运行。由于步骤在上一次已经完成，因此没有可以重新开始的中间地带，这就是需要重新运行步骤的原因所在。

> ■ **注意**　如果作业的 BatchStatus 为 COMPLETE，那么无论是否将所有步骤配置为 allowStartIfComplete(true)，都不能重新运行 JobInstance。

在配置批处理过程时，Spring Batch 提供了许多用于停止和重启的方法。在一些场景下，可以重新执行整个作业,也可以尝试执行给定的次数;但在其他一些场景下,则完全不会重启作业。开发人员必须设计批处理作业,使它们能够在具体的场景中安全运行。

6.6 本章小结

启动或停止程序并不是什么常见的话题。但是如你所见,有很多用于控制 Spring Batch 作业执行的方法。当考虑批处理过程必须支持的各种场景时,这些方法是十分有意义的。本书接下来将介绍 Spring Batch 框架的主要部分:ItemReader、ItemProcessor 和 ItemWriter。

第 7 章
ItemReader

阅读、写作和算术被认为是在校儿童必须学习的基础技能。如果仔细思考，你就会发现相同的概念也适用于软件开发。无论是 Web 应用、批处理作业还是其他任何程序，基础都是数据的输入、以某种方式对数据进行处理以及数据的输出。

在使用 Spring Batch 时，以上概念也是显而易见的。基于块的每个步骤都由一个 ItemReader、一个 ItemProcessor 和一个 ItemWriter 组成。但是，读取数据在任何系统中都不是你想象的那么简单。有很多不同格式的文件——平面文件、XML 文件以及各种数据库，它们都只是潜在的部分输入源。

Spring Batch 提供了无须编写代码即可处理大多数输入形式的标准方法，并为你提供了编写自己的读取器的能力，以便读取那些不被支持的格式，例如读取 Web 服务。本章将逐一介绍 Spring Batch 框架中的各个 ItemReader 接口实现所提供的不同特性。

7.1 ItemReader 接口

在本章之前，你已经大致了解了 ItemReader(条目读取器)的概念，但还没有研究 Spring Batch 用于定义输入操作的接口。org.springframework.batch.item.ItemReader<T>接口定义的 read()方法用于提供步骤的输入，代码清单 7-1 展示了这个接口。

代码清单 7-1　org.springframework.batch.item.ItemReader<T>接口

```
package org.springframework.batch.item;

public interface ItemReader<T> {

    T read() throws Exception, UnexpectedInputException, ParseException,
                NonTransientResourceException;
}
```

代码清单 7-1 中的 ItemReader 接口是策略接口。Spring Batch 根据将要处理的输入类型提供了许多实现，包括平面文件、数据库、JMS 资源和其他输入源在内的各种实现。你还可以通过实现 ItemReader 接口或其任何子接口来实现自己的输入读取器。

> **注意** Spring Batch 的 org.springframework.batch.item.ItemReader<T>接口与 JSR-352(JBatch)中的 javax.batch.api.chunk 接口不同，主要区别在于前者提供了泛型支持，后者则结合了 ItemStream 和 ItemReader 接口。本书将使用前者。

ItemReader 接口的 read()方法用于返回在 Spring Batch 中调用步骤时处理的单个条目。步骤会统计条目的数量，并且保存在块(chunk)中已经处理了多少个条目的信息。条目在作为块的一部分发送给 ItemWriter 之前，将会被传给任意已经配置的 ItemProcessor。

理解如何使用 ItemReader 接口的最好方法就是实战练习。下面先从 FlatFileItemReader 开始。

7.2 文件输入

在 Java 中，I/O 处理相关的 API 只是比使用 Java 语言处理日期的 API 略好一点(尽管近年来情况已经有了很大好转)。幸运的是，使用 Spring Batch 的开发人员已经提供了许多声明式的读取器，解决了大部分此类问题。有了这些读取器，只需要声明将要读取的格式即可，其余的事情将由它们来处理。在本节，你将看到 Spring Batch 提供的声明式读取器，以及如何配置它们来处理基于文件的 I/O。

7.2.1 平面文件

当谈到批处理中的平面文件(Flat File)时，指的是任何具有一条或多条记录的文件。每条记录可以占用一行或多行。平面文件和 XML 文件之间的区别在于，平面文件中的数据不是描述性的。换句话说，文件内部没有定义数据格式或含义的元信息。相反，XML 文件使用标签赋予数据一定的含义。

在真正地配置平面文件的 ItemReader 之前，我们先看看 Spring Batch 中用于读取文件的各个组件。图 7-1 展示了 FlatFileItemReader 的各个部分。org.springframework.batch.item.file.FlatFileItemReader 由两个主要部分组成：一部分是用来代表待读取文件的 Spring 资源，另一部分是 org.springframework.batch.item.file.LineMapper 接口的实现。LineMapper 的功能与 Spring JDBC 中的 RowMapper 类似。在 Spring JDBC 中使用 RowMapper 时，代表字段集合的 ResultSet 用于把字段映射到对象。

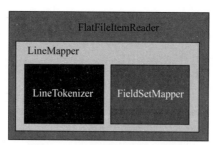

图 7-1 FlatFileItemReader 的各个部分

FlatFileItemReader 允许你配置与正在读取的文件相关的许多选项。表 7-1 展示了你可能会使用的一些选项，并解释了它们的含义。

表 7-1 FlatFileItemReader 的配置选项

选项	类型	默认值	描述
comments	String[]	null	此字符串数组指示了在文件解析期间哪些前缀将被视为行注释并被跳过
currentItemCount	int	0	正在读取的条目的索引，可在重启时使用
encoding	String	当前平台的默认字符集	文件的字符编码
lineMapper	LineMapper	null(必需的)	将文件的每一行作为字符串，并转换为将要处理的领域对象(条目)
linesToSkip	int	0	运行作业时，可以将 FlatFileItemReader 配置为在解析之前跳过文件开头的行。这个数字表示跳过多少行
maxItemCount	int	Integer.MAX_VALUE	指示要从文件读取的条目的最大数量
name	String	null	用于为持久化到 ExecutionContext 中的值创建唯一键
recordSeparatorPolicy	RecordSeparatorPolicy	DefaultRecordSeparatorPolicy	用于确定每条记录的结束。默认情况下，行结束字符表明记录已经结束；然而，这个选项也可以用于处理跨行引用的字符串之类的情况
resource	Resource	null(必需的)	要读取的资源
saveState	boolean	true	用于指示在处理每个块之后，应该保存 ItemReader 状态，以达到重启目的。如果要在多线程环境中使用，就应该设置为 false
skippedLinesCallback	LineCallbackHandler	null	为跳过的行调用的回调接口。跳过的每一行都将被传递给这个回调
strict	boolean	false	在严格(strict)模式下，如果找不到资源，将抛出异常

读取文件时，Spring Batch 会向 LineMapper 接口的实现提供一个字符串，以代表文件中的一条记录。最常用的 LineMapper 接口实现是 DefaultLineMapper。对于文件中的原始字符串，DefaultLineMapper 会通过两步将其转换成接下来要使用的领域对象。这两步由 LineTokenizer 和 FieldSetMapper 处理：

- LineTokenizer 的实现会将行解析为 org.springframework.batch.item.file.FieldSet。提供的字符串表示文件中的一整行。为了能够将每条记录的各个字段映射到领域对象，需要将行解析为字段的集合。Spring Batch 中的 FieldSet 表示单行的字段集合(类似于使用数据库时的 java.sql.ResultSet)。
- FieldSetMapper 的实现会将 FieldSet 映射到领域对象。把行划分为各个字段后，可以将每个输入字段映射到领域对象的字段，就像 RowMapper 将 ResultSet 行映射到领域对象一样。

这听起来很简单,不是吗?复杂的部分在于如何解析行,以及如何使用文件中的多条记录构建对象。下面首先看看如何读取固定宽度的文件。

1. 固定宽度的文件

在处理遗留的大型机系统时,通常必须使用固定宽度的文件(又称定宽文件),这种存储方式是由 COBOL、大数据以及其他技术决定的。因此,我们必须能够处理固定宽度的文件。

可以把客户文件作为固定宽度的文件使用,客户文件由客户的姓名和地址组成。表 7-2 概述了客户文件的格式。

表 7-2 客户文件的格式

字段	长度	描述
First Name	11	客户的名字
Middle Initial	1	客户的中间名的首字母
Last Name	10	客户的姓氏
Address Number	4	客户地址的街道编号
Street	20	客户地址的街道名称
City	16	客户所在的城市
State	2	包含两个字母的州的缩写
Zip Code	5	客户的邮政编码

为固定宽度的文件定义格式很重要。分隔符文件使用分隔符描述字段。由于给定的标签提供了元数据,因此 XML 文件或其他结构化文件是自描述的。数据库中包含用于描述数据的元数据。但是,固定宽度的文件与众不同,它们没有提供任何元数据来描述文件格式。代码清单 7-2 展示了前面描述的输入文件。

代码清单 7-2　customer.txt(固定宽度的文件)

```
Aimee     CHoover    7341Vel          Avenue Mobile    AL35928
Jonas     UGilbert   8852In St.       Saint Paul       MN57321
Regan     MBaxter    4851Nec Av.      Gulfport         MS3319
Sydnee    NRobinson  894 Ornare. Ave  Olathe           KS25606
```

为了读取以上输入文件,我们需要使用领域对象来表示这些记录。你在代码清单 7-3 中可以找到将要使用的 Customer 对象。

代码清单 7-3　Customer.java

```
...
public class Customer {

    private String firstName;
    private String middleInitial;
    private String lastName;
    private String addressNumber;
    private String street;
    private String city;
    private String state;
    private String zipCode;
```

第 7 章　ItemReader

```
    public Customer() {
    }

    public Customer(String firstName, String middleInitial, String lastName, String
    addressNumber, String street, String city, String state, String zipCode) {
        this.firstName = firstName;
        this. middleInitial = middleInitial;
        this.lastName = lastName;
        this.addressNumber = addressNumber;
        this.street = street;
        this.city = city;
        this.state = state;
        this.zipCode = zipCode;
    }

    // Getters and setters removed for brevity
    ...

    @Override
    public String toString() {
        return "Customer{" +
                "firstName='" + firstName + '\"' +
                ", middleInitial ='" + middleInitial + '\"' +
                ", lastName='" + lastName + '\"' +
                ", addressNumber='" + addressNumber + '\"' +
                ", street='" + street + '\"' +
                ", city='" + city + '\"' +
                ", state='" + state + '\"' +
                ", zipCode='" + zipCode + '\"' +
                '}';
    }
}
```

为了演示这些读取器如何工作，我们创建单步作业 copyJob，读取文件并将其写回。对于这个作业，可使用以下 Bean 创建 BatchConfiguration 配置类。

- customerReader：平面文件读取器(FlatFileItemReader)。
- outputWriter：平面文件写入器(FlatFileItemWriter)。
- copyStep：步骤的定义。
- copyJob：作业的定义。

customerReader 是 FlatFileItemReader 的实例。如前所述，FlatFileItemReader 主要由两部分组成：要读入的资源(在本例中为 customerFile)以及用来映射文件中每一行的方式(一个 LineMapper 实现)。

这个 LineMapper 实现将使用 Spring Batch 的 org.springframework.batch.item.file.DefaultLineMapper。DefaultLineMapper 将用于实现将行映射到先前讨论的领域对象的两步过程：首先将一行数据解析到 FieldSet，然后将 FieldSet 的字段映射到领域对象(在本例中为 Customer 对象)。

为了支持这个两步映射过程，DefaultLineMapper 使用了两个依赖项：一个 LineTokenizer 实现，用来把从文件中读取的字符串解析为 FieldSet；以及一个 FieldSetMapper 实现，用来将 FieldSet 中的字段映射到领域对象的字段。

这听起来可能需要很多代码。在 Spring Batch 4 之前确实如此。不过，Spring Batch 4 包含了一组构建器，可以简化常见用例的配置。我们将在本章的所有示例中使用构建器。代码清单 7-4 展示了通

145

过 FlatFileItemReaderBuilder 创建的 customerReader。

代码清单 7-4　BatchConfiguration 中的 customerReader

```
...
@Bean
@StepScope
public FlatFileItemReader<Customer> customerItemReader(
@Value("#{jobParameters['customerFile']}") Resource inputFile) {

    return new FlatFileItemReaderBuilder<Customer>()
                .name("customerItemReader")
                .resource(inputFile)
                .fixedLength()
                .columns(new Range[]{new Range(1,11), new Range(12, 12), new Range(13, 22),
                    new Range(23, 26), new Range(27,46), new Range(47,62),
                    new Range(63,64), new Range(65,69)})
                .names(new String[] {"firstName", "middleInitial", "lastName",
                    "addressNumber", "street", "city", "state","zipCode"})
                .targetType(Customer.class)
                .build();
}
...
```

代码清单 7-4 包含了用于配置读取固定宽度的文件所需的全部内容。然而，这里的构建器隐藏了许多实际发生的情况，因此我们下面逐步了解一下以上配置。

我们首先从 inputFile 方法的参数开始。这里把输入文件的路径作为作业参数传给了 Spring Boot 应用，Spring 会自动为其创建资源(Resource)并注入。

我们接下来创建了构建器，并使用构建器配置了名称(name)。我们将在本章后面详细介绍 ItemStream 接口，该接口要求我们提供名称(name)，以作为前缀添加到这个步骤的执行上下文 (ExecutionContext)的任何键。举个例子，这允许在同一步骤中使用两个 FlatFileItemReader 实例，而它们不会共享持久化状态。如果读取器的 saveState 配置被设置为 false，则不需要这个值(在这种情况下，不会在 ExecutionContext 中存储任何数据。因此，读取器在重新启动时会从头开始)。接下来，我们对构建器执行的操作是为了让它知道我们正在使用固定宽度的文件，也就是调用 fixedLength()方法。该调用将返回专用于构建 FixedLengthTokenizer 的构建器。FixedLengthTokenizer 是 LineTokenzier 接口的实现，读取器将使用该接口把每一行解析为 FieldSet。FixedLengthTokenizer 需要配置每条记录中每一列的名称以及一个 Range 对象的数组。每个 Range 实例代表了要解析的列的开始索引和结束索引。使用 FixedLengthTokenizer 进行配置时，也可采用用于创建 FieldSet 的 FieldSetFactory(默认提供了 DefaultFieldSetFactory)，或者使用 strict 标识，用于指示如何处理记录比定义的要解析的记录更多的情况(strict 标识的默认值为 true，表示在这种情况下会抛出异常)。在代码清单 7-4 中，由于默认值足够好，因此我们忽略了可选值，仅仅为输入文件配置了区间范围(range)和名称。

如果还记得的话，我们说过 FlatFileItemReader 使用 LineMapper 将文件记录转换为对象。代码清单 7-3 中的构建器使用了 DefaultLineMapper，其中有两个依赖项：我们刚刚配置的 LineTokenizer 和 FieldSetMapper(如果愿意，可以指定自己的构建器)。当调用.targetType(Class targetType)方法时，我们使用的构建器将创建一个新的 BeanWrapperFieldSetMapper。这个 BeanWrapperFieldSetMapper 将使用列名来设置配置的类实例的对应值。举例来说，BeanWrapperFieldSetMapper 将基于你在LineTokenizer 中配置的列名来调用 Customer#setFirstName、Customer#setMiddleInitial 等方法，从而设置对应的值。

■ **提示** FixedLengthTokenizer 不会裁剪(trim)字段中的任何前导字符或结尾字符(比如空格和 0)。为此，你必须实现自己的 LineTokenizer，或在 FieldSetMapper 中进行删除。

为了使用读取器，你需要配置步骤和作业，还需要配置写入器，以便让一切正常。第 9 章将深入介绍写入器。代码清单 7-5 展示了如何配置一个简单的写入器，以便将领域对象输出到标准输出。

代码清单 7-5　一个简单的写入器

```
...
@Bean
public ItemWriter<Customer> itemWriter() {
    return (items) -> items.forEach(System.out::println);
}
...
```

查看代码清单 7-5 中的写入器，ItemWriter 实际上是一个函数式接口，可返回一个匿名函数作为自身的实现。在这种情况下，对于传给 ItemWriter.write(List<T>items)方法的列表(List)中的每个条目，都会通过 System.out.println 调用.toString()方法，并在控制台中展示输出。

作业配置也非常简单。如代码清单 7-6 所示，只需要创建一个简单的步骤，它由一个读取器和一个写入器组成，并且提交次数为 10 次。

代码清单 7-6　copyFileStep 和 copyFileJob

```
...
@Bean
public Step copyFileStep() {
    return this.stepBuilderFactory.get("copyFileStep")
                .<Customer, Customer>chunk(10)
                .reader(customerItemReader(null))
                .writer(outputWriter(null))
                .build();
}

@Bean
public Job job() {
    return this.jobBuilderFactory.get("job")
                .start(copyFileStep())
                .build();
}
...
```

其中有趣的事情是，我们没有为领域对象(Customer 对象)编写应用代码。我们编写的代码仅用于配置应用。构建应用后，我们可以使用代码清单 7-7 所示的命令来执行应用。

代码清单 7-7　执行 copyJob

```
java -jar copyJob.jar customerFile=/path/to/customer/file.txt
```

输出结果将是基于我们为写入器配置的格式化字符串对输入文件进行格式化之后的内容，如代码清单 7-8 所示。

代码清单 7-8　执行 copyJob 后的输出结果

```
2019-01-28 16:11:44.089 INFO 54762 --- [           main] o.s.b.c.l.support.
SimpleJobLauncher        : Job: [SimpleJob: [name=job]] launched with the following
parameters: [{customerFile=/input/customerFixedWidth.txt}]
2019-01-28 16:11:44.159 INFO 54762 --- [           main] o.s.batch.core.job.
SimpleStepHandler        : Executing step: [copyFileStep]
Customer{firstName='Aimee', middleInitial='C', lastName='Hoover', addressNumber='7341',
street='Vel Avenue', city='Mobile', state='AL', zipCode='35928'}
Customer{firstName='Jonas', middleInitial='U', lastName='Gilbert', addressNumber='8852',
street='In St.', city='Saint Paul', state='MN', zipCode='57321'}
...
```

固定宽度的文件是许多企业为批处理提供的输入格式。如你所见，通过使用 FlatFileItemReader 和 FixedLengthTokenizer 将文件解析为对象，可以让这个过程变得容易一些。接下来将介绍分隔符文件，这种文件为我们提供了少量的元数据，以告诉我们如何解析文件。

2. 分隔符文件

在分隔符文件中，字符将充当记录中字段之间的分隔符。我们不必知道每个字段的定义。取而代之的是，文件本身就表明了字段的组成，方法是使用分隔符对每条记录进行分隔。

读取分隔符文件的过程与读取固定宽度的文件一样。记录将首先由 LineTokenizer 标记为 FieldSet。然后，FieldSet 被 FieldSetMapper 映射到领域对象。处理过程是相同的，你需要做的只是更新用于解析文件的 LineTokenizer 的实现。我们首先来看一下更新后的客户文件，它是分隔符文件而不是固定宽度的文件。代码清单 7-9 展示了新的输入文件。

代码清单 7-9　使用分隔符的客户文件

```
Aimee,C,Hoover,7341,Vel Avenue,Mobile,AL,35928
Jonas,U,Gilbert,8852,In St.,Saint Paul,MN,57321
Regan,M,Baxter,4851,Nec Av.,Gulfport,MS,33193
```

你将立刻注意到，新文件和旧文件之间有两处不同。首先，这里使用逗号来分隔字段。其次，你已经裁剪(trim)了所有字段。通常，当使用分隔符文件时，每个字段不会像固定宽度的文件中那样被填充为固定宽度。因此，与固定宽度的记录长度不同，这里的记录长度可以变化。

就像前面提到的那样，在使用新的文件格式时，唯一需要更改的配置是每条记录的解析方式。对于固定宽度的记录，可以使用 FixedLengthTokenizer 解析每一行。对于新的分隔符记录，可以使用 org.springframework.batch.item.file.transform.DelimitedLineTokenizer 将记录解析为 FieldSet。代码清单 7-10 展示了更新后的使用 DelimitedLineTokenizer 的读取器配置。

代码清单 7-10　使用 DelimitedLineTokenizer 的 customerFileReader

```
...
@Bean
@StepScope
public FlatFileItemReader<Customer> customerItemReader(@Value("#{jobParameters
    ['customerFile']}") Resource inputFile) {
        return new FlatFileItemReaderBuilder<Customer>()
            .name("customerItemReader")
            .delimited()
            .names(new String[] {"firstName",
```

```
                "middleInitial",
                "lastName",
                "addressNumber",
                "street",
                "city",
                "state",
                "zipCode"})
            .targetType(Customer.class)
            .resource(inputFile)
            .build();
}
...
```

DelimitedLineTokenizer 有两个非常有用的功能。首先是能够配置分隔符，默认使用的是逗号，但是也可以使用任何字符串。其次是能够配置引号(quote)字符的值，可指定字符代替默认的英文双引号作为引号字符。代码清单 7-11 给出的示例展示了如何使用#字符作为引号字符来解析字符串。

代码清单 7-11　使用引号字符解析分隔符文件

```
Michael,T,Minella,#123,4th Street#,Chicago,IL,60606

    Is parsed as

Michael
T
Minella
123,4th Street
Chicago
IL
60606
```

这就是处理分隔符文件所需的全部内容。前面的示例将地址编号和街道映射到了两个不同的字段。然而，如果想将它们一起映射到代码清单 7-12 中的领域对象所表示的单个字段，该怎么办呢？

代码清单 7-12　使用单一的街道地址字段的客户

```
package com.apress.springbatch.chapter7;

public class Customer {
    private String firstName;
    private String middleInitial;
    private String lastName;
    private String address;
    private String city;
    private String state;
    private String zip;

    // Getters & setters go here
    ...
}
```

为了使用新的对象格式，你需要更新将 FieldSet 映射到领域对象的方式。为此，创建自己的 org.springframework.batch.item.file.mapping.FieldSetMapper 接口实现。如代码清单 7-13 所示，FieldSetMapper

接口只有一个方法 mapFieldSet(FieldSet fieldSet)，该方法允许你把 LineTokenizer 返回的 FieldSet 映射到领域对象的字段。

代码清单 7-13　FieldSetMapper 接口

```java
package org.springframework.batch.item.file.mapping;

import org.springframework.batch.item.file.transform.FieldSet;
import org.springframework.validation.BindException;

public interface FieldSetMapper<T> {

    T mapFieldSet(FieldSet fieldSet) throws BindException;
}
```

为了创建自己的映射器(mapper)，你需要使用 Customer 类型来实现 FieldSetMapper 接口。然后，如代码清单 7-14 所示，将 FieldSet 中的每个字段映射到领域对象，并根据需求将 addressNumber 和 street 字段连接为 address 字段。

代码清单 7-14　把 FieldSet 中的字段映射到 Customer 对象

```java
...
public class CustomerFieldSetMapper implements FieldSetMapper<Customer> {

    public Customer mapFieldSet(FieldSet fieldSet) {
        Customer customer = new Customer();

        customer.setAddress(fieldSet.readString("addressNumber") +
                            " " + fieldSet.readString("street"));
        customer.setCity(fieldSet.readString("city"));
        customer.setFirstName(fieldSet.readString("firstName"));
        customer.setLastName(fieldSet.readString("lastName"));
        customer.setMiddleInitial(fieldSet.readString("middleInitial"));
        customer.setState(fieldSet.readString("state"));
        customer.setZipCode(fieldSet.readString("zipCode"));

        return customer;
    }
}
```

FieldSet 与 JDBC 领域的 ResultSet 非常相似。Spring 为每一种原生数据类型(包括 String(已裁剪或未裁剪)、BigDecimal 和 java.util.Date)提供了一个方法，其中每一个方法都有两个版本。其中一个版本采用整数(integer)作为参数，这里的整数表示要在记录中检索的字段的索引；代码清单 7-15 展示的另一个版本使用了字段的名称，尽管这种方式需要在作业配置中为字段命名，但从长远看，这是一种更易于维护的模型。

代码清单 7-15　FieldSet 接口

```java
package org.springframework.batch.item.file.transform;
import java.math.BigDecimal;
import java.sql.ResultSet;
import java.util.Date;
import java.util.Properties;
```

```java
public interface FieldSet {

    String[] getNames();
    boolean hasNames();
    String[] getValues();
    String readString(int index);
    String readString(String name);
    String readRawString(int index);
    String readRawString(String name);
    boolean readBoolean(int index);
    boolean readBoolean(String name);
    boolean readBoolean(int index, String trueValue);
    boolean readBoolean(String name, String trueValue);
    char readChar(int index);
    char readChar(String name);
    byte readByte(int index);
    byte readByte(String name);
    short readShort(int index);
    short readShort(String name);
    int readInt(int index);
    int readInt(String name);
    int readInt(int index, int defaultValue);
    int readInt(String name, int defaultValue);
    long readLong(int index);
    long readLong(String name);
    long readLong(int index, long defaultValue);
    long readLong(String name, long defaultValue);
    float readFloat(int index);
    float readFloat(String name);
    double readDouble(int index);
    double readDouble(String name);
    BigDecimal readBigDecimal(int index);
    BigDecimal readBigDecimal(String name);
    BigDecimal readBigDecimal(int index, BigDecimal defaultValue);
    BigDecimal readBigDecimal(String name, BigDecimal defaultValue);
    Date readDate(int index);
    Date readDate(String name);
    Date readDate(int index, Date defaultValue);
    Date readDate(String name, Date defaultValue);
    Date readDate(int index, String pattern);
    Date readDate(String name, String pattern);
    Date readDate(int index, String pattern, Date defaultValue);
    Date readDate(String name, String pattern, Date defaultValue);
    int getFieldCount();
    Properties getProperties();
}
```

■ **注意** 与 JDBC 中的 ResultSet 的索引列从 1 开始索引不同，Spring Batch 中的 FieldSet 使用的索引从 0 开始编号。

为了使用 CustomerFieldSetMapper，你还需要更新配置。把 BeanWrapperFieldSetMapper 替换为已经编写的 Bean 引用，如代码清单 7-16 所示。

代码清单 7-16　使用 CustomerFieldSetMapper 的 customerFileReader

```
...
@Bean
@StepScope
public FlatFileItemReader<Customer> customerItemReader(@Value("#{jobParameters
['customerFile']}")Resource inputFile) {
    return new FlatFileItemReaderBuilder<Customer>()
            .name("customerItemReader")
            .delimited()
            .names(new String[] {"firstName",
                    "middleInitial",
                    "lastName",
                    "addressNumber",
                    "street",
                    "city",
                    "state",
                    "zip"})
            .fieldSetMapper(new CustomerFieldSetMapper())
            .resource(inputFile)
            .build();
}
...
```

注意，当使用新的 CustomerFieldSetMapper 时，无须对 Customer Bean 的引用进行配置。

如你所见，使用标准的 Spring Batch 解析器解析文件时只需要几行 Java 代码。但是，并不是所有文件中包含的 Unicode 字符对 Java 来说都是易于理解的。在处理遗留系统时，你经常会遇到需要自定义记录解析的数据存储技术。接下来，我们将研究如何实现自己的 LineTokenizer，以便能够处理自定义的文件格式。

3. 自定义记录解析

前面探讨了如何通过创建自定义的 FieldSetMapper 实现，将文件中的字段映射到领域对象的字段。但是，这不是唯一选择。相反，你也可以创建自己的 LineTokenizer 实现，这样就可以按需解析每一条记录。

就像 FieldSetMapper 接口一样，org.springframework.batch.item.file.transform.LineTokenizer 接口也只有 tokenize 方法。代码清单 7-17 展示了 LineTokenizer 接口。

代码清单 7-17　LineTokenizer 接口

```
package org.springframework.batch.item.file.transform;

public interface LineTokenizer {

    FieldSet tokenize(String line);
}
```

在这种方式下，你将使用与以前一样的分隔符文件。但是，由于领域对象将地址编号和街道连接为单个字段，因此你需要把这两个标记(token)合并到 FieldSet 中的一个字段中。代码清单 7-18 展示了 CustomerFileLineTokenizer。

第 7 章 ItemReader

代码清单 7-18　CustomerFileLineTokenizer

```java
...
public class CustomerFileLineTokenizer implements LineTokenizer {

    private String delimiter = ",";
    private String[] names = new String[] {"firstName",
                                            "middleInitial",
                                            "lastName",
                                            "address",
                                            "city",
                                            "state",
                                            "zipCode"};

    private FieldSetFactory fieldSetFactory = new DefaultFieldSetFactory();

    public FieldSet tokenize(String record) {

        String[] fields = record.split(delimiter);

        List<String> parsedFields = new ArrayList<>();

        for (int i = 0; i < fields.length; i++) {
            if (i == 4) {
                parsedFields.set(i - 1,
                    parsedFields.get(i - 1) + " " + fields[i]);
            } else {
                parsedFields.add(fields[i]);
            }
        }

        return fieldSetFactory.create(parsedFields.toArray(new String [0]),
                    names);
    }
}
```

CustomerFileLineTokenizer 的 tokenize(String line)方法接收每一条记录,并使用配置好的分隔符对记录进行拆分。遍历这些字段,将第三和第四个字段组合在一起,使它们成为一个字段。然后,使用 DefaultFieldSetFactory 创建一个 FieldSet,向其传递一个必要参数(一个作为字段值的数组)和一个可选参数(一个作为字段名称的数组)。CustomerFileLineTokenizer 给每个字段添加了名称,以便使用 BeanWrapperFieldSetMapper 把 FieldSet 映射到领域对象,并且无需任何额外的代码。

配置 CustomerFileLineTokenizer 的方式与配置自定义的 FieldSetMapper 类似:删除其他的配置并替换为单个方法调用。代码清单 7-19 展示了更新后的配置。

代码清单 7-19　配置 CustomerFileLineTokenizer

```java
...
@Bean
@StepScope
public FlatFileItemReader<Customer> customerItemReader(@Value("#{jobParameters['customer File']}")Resource inputFile) {
    return new FlatFileItemReaderBuilder<Customer>()
            .name("customerItemReader")
```

153

```
                    .lineTokenizer(new CustomerFileLineTokenizer())
                    .fieldSetMapper(new CustomerFieldSetMapper())
                    .resource(inputFile)
                    .build();
    }
    ...
```

使用自己的 LineTokenizer 和 FieldSetMapper 时是没有限制的。自定义 LineTokenizer 的其他用途可能包括：

- 解析非常规的文件格式。
- 解析第三方文件格式，例如微软的 Excel 工作表。
- 处理特殊类型的转换需求

然而，并非所有文件都像你使用过的客户文件那样简单。如果文件中包含多种记录格式，怎么办？接下来我们将讨论 Spring Batch 如何选择合适的 LineTokenizer 来解析遇到的每条记录。

4. 多种记录格式

到目前为止，你一直在处理包含客户记录集合的客户文件。客户文件中的每条记录都具有完全相同的格式。可是，如果收到同时包含客户信息和交易信息的文件怎么办？可以自定义 LineTokenizer。但是，这样做存在如下两个问题。

- 复杂性：如果文件具有三种、四种、五种或更多种行格式(每种都有大量字段)，那么这个类可能很快就失控了。
- 关注点分离：LineTokenizer 用于解析记录，仅此而已，在解析之前，不需要决定记录的类型。

为此，Spring Batch 提供了另一个 LineMapper 实现：org.springframework.batch.item.file.mapping.PatternMatchingCompositeLineMapper。前面的示例使用了 DefaultLineMapper，从而让你有了使用单个 LineTokenizer 和单个 FileSetMapper 的能力。通过使用 PatternMatchingCompositeLineMapper，就可以定义 LineTokenizer 的映射(Map)以及对应的 FieldSetMapper 的映射。每个映射的键都是一种模式，由 LineMapper 用来标识在解析每条记录时使用哪个 LineTokenizer。

让我们看看这个示例，在该例中，客户记录没有变化。然而，每条客户记录之间夹杂着随机数量的交易记录。为了帮助识别每条记录的类型，这里为每条记录添加了前缀。代码清单 7-20 展示了更新后的客户文件。

代码清单 7-20　更新后的客户文件

```
CUST,Warren,Q,Darrow,8272 4th Street,New York,IL,76091
TRANS,1165965,2011-01-22 00:13:29,51.43
CUST,Ann,V,Gates,9247 Infinite Loop Drive,Hollywood,NE,37612
CUST,Erica,I,Jobs,8875 Farnam Street,Aurora,IL,36314
TRANS,8116369,2011-01-21 20:40:52,-14.83
TRANS,8116369,2011-01-21 15:50:17,-45.45
TRANS,8116369,2011-01-21 16:52:46,-74.6
TRANS,8116369,2011-01-22 13:51:05,48.55
TRANS,8116369,2011-01-21 16:51:59,98.53
```

在代码清单 7-20 所示的客户文件中，有两种以逗号分隔的格式。其中，第一种包含到目前为止我们一直使用的标准客户格式，比如连接起来的地址编号和街道。这些记录以前缀 CUST 作为标志。

其他记录是交易记录；其中每一条记录都以 TRANS 为前缀，也用逗号分隔，并且具有以下三个字段。
- accountNumber：客户的账号。
- Date：交易发生的日期。交易可能会也可能不会以日期排序。
- amount：以美元为单位的交易金额。负值表示借方，正值表示贷方。

代码清单 7-21 展示了 Transaction 领域对象的代码。

代码清单 7-21　Transaction 领域对象的代码

```
...
public class Transaction {

    private String accountNumber;
    private Date transactionDate;
    private Double amount;

    private DateFormat formatter = new SimpleDateFormat("mm/dd/yyyy");

    // Getters and setters are omitted
      @Override
      public String toString() {
              return "Transaction{" +
                      "accountNumber='" + accountNumber + '\"' +
                      ", transactionDate=" + transactionDate +
                      ", amount=" + amount +
                      '}';
    }
}
```

确定了记录的格式后，就可以查看读取器。代码清单 7-22 展示了更新后的 customerFileReader 的配置。如前所述，使用 PatternMatchingCompositeLineMapper 可以映射 DelimitedLineTokenizer 的两个实例，其中每个实例都配置了正确的记录格式。你会注意到，每个 LineTokenizer 都有一个名为 prefix 的附加字段，用于指明每条记录开头的标志字符串（CUST 和 TRANS）。Spring Batch 会解析前缀，并在 FieldSet 中对它们进行命名；然而，由于这里的两个领域对象都没有前缀字段，因此在映射中将忽略前缀。

代码清单 7-22　使用多种记录格式配置 customerFileReader

```
...
@Bean
@StepScope
public FlatFileItemReader customerItemReader(
      @Value("#{jobParameters['customerFile']}")Resource inputFile) {

    return new FlatFileItemReaderBuilder<Customer>()
            .name("customerItemReader")
            .lineMapper(lineTokenizer())
            .resource(inputFile)
            .build();
}

@Bean
public PatternMatchingCompositeLineMapper lineTokenizer() {
```

```java
        Map<String, LineTokenizer> lineTokenizers = new HashMap<>(2);

        lineTokenizers.put("CUST*", customerLineTokenizer());
        lineTokenizers.put("TRANS*", transactionLineTokenizer());

        Map<String, FieldSetMapper> fieldSetMappers = new HashMap<>(2);

        BeanWrapperFieldSetMapper<Customer> customerFieldSetMapper =
          new BeanWrapperFieldSetMapper<>();
        customerFieldSetMapper.setTargetType(Customer.class);

        fieldSetMappers.put("CUST*", customerFieldSetMapper);
        fieldSetMappers.put("TRANS*", new TransactionFieldSetMapper());

        PatternMatchingCompositeLineMapper lineMappers =
          new PatternMatchingCompositeLineMapper();

        lineMappers.setTokenizers(lineTokenizers);
        lineMappers.setFieldSetMappers(fieldSetMappers);

        return lineMappers;
    }

    @Bean
    public DelimitedLineTokenizer transactionLineTokenizer() {
        DelimitedLineTokenizer lineTokenizer = new DelimitedLineTokenizer();

        lineTokenizer.setNames("prefix", "accountNumber", "transactionDate", "amount");

        return lineTokenizer;
    }

    @Bean
    public DelimitedLineTokenizer customerLineTokenizer() {
        DelimitedLineTokenizer lineTokenizer = new DelimitedLineTokenizer();

        lineTokenizer.setNames("firstName",
          "middleInitial",
          "lastName",
          "address",
          "city",
          "state",
          "zip");

        lineTokenizer.setIncludedFields(1, 2, 3, 4, 5, 6, 7);

        return lineTokenizer;
    }
    ...
```

customerFileReader 读取器的配置开始变得有些冗长。让我们来看看执行这个读取器时实际发生的情况。查看图 7-2，你可以在流程中看到 customerFileReader 如何处理每一行。

第 7 章 ItemReader

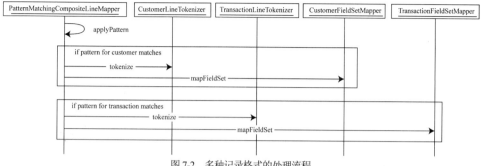

图 7-2 多种记录格式的处理流程

如图 7-2 所示，PatternMatchingCompositeLineMapper 将查看文件中的每条记录并把模式应用于记录。如果记录以 CUST,*开头(其中*是零个或多个字符)，就把记录传递给 customerLineTokenizer 进行解析。一旦将记录解析为 FieldSet 之后，记录将被传递给 FieldSetMapper。在本例中，我们使用的是 Spring Batch 框架中的 BeanWrapperFieldSetMapper。出于这个原因，并且因为我们的 Customer 领域对象没有前缀字段，所以我们希望分词器跳过前缀字段。为此，你需要做两件事。首先，打开我们在 DelimitedLineTokenizer 中配置的字段名称列表，将前缀名称排除。其次，提供要包括的字段的索引(从 0 开始)列表。在本例中，我们希望包括除前缀外的所有字段。

如果记录以 TRANS,*开头，记录将被传给 transactionLineTokenizer 进行解析，解析结果 FieldSet 将被传给自定义的 transactionFieldSetMapper。

但是，为什么需要自定义 FieldSetMapper 呢？自定义类型转换是必需的。默认情况下，BeanWrapperFieldSetMapper 并不执行任何特殊的类型转换。Transaction 领域对象包含的 accountNumber 字段是一个字符串；但是，其他两个字段——transactionDate 和 amount 分别是 java.util.Date 和 Double 类型。因此，我们需要自定义 FieldSetMapper 以进行必要的类型转换。代码清单 7-23 展示了 TransactionFieldSetMapper。

代码清单 7-23　TransactionFieldSetMapper

```
package com.apress.springbatch.chapter7;

import org.springframework.batch.item.file.mapping.FieldSetMapper;
import org.springframework.batch.item.file.transform.FieldSet;
import org.springframework.validation.BindException;

public class TransactionFieldSetMapper implements FieldSetMapper<Transaction> {

    public Transaction mapFieldSet(FieldSet fieldSet) {
        Transaction trans = new Transaction();

        trans.setAccountNumber(fieldSet.readString("accountNumber"));
        trans.setAmount(fieldSet.readDouble("amount"));
        trans.setTransactionDate(fieldSet.readDate("transactionDate",
                                    "yyyy-mm-dd hh:mm:ss"));
        return trans;
    }
}
```

Spring Batch 权威指南

如你所见，FieldSet 接口类似于 JDBC 世界里的 ResultSet 接口，它们都为每种数据类型提供了自定义方法。对于 Transaction 领域对象，可以使用 readDouble 方法将文件中的字符串转换为 java.lang.Double 类型，然后使用 readDate 方法将文件中包含的字符串解析为 java.util.Date 类型。对于日期转换，不仅要指定字段名称，还要指定解析时使用的日期格式。

当执行作业时，就能够读取两种不同的记录格式，并且将它们解析为各自的领域对象。代码清单 7-24 展示了 copyJob 的运行结果。

代码清单 7-24 处理多种记录格式的 copyJob 的运行结果

```
2019-01-28 22:41:09.812 INFO 60498 --- [           main] o.s.batch.core.job.
SimpleStepHandler        : Executing step: [copyFileStep]
Customer{firstName='Warren', middleInitial='Q', lastName='Darrow', address='8272 4th
Street', city='New York', state='IL', zipCode='76091'}
Transaction{accountNumber='1165965', transactionDate=Sat Jan 22 00:13:29 CST 2011,
amount=51.43}
Customer{firstName='Ann', middleInitial='V', lastName='Gates', address='9247 Infinite Loop
Drive', city='Hollywood', state='NE', zipCode='37612'}
Customer{firstName='Erica', middleInitial='I', lastName='Jobs', address='8875 Farnam
Street', city='Aurora', state='IL', zipCode='36314'}
Transaction{accountNumber='8116369', transactionDate=Fri Jan 21 20:40:52 CST 2011,
amount=-14.83}
Transaction{accountNumber='8116369', transactionDate=Fri Jan 21 15:50:17 CST 2011,
amount=-45.45}
Transaction{accountNumber='8116369', transactionDate=Fri Jan 21 16:52:46 CST 2011,
amount=-74.6}
Transaction{accountNumber='8116369', transactionDate=Sat Jan 22 13:51:05 CST 2011,
amount=48.55}
Transaction{accountNumber='8116369', transactionDate=Fri Jan 21 16:51:59 CST 2011,
amount=98.53}
```

处理单个文件中的多种记录格式是批处理中的常见需求。然而，此例假定不同记录之间没有实际关系。如果有，那该怎么办？接下来将介绍如何将多行记录读入单个条目。

5. 多行记录

上一个示例研究了如何将两种不同的记录格式处理为两个不同的、不相关的条目。然而，如果仔细查看使用的文件格式，就可以看到读取的记录实际上是相关联的。尽管它们不是通过文件中的字段相关联，但交易记录其实属于上面的客户记录。让一个 Customer 对象具有一个 Transaction 对象的集合，而不是单独处理每条记录，是不是更有意义？

为了做到这一点，需要使用一些技巧。Spring Batch 提供的示例使用尾记录来标识记录的真正结尾。尽管很方便，但是批处理中的很多文件都没有尾记录。对于这里的文件格式，你会遇到一个问题，即不读取下一行就不知道记录什么时候结束。为了解决这个问题，你可以实现自己的 ItemReader，在之前配置的 customerFileReader 的前后增加一些逻辑。图 7-3 展示了我们将在自定义的 ItemReader 中使用的处理流程。

如图 7-3 所示，首先确定是否读取了 Customer 对象。如果还没有，就尝试从 FlatFileItemReader 中读取。假设读入一条记录(一旦到达文件末尾，就不要再读取了)，可在 Customer 对象上初始化交易列表。当读入的下一条记录是交易记录时，将其添加到 Customer 对象中。

第 7 章 ItemReader

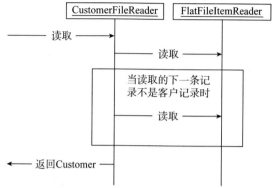

图 7-3 CustomerFileReader 的处理流程

在介绍自定义 ItemReader 的实现代码之前，对领域对象需要进行稍许修改。我们在新的配置中使用的 Customer 对象将包含类型为 List <Transaction>的交易列表而不是两个独立的领域对象 Customer 和 Transaction。代码清单 7-25 展示了更新后的 Customer 领域对象。

代码清单 7-25　更新后的 Customer 领域对象

```
...
public class Customer {

    private String firstName;
    private String middleInitial;
    private String lastName;
    private String address;
    private String city;
    private String state;
    private String zipCode;

    List<Transaction> transactions;

    public Customer() {
    }

    // Getters and setters removed for brevity
    ...
}
```

有了更新后的领域对象后，我们下面看看代码清单 7-26 所示的 CustomerFileReader 的实现。

代码清单 7-26　CustomerFileReader

```
public class CustomerFileReader implements ItemStreamReader<Customer> {

    private Object curItem = null;

    private ItemStreamReader<Object> delegate;

    public CustomerFileReader(ItemStreamReader<Object> delegate) {
        this.delegate = delegate;
    }
```

```java
    public Customer read() throws Exception {
        if(curItem == null) {
            curItem = delegate.read();
        }

        Customer item = (Customer) curItem;
        curItem = null;

        if(item != null) {
            item.setTransactions(new ArrayList<>());

            while(peek() instanceof Transaction) {
                item.getTransactions().add((Transaction) curItem);
                curItem = null;
            }
        }

        return item;
    }

    private Object peek() throws Exception {
        if (curItem == null) {
            curItem = delegate.read();
        }
        return curItem;
    }

    public void close() throws ItemStreamException {
        delegate.close();
    }

    public void open(ExecutionContext arg0) throws ItemStreamException {
        delegate.open(arg0);
    }

    public void update(ExecutionContext arg0) throws ItemStreamException {
        delegate.update(arg0);
    }
}
```

CustomerFileReader 有两个关键方法。第一个方法是 read 方法。read 方法负责实现读取和组装单个 Customer 条目(及其子交易记录)时涉及的逻辑,可通过从正在读取的文件中读取客户记录来实现。然后读取相关的交易记录,直到下一条记录是客户记录为止。一旦找到下一条客户记录,ItemReader 就会认为已经完成对当前客户的读取。这种逻辑称为控制中断逻辑。

第二个方法是 peak 方法。peak 方法用于在处理当前客户时,提前读取下一条记录,从而缓存当前记录。如果记录已被读取但未处理,就返回当前记录。如果记录已被处理(通过将 curItem 设置为 null),那就读取下一条记录[1]。

注意,这个自定义的 ItemReader 并没有实现 ItemReader 接口,而是实现了 ItemReader 的子接口

---

[1] 需要注意的是,ItemReader 有一个名为 org.springframework.batch.item.PeekableItemReader<T>的子接口。由于 CustomerFileReader 不完全符合这个子接口的定义,因此这里没有实现这个子接口。

第 7 章 ItemReader

ItemStreamReader。原因在于，当使用 Spring Batch 中的某个 ItemReader 实现时，它们负责处理正在读取的资源的打开和关闭，并在读取记录时维护执行上下文(ExecutionContext)。然而，如果自行实现，那么需要自己进行管理。由于这里只是在封装 Spring Batch 的 ItemReader(FlatFileItemReader)，因此可以用它来维护那些资源。

为了配置 CustomerFileReader，这里唯一的依赖项就是这个委托。本例中的委托是一个读取器，用于完成实际的读取和解析工作。代码清单 7-27 展示了 CustomerFileReader 的配置。

代码清单 7-27　CustomerFileReader 的配置

```java
...
@Bean
@StepScope
public FlatFileItemReader customerItemReader(@Value("#{jobParameters['customerFile']}")
Resource inputFile) {
    return new FlatFileItemReaderBuilder()
            .name("customerItemReader")
            .lineMapper(lineTokenizer())
            .resource(inputFile)
            .build();
}

@Bean
public CustomerFileReader customerFileReader() {
    return new CustomerFileReader(customerItemReader(null));
}

@Bean
public PatternMatchingCompositeLineMapper lineTokenizer() {
    Map<String, LineTokenizer> lineTokenizers = new HashMap<>(2);

    lineTokenizers.put("CUST*", customerLineTokenizer());
    lineTokenizers.put("TRANS*", transactionLineTokenizer());

    Map<String, FieldSetMapper> fieldSetMappers = new HashMap<>(2);

    BeanWrapperFieldSetMapper<Customer> customerFieldSetMapper = new
        BeanWrapperFieldSetMapper<>();
    customerFieldSetMapper.setTargetType(Customer.class);

    fieldSetMappers.put("CUST*", customerFieldSetMapper);
    fieldSetMappers.put("TRANS*", new TransactionFieldSetMapper());

    PatternMatchingCompositeLineMapper lineMappers = new
        PatternMatchingCompositeLineMapper();

    lineMappers.setTokenizers(lineTokenizers);
    lineMappers.setFieldSetMappers(fieldSetMappers);

    return lineMappers;
}

@Bean
public DelimitedLineTokenizer transactionLineTokenizer() {
    DelimitedLineTokenizer lineTokenizer = new DelimitedLineTokenizer();
```

```
        lineTokenizer.setNames("prefix", "accountNumber", "transactionDate", "amount");

        return lineTokenizer;
    }

    @Bean
    public DelimitedLineTokenizer customerLineTokenizer() {
        DelimitedLineTokenizer lineTokenizer = new DelimitedLineTokenizer();

        lineTokenizer.setNames("prefix", "firstName", "middleInitial", "lastName",
            "address","city", "state", "zip");
        return lineTokenizer;
    }
...
```

代码清单7-27 中的配置你应该很熟悉，实质上它们与用于多种记录格式的配置完全相同(参见代码清单 7-19)。粗体部分展示了我们新添加的内容——用于 CustomerFileReader 的配置，这里引用了旧的 ItemReader 并重新进行了命名。

在使用封装了旧 ItemReader 的新 CustomerFileReader 完成处理后，我们需要更新步骤，以引用 CustomerFileReader 并作为条目读取器来完成处理。代码清单 7-28 展示了更新后的步骤的配置。

代码清单 7-28　copyFileStep

```
...
@Bean
public Step copyFileStep() {
return this.stepBuilderFactory.get("copyFileStep")
.<Customer, Customer>chunk(10)
.reader(customerFileReader())
.writer(itemWriter())
.build();
}
...
```

对于每个 Customer 对象，我们想要打印每个用户的交易数量，从而提供足够的用于验证读取是否工作的详情。鉴于我们一直在使用简单的 ItemWriter，只需要重写 Customer 对象的 toString()方法即可格式化输出。代码清单 7-29 展示了更新后的 toString()方法。

代码清单 7-29　更新后的 toString()方法

```
...
    @Override
    public String toString() {
        StringBuilder output = new StringBuilder();

        output.append(firstName);
        output.append(" ");
        output.append(middleInitial);
        output.append(". ");
        output.append(lastName);

        if(transactions != null&& transactions.size() > 0) {
            output.append(" has ");
```

```
            output.append(transactions.size());
            output.append(" transactions.");
        } else {
            output.append(" has no transactions.");
        }
        return output.toString();
    }
...
```

只要运行一次作业，就可以看到每个客户以及读入的交易记录的数量。需要注意的是，以这种方式读取记录时，客户记录和所有后续的交易记录都将视为单个条目。这样做的原因是：Spring Batch 认为条目就是 ItemReader 返回的对象。在本例中，Customer 对象就是 ItemReader 返回的对象，因而也就是用于诸如提交计数等事项的条目。每个 Customer 对象将由已配置的某个 ItemProcessor 处理一次，再由已配置的某个 ItemWriter 处理一次。代码清单 7-30 展示了使用新的条目读取器配置的多行作业的输出。

代码清单 7-30　多行作业的输出

```
2019-01-28 23:32:17.635 INFO 61271 --- [           main] o.s.batch.core.job.
SimpleStepHandler        : Executing step: [copyFileStep]
Warren Q. Darrow has 1 transactions.
Ann V. Gates has no transactions.
Erica I. Jobs has 5 transactions.
```

多行记录是批处理中的常见元素。尽管它们相比基本的记录处理要复杂一些，但是从本例可以看出，实际上仍然只需要编写极少量的代码，就可以健壮地处理这些情况。

6．多个数据源

到目前为止，所有的示例都基于单个客户文件，其中包含了每个客户的交易信息。许多公司都有多个部门或多个销售产品的地点。以一家在全国范围内设有店铺的连锁餐厅为例，每个位置都可能贡献具有相同格式的文件以进行处理。如果像现在一样，对于每个文件都使用一个读取器，那么无论是性能还是可维护性都将存在许多问题。为了读取相同格式的多个输入文件，Spring Batch 是如何提供这种能力的呢？

与在多行记录示例中使用的模式类似，Spring Batch 提供了名为 MultiResourceItemReader 的 ItemReader。这个读取器封装了另一个 ItemReader，就像 CustomerFileItemReader 一样。然而，不是把想要读取的资源定义为子条目读取器的一部分，而是定义一种文件模式作为 MultiResourceItemReader 的依赖项。

你可以使用与多行记录示例中相同的文件格式(如代码清单 7-20 所示)，从而创建与多行记录示例中相同的 ItemReader 配置。但是，如果你有五个文件，它们分别是 customerFile1.csv、customerFile2.csv、customerFile3.csv、customerFile4.csv 和 customerFile5.csv，那么需要进行两处小的修改。首先需要调整配置，让 MultiResourceItemReader 使用正确的资源模式。代码清单 7-31 展示了更新后的配置。

代码清单 7-31　处理多个客户文件的配置

```
...
@Bean
@StepScope
```

```java
    public MultiResourceItemReader multiCustomerReader(@Value("#{jobParameters
['customerFile']}")Resource[] inputFiles) {
        return new MultiResourceItemReaderBuilder<>()
                .name("multiCustomerReader")
                .resources(inputFiles)
                .delegate(customerFileReader())
                .build();
    }

    @Bean
    public CustomerFileReader customerFileReader() {
        return new CustomerFileReader(customerItemReader());
    }

    @Bean
    public FlatFileItemReader customerItemReader() {
        return new FlatFileItemReaderBuilder()
                .name("customerItemReader")
                .lineMapper(lineTokenizer())
                .build();
    }

    @Bean
    public PatternMatchingCompositeLineMapper lineTokenizer() {
        Map<String, LineTokenizer> lineTokenizers = new HashMap<>(2);

        lineTokenizers.put("CUST*", customerLineTokenizer());
        lineTokenizers.put("TRANS*", transactionLineTokenizer());

        Map<String, FieldSetMapper> fieldSetMappers = new HashMap<>(2);

        BeanWrapperFieldSetMapper<Customer> customerFieldSetMapper = new
BeanWrapperFieldSetMapper<>();
        customerFieldSetMapper.setTargetType(Customer.class);

        fieldSetMappers.put("CUST*", customerFieldSetMapper);
        fieldSetMappers.put("TRANS*", new TransactionFieldSetMapper());

        PatternMatchingCompositeLineMapper lineMappers = new PatternMatchingCompositeLineMapper();

        lineMappers.setTokenizers(lineTokenizers);
        lineMappers.setFieldSetMappers(fieldSetMappers);

        return lineMappers;
    }

    @Bean
    public DelimitedLineTokenizer transactionLineTokenizer() {
        DelimitedLineTokenizer lineTokenizer = new DelimitedLineTokenizer();

        lineTokenizer.setNames("prefix", "accountNumber", "transactionDate", "amount");

        return lineTokenizer;
    }

    @Bean
```

```
public DelimitedLineTokenizer customerLineTokenizer() {
    DelimitedLineTokenizer lineTokenizer = new DelimitedLineTokenizer();

    lineTokenizer.setNames("prefix", "firstName", "middleInitial", "lastName", "address",
        "city", "state", "zip");

    return lineTokenizer;
}
...
```

MultiResourceItemReader 包含三个主要部分。第一部分是读取器的名称(name)，与到目前为止我们研究的其他 ItemReader 实现一样，它们都是有状态的。第二部分是 Resource 类型的资源对象数组，这些是要读取的文件，我们可以使用 Spring 的 SpEL(Spring 表达式语言)解析该数组。第三部分是完成实际工作的委托，在此例中，使用的委托是自定义的 CustomerFileReader。

为了使 MultiResourceItemReader 能完成工作，你还需要更改 FlatFileItemReader 的配置。在前面的代码清单 7-31 中，我们配置了要读取的资源。然而，在本例中，MultiResourceItemReader 将遍历 Resource 类型的资源对象数组，并在读取完每个对象时注入另一个新的对象。因此，你需要删除 customerItemReader Bean 的资源配置，也就是代码清单 7-31 中的粗体部分。

需要更改的另一个地方是 CustomerFileReader 的代码。之前，可以把 ItemStreamReader 作为要实现的接口和委托的类型。但是，在这里仅仅这么做还不够，你还需要使用 ItemStreamResource 的一个子接口——ResourceAwareItemReaderItemStream 子接口适用于从资源对象数组中读取输入的任何 ItemReader。修改这两个地方的原因是，你需要能够将多个资源对象注入 ItemReader 中。

为了实现 org.springframework.batch.item.file.ResourceAwareItemReaderItemStream，你需要添加一个新的方法 setResource(Resource resource)。就像 ItemStream 接口的 open、close 和 update 方法，你只需要在实现中的委托上调用 setResource 方法即可。另一处更改是使委托类型为 ResourceAwareItemReaderItemStream。由于这里使用 FlatFileItemReader 作为委托，因此不需要使用其他 ItemReader 作为委托。代码清单 7-32 展示了更新后的代码。

代码清单 7-32　CustomerFileReader

```
public class CustomerFileReader implements ResourceAwareItemReaderItemStream
<Customer> {

    private Object curItem = null;

    private ResourceAwareItemReaderItemStream<Object> delegate;

    public CustomerFileReader(ResourceAwareItemReaderItemStream<Object> delegate) {
        this.delegate = delegate;
    }

    public Customer read() throws Exception {
        if(curItem == null) {
            curItem = delegate.read();
        }

        Customer item = (Customer) curItem;
        curItem = null;
```

```java
        if(item != null) {
            item.setTransactions(new ArrayList<>());

            while(peek() instanceof Transaction) {
                item.getTransactions().add((Transaction) curItem);
                curItem = null;
            }
        }

        return item;
    }

    private Object peek() throws Exception {
        if (curItem == null) {
            curItem = delegate.read();
        }
        return curItem;
    }

    public void close() throws ItemStreamException {
        delegate.close();
    }

    public void open(ExecutionContext arg0) throws ItemStreamException {
        delegate.open(arg0);
    }

    public void update(ExecutionContext arg0) throws ItemStreamException {
        delegate.update(arg0);
    }

    @Override
    public void setResource(Resource resource) {
        this.delegate.setResource(resource);
    }
}
```

从处理的角度看，代码清单 7-32 中展示的内容与你最初在代码清单 7-26 中看到的内容之间的唯一区别，就在于多了注入资源对象的功能。这使得 Spring Batch 可以按需创建每个文件，并将它们注入 ItemReader，而不是由 ItemReader 本身负责文件的管理。

当使用命令 java -jar copyJob.jar customerFile=/input/customerMulitFormat* 运行这个示例作业时，Spring Batch 会迭代所有能满足指定模式的资源，然后针对每个文件执行读取器。这个作业的输出只是比多行记录示例的输出更多而已。

代码清单 7-33 使用多个输入文件的作业的输出

```
Warren Q. Darrow has 1 transactions.
Ann V. Gates has no transactions.
Erica I. Jobs has 5 transactions.
Joseph Z. Williams has 2 transactions.
Estelle Y. Laflamme has 3 transactions.
Robert X. Wilson has 1 transactions.
Clement A. Blair has 1 transactions.
```

```
Chana B. Meyer has 1 transactions.
Kay C. Quinonez has 1 transactions.
Kristen D. Seibert has 1 transactions.
Lee E. Troupe has 1 transactions.
Edgar F. Christian has 1 transactions.
```

需要特别注意的是，在处理多个这样的文件时，Spring Batch 不会为重启等情况提供额外的安全性。因此，在此例中，如果作业以文件 customerFile1.csv、customerFile2.csv 和 customerFile3.csv 启动，但是在处理了 customerFile2.csv 之后失败，然后在重启作业之前加入 customerFile4.csv，那么即使在第一次执行作业时 customerFile4.csv 不存在，在第二次执行时也会将其作为作业的一部分进行处理。为了避免这种情况，通常的做法是为批处理的每一次运行创建一个目录，使一次运行中将要处理的所有文件都进入合适的目录并得到处理。由于任何新的文件都将进入新目录，因此它们对当前的执行没有影响。

前面已经讨论了许多涉及平面文件的场景——从固定宽度的记录、分隔符记录、多行记录到来自多个文件的输入。但是，平面文件并不是你可能会看到的唯一文件类型。XML 虽然不是最流行的输入格式，但却代表了企业中大量的基于文件的输入。让我们看看在面对 XML 文件时，Spring Batch 能为你做什么。

7.2.2 XML 文件

本章在开头讨论基于文件的处理时，介绍了不同的文件格式如何具有不同数量的用于描述文件格式的元数据。固定宽度的记录的元数据量最少，你需要提前知道尽可能多的与记录格式相关的信息。XML 文件则是另一个极端。XML 使用标记来描述文件中的数据，从而提供对所包含数据的完整描述。

XML 解析器通常有两种：DOM 和 SAX。DOM 解析器将整个文件以树状结构加载到内存中，用于节点的导航。由于性能方面的影响，这种方法对批处理没有用，因此只能使用 SAX 解析器。SAX 是基于事件的解析器，当找到某些元素时就触发事件。

在 Spring Batch 中，我们使用的是 StAX 解析器。尽管是类似于 SAX 的基于事件的解析器，但 StAX 解析器的优势在于能够独立地解析文档的各个小节。这与面向条目的读取直接相关。SAX 解析器将在一次运行中解析整个文件，而 StAX 解析器可以读取文件的每个小节，每个小节代表一次要处理的条目。

在介绍如何使用 Spring Batch 解析 XML 文件之前，我们先看一个示例。为了了解 Spring Batch 解析 XML 文件的方式，下面将使用相同的客户文件。但是，你将使用 XML 文件而不是平面文件中的数据。代码清单 7-34 展示了这个示例。

代码清单 7-34　使用 XML 格式的客户文件

```xml
<customers>
    <customer>
            <firstName>Laura</firstName>
            <middleInitial>O</middleInitial>
            <lastName>Minella</lastName>
            <address>2039 Wall Street</address>
            <city>Omaha</city>
            <state>IL</state>
            <zipCode>35446</zipCode>
```

```xml
            <transactions>
                <transaction>
                    <accountNumber>829433</accountNumber>
                    <transactionDate>2010-10-14 05:49:58</transactionDate>
                    <amount>26.08</amount>
                </transaction>
            </transactions>
        </customer>
        <customer>
            <firstName>Michael</firstName>
            <middleInitial>T</middleInitial>
            <lastName>Buffett</lastName>
            <address>8192 Wall Street</address>
            <city>Omaha</city>
            <state>NE</state>
            <zipCode>25372</zipCode>
            <transactions>
                <transaction>
                    <accountNumber>8179238</accountNumber>
                    <transactionDate>2010-10-27 05:56:59</transactionDate>
                    <amount>-91.76</amount>
                </transaction>
                <transaction>
                    <accountNumber>8179238</accountNumber>
                    <transactionDate>2010-10-06 21:51:05</transactionDate>
                    <amount>-25.99</amount>
                </transaction>
            </transactions>
        </customer>
</customers>
```

客户文件被结构化为 customer(客户)小节的集合。其中每个 customer 小节又都包含 transaction (交易)小节的集合。Spring Batch 会将平面文件中的行解析为一个个 FieldSet。在使用 XML 时，Spring Batch 会将 XML 片段解析为领域对象。什么是 XML 片段？如图 7-4 所示，XML 片段是从开始标签到关闭标签的 XML 块。当指定的 XML 片段存在于文件中时，XML 片段将被视为一条记录，并被转换为要处理的条目。

在客户文件中，你将在客户级别具有相同的数据。对于每个客户，还有一些交易元素的集合，这些交易元素代表了前面的多行记录示例中的交易列表。

为了解析 XML 输入文件，我们需要使用 Spring Batch 提供的 org.springframework.batch.item.xml. StaxEventItemReader。为此，定义根元素，用于标识 XML 中被视为条目的 XML 片段。在本例中，根元素是 customer 标签。这里需要使用的资源与前面示例中的 customerFile Bean 相同。最后，我们还需要接收 org.springframework.oxm.Unmarshaller 的一个实现，它被用于把 XML 转换为领域对象。代码清单 7-35 展示了使用 StaxEventItemReader 实现的 customerFileReader 的配置。

代码清单 7-35　使用 StaxEventItemReader 配置的 customerFileReader

```
...
@Bean
@StepScope
public StaxEventItemReader<Customer> customerFileReader(
        @Value("#{jobParameters['customerFile']}") Resource inputFile) {
```

第 7 章 ItemReader

```
        return new StaxEventItemReaderBuilder<Customer>()
                .name("customerFileReader")
                .resource(inputFile)
                .addFragmentRootElements("customer")
                .unmarshaller(customerMarshaller())
                .build();
}
...
```

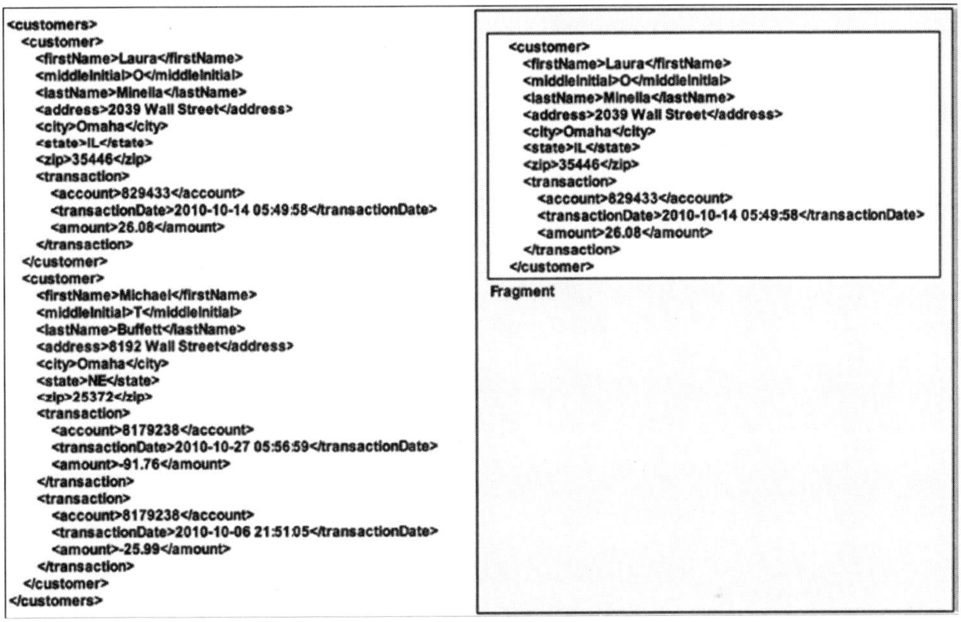

图 7-4　Spring Batch 看到的 XML 片段

Spring Batch 不会挑剔你选择使用的 XML 绑定技术。Spring 提供的 Unmarshaller 实现会使用各个 oxm 包里的 Castor、JAXB、JiBX、XMLBeans 和 XStream 等依赖项。本例将使用 JAXB 依赖项。

对于 customerMarshaller，可以使用 Spring 提供的 org.spriingframework.oxm.jaxb.Jaxb2Marshaller 实现。为此，需要在项目中添加一些依赖项。代码清单 7-36 展示了需要在类路径中添加的 JAXB 依赖项。

代码清单 7-36　JAXB 依赖项

```
...
<dependency>
        <groupId>org.springframework</groupId>
        <artifactId>spring-oxm</artifactId>
</dependency>
<dependency>
        <groupId>javax.xml.bind</groupId>
        <artifactId>jaxb-api</artifactId>
        <version>2.2.11</version>
</dependency>
<dependency>
        <groupId>com.sun.xml.bind</groupId>
        <artifactId>jaxb-core</artifactId>
```

169

```xml
        <version>2.2.11</version>
</dependency>
<dependency>
        <groupId>com.sun.xml.bind</groupId>
        <artifactId>jaxb-impl</artifactId>
        <version>2.2.11</version>
</dependency>
<dependency>
        <groupId>javax.activation</groupId>
        <artifactId>activation</artifactId>
        <version>1.1.1</version>
</dependency>
...
```

有了用于 JAXB 以及使用 JAXB 的 Spring 组件的依赖项之后(通过 Spring OXM 模块)，你还需要配置应用以解析 XML。首先需要将 JAXB 注解添加到领域对象中，包括客户对象和交易对象。为了使 JAXB 理解它们与 XML 中的标签之间的映射，我们需要给类添加注解并配置映射。对于 Transaction 类，只需要为类添加@XmlType(name ="transaction")。但是，对于 Customer 类，我们不仅需要告诉 JAXB 期望的元素(通过@XmlRootElement 注解)，而且需要向解析器解释如何构造交易集合(交易集合由<transactions>元素包裹，并且集合中的每个元素都包含一个<transaction>块)。代码清单 7-37 展示了应用注解后的 Customer 类。

代码清单 7-37　用于 Customer 类的 JAXB 注解

```java
...
@XmlRootElement
public class Customer {

    private String firstName;
    private String middleInitial;
    private String lastName;
    private String address;
    private String city;
    private String state;
    private String zipCode;

    private List<Transaction> transactions;

    public Customer() {
    }

    // Other getters and setters were removed for brevity
    // No change to them is required

    @XmlElementWrapper(name = "transactions")
    @XmlElement(name = "transaction")
    public void setTransactions(List<Transaction> transactions) {
        this.transactions = transactions;
    }

    @Override
    public String toString() {
        StringBuilder output = new StringBuilder();
```

第 7 章 ItemReader

```
        output.append(firstName);
        output.append(" ");
        output.append(middleInitial);
        output.append(". ");
        output.append(lastName);

        if(transactions != null&& transactions.size() > 0) {
            output.append(" has ");
            output.append(transactions.size());
            output.append(" transactions.");
        } else {
            output.append(" has no transactions.");
        }

        return output.toString();
    }
}
```

一旦为领域对象配置了适当的映射关系,就可以配置 StaxEventItemReader 使用的用于解析每个块的 Unmarshaller 的具体实现。由于领域对象上的注解能够处理大多数的配置,因此实际上我们只需要告诉 Jaxb2Marshaller[1]需要感知哪些类。

代码清单 7-38 展示了配置 Unmarshaller 所需的代码。

代码清单 7-38　Jaxb2Marshaller 的配置

```
...
@Bean
public Jaxb2Marshaller customerMarshaller() {
    Jaxb2Marshaller jaxb2Marshaller = new Jaxb2Marshaller();

    jaxb2Marshaller.setClassesToBeBound(Customer.class,
                Transaction.class);
    return jaxb2Marshaller;
}
...
```

最后是配置步骤,将这个新的条目读取器作为输入源。代码清单 7-39 展示了更新后的步骤。

代码清单 7-39　copyFileStep

```
...
@Bean
public Step copyFileStep() {
    return this.stepBuilderFactory.get("copyFileStep")
                .<Customer, Customer>chunk(10)
                .reader(customerFileReader(null))
                .writer(itemWriter())
                .build();
}
...
```

以上就是在 Spring Batch 中把 XML 解析为条目所需的一切!运行作业后,就能得到与多行记录

---

1 Jaxb2Marshaller 也实现了 Unmarshaller 接口。

171

示例中相同的输出。

尽管许多企业仍在使用 XML 文件，但 XML 并不是当今首选的序列化格式。JSON 已在许多方面取代 XML，成为数据的首选存储格式。如果需要读取 JSON，那么也可以使用 Spring Batch。

7.3 JSON

XML 由于冗长，正逐渐失宠，XML 的替代者 JSON 则发展非常迅速。JSON 作为一种数据格式不仅不那么冗长，而且与 XML 一样灵活。基于 JavaScript 的复杂 Web 前端的兴起，引发了对在后端和前端之间传递数据的通用机制的需求。一旦后端开始使用 JSON 进行通信，信息就会迅速传播到许多其他应用中。正因为如此，你可能会遇到在批处理中需要读取 JSON 的情况。幸运的是，Spring Batch 的 ItemReader 正好能满足这样的需求。

JsonItemReader 与 StaxEventItemReader 拥有相似的理念：读取 JSON 格式的块，并将它们解析为对象。JSON 文档应该是包含单个对象数组的完整文档。JsonItemReader 把解析委托给了 JsonObjectReader 接口的实现。该接口将真正把 JSON 文本解析成对象，类似于 Unmarshaller 在 StaxEventItemReader 中将 XML 解析为对象。Spring Batch 提供了 JsonObjectReader 接口的两个开箱即用的实现，它们分别使用 Jackson 和 Gson 作为解析引擎。对于本例，我们将使用 Jackson 实现。

在查看代码之前，我们先看一下将要读取的输入文件。实际上，这里使用的数据与我们之前从 customer.xml 文件中读取的数据相同，但使用的是 JSON 格式。代码清单 7-40 展示了输入文件。

代码清单 7-40　customer.json 输入文件

```json
[
  {
    "firstName": "Laura",
    "middleInitial": "O",
    "lastName": "Minella",
    "address": "2039 Wall Street",
    "city": "Omaha",
    "state": "IL",
    "zipCode": "35446",
    "transactions": [
      {
        "accountNumber": 829433,
        "transactionDate": "2010-10-14 05:49:58",
        "amount": 26.08
      }
    ]
  },
  {
    "firstName": "Michael",
    "middleInitial": "T",
    "lastName": "Buffett",
    "address": "8192 Wall Street",
    "city": "Omaha",
    "state": "NE",
    "zipCode": "25372",
    "transactions": [
      {
```

```
      "accountNumber": 8179238,
      "transactionDate": "2010-10-27 05:56:59",
      "amount": -91.76
    },
    {
      "accountNumber": 8179238,
      "transactionDate": "2010-10-06 21:51:05",
      "amount": -25.99
    }
  ]
 }
]
```

为了配置条目读取器 JsonItemReader，我们将使用 Spring Batch 提供的构建器进行配置。对于 customer.json 文件，我们需要配置三个依赖项：用于重启的名称(name)、将要使用的 JsonObjectReader 以及将要读取的资源(Resource)。用于这个条目读取器的其他配置选项包括：标识 strict，用于指示输入是否必须存在(默认为 true)；标识 saveState，用于指示是否应保存状态(默认为 true)；以及要读取的最大条目数 maxItemCount(默认情况下为 Integer.MAX_VALUE)和当前条目数 currentItemCount(用于重新启动)。代码清单 7-41 展示了用于读入 JSON 文件的 JsonItemReader 的配置。

代码清单 7-41　JsonItemReader 的配置

```
...
@Bean
@StepScope
public JsonItemReader<Customer> customerFileReader(
            @Value("#{jobParameters['customerFile']}") Resource inputFile) {

    ObjectMapper objectMapper = new ObjectMapper();
    objectMapper.setDateFormat(new SimpleDateFormat("yyyy-mm-dd hh:mm:ss"));

    JacksonJsonObjectReader<Customer> jsonObjectReader =
        new JacksonJsonObjectReader<>(Customer.class);
    jsonObjectReader.setMapper(objectMapper);

    return new JsonItemReaderBuilder<Customer>()
                .name("customerFileReader")
                .jsonObjectReader(jsonObjectReader)
                .resource(inputFile)
                .build();
}
...
```

代码清单 7-41 首先创建了一个 ObjectMapper 实例，ObjectMapper 是 Jackson 用来读写 JSON 的主类。在许多情况下，无须执行此步骤；然而在本例中，你需要在输入文件中指定日期的格式，这意味着需要自定义将要使用的 ObjectMapper 实例。一旦创建了 ObjectMapper 实例并为输入文件中的 transactionDate 字段配置了格式，就可以创建 JacksonJsonObjectReader。JacksonJsonObjectReader 类有两个依赖项：其一是将要返回的类(在本例中为 Customer 类)，其二是我们刚刚自定义的 ObjectMapper 实例。最后，我们配置了 JsonItemReader 实例：我们创建了新的 JsonItemReaderBuilder 实例，配置了名称、JsonObjectReader 实例以及要读取的资源，并通过调用 build()方法构造了 JsonItemReader 实例。

以上就是需要对基于 XML 的示例进行的唯一更改[1]。如果使用命令 java -jar copyJob.jar customerFile=/path/to/customer/customer.json 运行作业，你将看到与运行 XML 示例时相同的输出，如代码清单 7-42 所示。

代码清单 7-42　JsonItemReader 作业的输出

```
2019-01-30 23:50:27.012 INFO 10451 --- [main] o.s.b.a.b.JobLauncherCommandLineRunner:
Running default command line with: [customerFile=/input/customer.json]
2019-01-30 23:50:27.153 INFO 10451 --- [main] o.s.b.c.l.support.SimpleJobLauncher:
Job: [SimpleJob: [name=job]] launched with the following parameters:
[{customerFile=/input/customer.json}]
2019-01-30 23:50:27.222 INFO 10451 --- [main] o.s.batch.core.job.SimpleStepHandler:
Executing step: [copyFileStep]
Laura O. Minella has 1 transactions.
Michael T. Buffett has 2 transactions.
2019-01-30 23:50:27.355 INFO 10451 --- [main] o.s.b.c.l.support.SimpleJobLauncher:
Job: [SimpleJob: [name=job]] completed with the following parameters: [{customerFile
=/input/customer.json}] and the following status: [COMPLETED]
```

本节介绍了各种各样的基于文件的输入格式。如你所见，固定宽度的文件、分隔符文件、各种记录的配置以及 XML 和 JSON 文件等都可以通过 Spring Batch 进行处理，同时无须编码或者只需要编写非常有限的代码。但是，并非所有的输入都来自文件。关系数据库将为批处理过程提供大量输入。7.4 节将介绍 Spring Batch 为数据库输入提供的设施。

7.4　数据库输入

数据库可以作为批处理过程的重要输入源，原因有很多。它们提供了内置的事务性，通常性能更高，并且伸缩性比文件好。它们还提供了相比其他大多数输入格式更好的恢复特性。在考虑了所有这些因素，以及大多数企业的数据存储在关系数据库中这一事实之后，批处理过程将需要能够处理来自数据库的输入。在本节中，你将查看 Spring Batch 提供的用于从数据库读取数据的开箱即用的设施，包括 JDBC、Hibernate 和 JPA 等。

7.4.1　JDBC

在 Java 世界里，数据库连接始于 JDBC。在学习 JDBC 时，编写代码让我们尝尽苦头，然而，当意识到大多数框架已经处理了诸如数据库连接的事情时，我们便会很快忘记这些。Spring 框架的优势之一就是封装了 JDBC 之类的痛点，使开发人员仅专注于业务的细节。

按照以上传统，Spring Batch 框架的开发人员也使用批处理世界中所需的特性扩展了 Spring 框架的 JDBC 功能。但是，这些特性是什么？Spring Batch 又是如何实现这些功能的？

在使用批处理时，通常需要处理大量的数据。如果一个查询返回数百万条记录，那么你可能不希望一次性将所有数据加载到内存中。但是，如果使用 Spring 的 JdbcTemplate，那么确实会发生这样的后果。JdbcTemplate 会遍历整个 ResultSet，把每一行映射成内存中的领域对象。

---

[1] Jackson 已经包含在基于 Spring Batch 的应用中，所以不需要更新依赖项。如果想要使用 Gson，那么需要先导入它。

第 7 章 ItemReader

相反，Spring Batch 提供了两种不同的方法，用于在处理记录时一次性加载部分记录：游标和分页。游标是通过标准的 java.sql.ResultSet 实现的。当打开 ResultSet 时，对 next()方法的每一次调用，都会从数据库中返回一批记录。这允许你按需从数据库中流式地传输记录，这是使用游标时的行为。

分页指的是从数据库中检索一块又一块的记录(每一块叫作一"页")。每一页可使用独立的 SQL 查询来创建。当浏览每一页时，将通过新的查询从数据库中读取新页。图 7-5 展示了游标和分页这两种方法之间的区别。

图 7-5　游标和分页之间的区别

如图 7-5 所示，游标中的第一次读取将返回一条记录，并将指向的记录前移到下一条记录，每次流出一条记录；而在分页方法中，将会一次性从数据库接收 10 条记录。下面将针对每种数据库技术分别介绍这两种法。让我们从简单的 JDBC 开始。

1. JDBC 游标处理

本例将使用 CUSTOMER 数据表。使用到目前为止我们一直在使用的相同字段，创建新的 CUSTOMER 数据表以保存数据。图 7-6 展示了新的客户数据模型。

为了实现 JDBC 读取器(基于游标或基于页面的读取器)，需要做两件事：配置读取器以执行所需的查询；创建 RowMapper 的实现，就像 Spring 的 JdbcTemplate 需要将 ResultSet 映射到领域对象一样。在介绍新的组件之前，让我们重新查看领域对象 Customer，并找到需要调整的地方，从而与现在使用的数据表兼容。代码清单 7-43 展示了更新后的 Customer 领域对象。

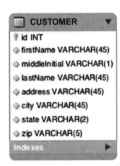

图 7-6　新的客户数据模型

代码清单 7-43　Customer 领域对象

```
...
public class Customer {

    private Long id;

    private String firstName;
    private String middleInitial;
    private String lastName;
    private String address;
    private String city;
    private String state;
    private String zipCode;

    public Customer() {}

    // Getters and setters removed
```

175

```
    @Override
    public String toString() {
        return "Customer{" +
                "id=" + id +
                ", firstName='" + firstName + '\"' +
                ", middleInitial='" + middleInitial + '\"' +
                ", lastName='" + lastName + '\"' +
                ", address='" + address + '\"' +
                ", city='" + city + '\"' +
                ", state='" + state + '\"' +
                ", zipCode='" + zipCode + '\"' +
                '}';
    }
}
```

Customer 对象删除了所有与交易相关的代码和所有的 JAXB 注解。我们还在数据表中为主键添加了 id 字段。正确地定义了领域对象后，我们下面看看 RowMapper 的实现。

RowMapper 确实听起来像是核心 Spring 框架的 JDBC 支持的标准部分。RowMapper 从 ResultSet 中获取一行，并将字段映射到领域对象。在本例中，是把 CUSTOMER 表的字段映射到 Customer 领域对象。代码清单 7-44 展示了将用于 JDBC 实现的 CustomerRowMapper。

代码清单 7-44　CustomerRowMapper

```
...
public class CustomerRowMapper implements RowMapper<Customer> {

    @Override
    public Customer mapRow(ResultSet resultSet, int rowNumber) throws
            SQLException {
        Customer customer = new Customer();
        customer.setId(resultSet.getLong("id"));
        customer.setAddress(resultSet.getString("address"));
        customer.setCity(resultSet.getString("city"));
        customer.setFirstName(resultSet.getString("firstName"));
        customer.setLastName(resultSet.getString("lastName"));
        customer.setMiddleInitial(resultSet.getString("middleInitial"));
        customer.setState(resultSet.getString("state"));
        customer.setZipCode(resultSet.getString("zipCode"));

        return customer;
    }
}
```

在能够将查询结果映射到领域对象之后，还需要通过打开游标执行查询才能按需返回结果，这可通过使用 Spring Batch 的 org.springframework.batch.item.database.JdbcCursorItemReader 条目读取器来完成。每当 Spring Batch 调用 read 方法时，这个条目读取器就会打开一个游标(通过创建 ResultSet)，并将一行映射到一个领域对象。为了配置 JdbcCursorItemReader，至少需要提供三个依赖项：一个数据源(DataSource)、将要运行的查询以及 RowMapper 的一个实现。代码清单 7-45 展示了 customerItemReader 的配置。

代码清单 7-45　基于 JDBC 游标的 customerItemReader

```
...
@Bean
public JdbcCursorItemReader<Customer> customerItemReader(DataSource dataSource) {
    return new JdbcCursorItemReaderBuilder<Customer>()
            .name("customerItemReader")
            .dataSource(dataSource)
            .sql("select * from customer")
            .rowMapper(new CustomerRowMapper())
            .build();
}
...
```

必须指出的是，虽然不需要修改这个作业的其他配置(原来的 ItemWriter 可以正常工作)，但是需要修改 copyFileStep 中的读取器引用，把 customerFileReader 改为 customerItemReader。

使用现在的配置，Spring Batch 在每次调用 JdbcCursorItemReader 的 read 方法时，数据库将返回一行记录，以便映射到领域对象进行处理。

> **注意**　默认情况下，并非所有数据库都把数据以流的方式传入 ResultSet。有些数据库会尝试一次性将所有行加载到内存中，这在较大的数据集中可能会出现问题。此时可能需要进行特殊的配置。有关详细信息，请参考相应的数据库文档。

要执行作业，可使用以下命令：java –jar copyJob。执行作业后，将生成与先前示例中相同的输出。

此例尽管很好，但却存在如下硬伤：其中的 SQL 是硬编码的。我们几乎想不到 SQL 不需要参数的情况。在使用 JdbcCursorItemReader 时，也可以在 SQL 中设置参数，这与使用 JdbcTemplate 和 PreparedStatement 设置参数是相同的。为此，我们需要使用一个 org.springframework.jdbc.core.PreparedStatementSetter 实现。PreparedStatementSetter 与 RowMapper 类似；但是，前者不是将 ResultSet 行映射到领域对象，而是将参数映射到 SQL 语句。你也可以自行实现，但是 Spring 已经为我们提供了一些有用的实现。我们将要使用的是 Spring 框架中的 ArgumentPreparedStatementSetter，从而接收对象数组。如果这些对象的类型不是 SqlParameterValue，那就把对象设置为 PreaparedStatement 中的值(按数组顺序)。如果这些值是 SqlParameterValue 实例，那么 Spring 框架会提供有关如何处理这些值的更多元数据(比如设置到哪些索引上，它们是什么类型，等等)。代码清单 7-46 展示了更新后的构建器和 ArgumentPreparedStatementSetter 配置。

代码清单 7-46　仅处理给定城市的客户

```
...
@Bean
public JdbcCursorItemReader<Customer> customerItemReader(DataSource dataSource) {
    return new JdbcCursorItemReaderBuilder<Customer>()
            .name("customerItemReader")
            .dataSource(dataSource)
            .sql("select * from customer where city = ?")
            .rowMapper(new CustomerRowMapper())
            .preparedStatementSetter(citySetter(null))
            .build();
}
```

```
@Bean
@StepScope
public ArgumentPreparedStatementSetter citySetter(
        @Value("#{jobParameters['city']}") String city) {

    return new ArgumentPreparedStatementSetter(new Object [] {city});
}
...
```

可使用包含作业参数 city=Chicago 的命令执行作业，完整的命令为 java -jar copyJob.jar city=Chicago。执行结果只会展示位于芝加哥的客户，如代码清单 7-47 所示。

代码清单 7-47　位于芝加哥的客户

```
2019-01-31 22:31:41.939 INFO 33800 --- [           main] o.s.b.c.l.support.
SimpleJobLauncher        : Job: [SimpleJob: [name=job]] launched with the following
parameters: [{city=Chicago}]
2019-01-31 22:31:41.995 INFO 33800 --- [           main] o.s.batch.core.job.
SimpleStepHandler        : Executing step: [copyFileStep]
Customer{id=297, firstName='Hermione', middleInitial='K', lastName='Kirby',
address='599-9125 Et St.', city='Chicago', state='IL', zipCode='95546'}
Customer{id=831, firstName='Oren', middleInitial='Y', lastName='Benson', address='P.O. Box
201, 1204 Sed St.', city='Chicago', state='IL', zipCode='91416'}
2019-01-31 22:31:42.063 INFO 33800 --- [           main] o.s.b.c.l.support.
SimpleJobLauncher        : Job: [SimpleJob: [name=job]] completed with the following
parameters: [{city=Chicago}] and the following status: [COMPLETED]
...
```

不仅能够从数据库读取数据流，而且可以将参数注入查询的能力，在现实世界中非常有用。这种方法有好有坏。在某些情况下，流式记录是一件好事；但是，当处理一百万行时，每个请求的网络开销可能会累加起来。另外，ResultSet 不是线程安全的，这意味着不能在多线程环境中使用这种方法。所有这些都会让你倾向于使用分页方法。

2. JDBC 分页处理

当使用分页方法时，Spring Batch 以块的形式返回的结果集称为页。每一页是数据库返回的包含预定义数量的记录。需要注意的是，在使用页时，作业仍将单独地处理每个条目，这很重要。记录的处理是没有区别的。区别在于从数据库中检索记录的方式有所不同。进行分页处理时，不是执行单个 SQL 查询以检索出所有的记录，而是为每一页执行一次查询，这意味着只会把这一页的记录加载到内存中。接下来我们将更新配置，返回包含 10 条记录的页。

为了使分页生效，你需要能够基于页的大小和页码(要返回的记录数和当前正在处理的页号)进行查询。举例来说，如果记录数为 10 000，并且页的大小为 100 条记录，那么你需要能够查询包含 100 条记录的第 20 页。为此，需要为 JdbcPagingItemReader 提供 org.springframework.batch.item.database.PagingQueryProvider 接口的实现。PagingQueryProvider 接口提供了用于在分页的 ResultSet 中进行导航所需的所有功能。

遗憾的是，每个数据库都提供了自己的分页实现。因此，你有以下两种选择：

1) 配置特定于数据库的 PagingQueryProvider 实现。在撰写本书时，Spring Batch 提供了针对 DB2、Derby、H2、HSQL、MySQL、Oracle、PostgreSQL、SQL Server 和 Sybase 的实现。

2) 配置读取器以使用工厂 org.springframework.batch.item.database.support.SqlPagingQueryProvider-

FactoryBean。该工厂会检测出使用了哪个数据库。

SqlPagingQueryProviderFactoryBean 通常会自动检测我们正在使用的数据库平台，并返回适当的 PagingQueryProvider，以便提供我们所需的信息。

为了配置 JdbcPagingItemReader，我们需要四个依赖项：一个数据源(DataSource)、PagingQueryProvider 的一个实现、RowMapper 的一个实现以及页的大小。此时，还可以配置由 Spring 注入的 SQL 语句的参数。代码清单 7-48 展示了 JdbcPagingItemReader 的配置。

代码清单 7-48　JdbcPagingItemReader 的配置

```
...
@Bean
@StepScope
public JdbcPagingItemReader<Customer> customerItemReader(DataSource dataSource,
        PagingQueryProvider queryProvider,
        @Value("#{jobParameters['city']}") String city) {

    Map<String, Object> parameterValues = new HashMap<>(1);
    parameterValues.put("city", city);

    return new JdbcPagingItemReaderBuilder<Customer>()
            .name("customerItemReader")
            .dataSource(dataSource)
            .queryProvider(queryProvider)
            .parameterValues(parameterValues)
            .pageSize(10)
            .rowMapper(new CustomerRowMapper())
            .build();
}

@Bean
public SqlPagingQueryProviderFactoryBean pagingQueryProvider(DataSource dataSource) {
    SqlPagingQueryProviderFactoryBean factoryBean = new SqlPagingQueryProviderFactoryBean();

    factoryBean.setDataSource(dataSource);
    factoryBean.setSelectClause("select *");
    factoryBean.setFromClause("from Customer");
    factoryBean.setWhereClause("where city = :city");
    factoryBean.setSortKey("lastName");

    return factoryBean;
}
```

如你所见，为了配置 JdbcPagingItemReader，以上代码提供了数据源(DataSource)、PagingQueryProvider、要注入的 SQL 语句的参数、每个页的大小以及用于映射结果的 RowMapper 实现。

在 PagingQueryProvider 的配置中，代码提供了五部分信息。前三部分信息是 SQL 语句的不同部分：select 子句、from 子句和 where 子句。第四部分信息是排序键。在分页时对结果进行排序很重要，因为采用分页时通常会为每一页执行查询，而不是执行单个查询并返回结果流。为了保证在多次查询执行之间记录的顺序正确，order by 是必需的，sortKey(排序键)中列出的任何字段都被用于生成 SQL 语句中的 order by 子句。需要注意的是，排序键在 ResultSet 中必须是唯一的。原因在于 Spring Batch 使用排序键创建用于执行的 SQL 查询。最后一部分信息是对数据源的引用。你可能想知道，为什么需要同时在 SqlPagingQueryProviderFactoryBean 和 JdbcPagingItemReader 中配置数据源。SqlPagingQueryProviderFactoryBean 使用数据源来确定当前数据库的类型。如果需要，也可以通过 setDatabaseType(String databaseType) 方法显式地配置数据库的类型。在此之上，我们才能提供合适的用于读取器的 PagingQueryProvider 实现。

我们在分页上下文中使用参数的方式与游标示例不同。在这里，不是创建带有问号作为参数占位符的单个 SQL 语句，而是分段地构建 SQL 语句。在 whereClause 字符串中，可以选择使用标准的问号占位符，也可以像代码清单 7-48 中的 customerItemReader 那样使用命名参数。然后，可以把值作为映射注入配置中。在本例中，parameterValues 中的城市条目将被映射到 whereClause 字符串中的命名参数 city。如果要使用问号而不是名称，那么可以将问号的编号作为每个参数的键。以上所有五部分信息都准备就绪后，Spring Batch 将在每次需要时为每一页构造适当的查询。

如你所见，通过与数据库直接进行 JDBC 交互来读取要处理的条目实际上是非常简单的。仅仅使用几行 Java 代码，就可以使用高效的条目读取器，以便将数据输入作业中。然而，JDBC 不是访问数据库记录的唯一方法，像 Hibernate 和 MyBatis 这样的对象关系映射(ORM)技术已经成为流行的数据访问选择，因为它们给出了将关系数据表映射到领域对象的良好解决方案。接下来，你将了解如何使用 Hibernate 访问数据。

7.4.2　Hibernate

Hibernate 是当今 Java 世界中领先的 ORM 技术。Hibernate 由 Gaven King 早在 2001 年发布，为你提供了将应用中的面向对象模型映射到关系数据库的能力。Hibernate 使用 XML 文件或注解来配置对象到数据表的映射，此外还提供了一个通过对象查询数据库的框架。这让你有了基于对象结构编写查询的能力，即使你对底层数据库结构的了解很少或根本不知道。在本节中，你将研究使用 Hibernate 从数据库读取条目的方法。

在批处理中使用 Hibernate 并不像在 Web 应用中那样直截了当。对于 Web 应用，典型的场景是使用视图会话模式。在这种模式下，会话将在请求进入服务器时打开，所有的处理都在同一个会话中完成，然后在视图返回给客户端时关闭。尽管这对于典型的具有较小的独立交互的 Web 应用非常有效，但批处理却与之完全不同。

对于批处理，如果非常天真地使用 Hibernate，你将使用有状态的会话实现，在处理条目时读取会话，并在完成处理时，当步骤结束后写入并关闭会话。然而，就像之前提到的那样，Hibernate 中的标准会话是有状态的。如果正在读取一百万个条目，并且在对它们进行处理后再写入相同的一百万个条目，那么 Hibernate 会话将在读取它们时缓存这些条目，并且抛出 OutOfMemoryException 异常。

将 Hibernate 用作批处理的持久化框架的另一个问题是，与直接使用 JDBC 相比，Hibernate 产生的开销更大。在处理数百万条记录时，每一毫秒的差异都会产生很大的不同[1]。

开发基于 Hibernate 的 Spring Batch 的条目读取器的目的是希望做正确的事。与基于 Web 的 Hibernate 相比，它们会在提交时执行诸如刷新会话和其他更多的与批处理相关的功能。在已经将 Hibernate 对象映射到数据表的系统中，这可能是进行开发并运行的有效方法。Hibernate 确实也以一种非常健壮的方式解决了将对象映射到数据表的根本问题。需求决定了 Hibernate 或其他任何 ORM

---

[1] 在一百万个条目中，每个条目每增加一毫秒，就会使单个步骤的处理时间增加 15 分钟以上。

工具是否适合作业。

1. 使用 Hibernate 进行游标处理

为了将 Hibernate 与游标一起使用，需要配置 SessionFactory、Customer 映射和 HibernateCursorItemReader，并将 Hibernate 依赖项添加到 pom.xml 文件中。下面首先从更新的 pom.xml 文件开始。

在作业中使用 Hibernate 时，需要添加依赖项 spring-boot-starter-jpa。现在，该例不会直接使用 JPA 进行数据访问(我们将把 JPA 注解用于数据映射)。但是，由于这个启动包由 Hibernate 支持，因此我们将获得特定于 Hibernate 的所有依赖项以及 Spring Data JPA 提供的所有其他功能(注册自定义类型转换器等)。在 7.4.3 节中，我们将介绍如何使用 JPA。代码清单 7-49 展示了 pom.xml 中的启动包。

代码清单 7-49　pom.xml 中的启动包

```xml
...
<dependency>
    <groupId>org.springframework.boot</groupId>
    <artifactId>spring-boot-starter-data-jpa</artifactId>
</dependency>
...
```

在将 Hibernate 框架添加到项目中之后，就可以将 Customer 对象映射到数据库中的 CUSTOMER 表。为了简单起见，这里将使用 Hibernate 注解来配置映射。代码清单 7-50 展示了更新后的映射到 CUSTOMER 表的 Customer 对象。

代码清单 7-50　通过 Hibernate 注解映射到 CUSTOMER 表的 Customer 对象

```java
...
@Entity
@Table(name = "customer")
public class Customer {

    @Id
    private Long id;

    @Column(name = "firstName")
    private String firstName;
    @Column(name = "middleInitial")
    private String middleInitial;
    @Column(name = "lastName")
    private String lastName;
    private String address;
    private String city;
    private String state;
    private String zipCode;

    public Customer() {
    }

    @Override
    public String toString() {
        return "Customer{" +
                "id=" + id +
                ", firstName='" + firstName + '\" +
```

```
            ", middleInitial='" + middleInitial + '\" +
            ", lastName='" + lastName + '\" +
            ", address='" + address + '\" +
            ", city='" + city + '\" +
            ", state='" + state + '\" +
            ", zipCode='" + zipCode + '\" +
            '}';
    }
}
```

Customer 类的映射包括使用 JPA 注解@Entity 将对象标识为实体，使用@Table 注解指定要将实体映射到的数据表，以及使用@Id 注解指定数据表的 ID。由于数据表中的列被命名为与对象中的属性相同，因此 Hibernate 将自动映射 Customer 对象中的所有其他属性。然而，我们确实需要在 Spring Boot 中指定两个属性才能应用这种映射。由于此例使用的是驼峰命名而不是约定的下画线命名，因此需要为 Hibernate 使用正确的命名策略。为此，更新 application.yml 并设置以下属性。

- spring.jpa.hibernate.naming.implicit-strategy: "org.hibernate.boot.model.naming.ImplicitNamingStrategyLegacyJpaImpl"
- spring.jpa.hibernate.naming.physical-strategy:"org.hibernate.boot.model.naming.PhysicalNamingStrategyStandardImpl"

在映射了数据对象后，需要自定义批处理作业使用的 TransactionManager(事务管理器)。默认情况下，Spring Batch 将提供 DataSourceTransactionManager，但我们需要的是能够在常规数据源连接和 Hibernate 会话之间工作的 TransactionManager。Spring 提供的 HibernateTransactionManager 正好可用于此。另外，你还需要通过 BatchConfigurer 的自定义实现对其进行配置。我们需要做的事情就是重写 DefaultBatchConfigurer.getTransactionManager()方法。代码清单 7-51 展示了新的 HibernateBatchConfigurer。

代码清单 7-51　HibernateBatchConfigurer

```
...
@Component
public class HibernateBatchConfigurer extends DefaultBatchConfigurer {

    private DataSource dataSource;
    private SessionFactory sessionFactory;
    private PlatformTransactionManager transactionManager;

    public HibernateBatchConfigurer(DataSource dataSource,
                EntityManagerFactory entityManagerFactory) {

        super(dataSource);
        this.dataSource = dataSource;
        this.sessionFactory = entityManagerFactory.unwrap(SessionFactory.class);
        this.transactionManager = new HibernateTransactionManager(this.sessionFactory);
    }

    @Override
    public PlatformTransactionManager getTransactionManager() {
        return this.transactionManager;
    }
}
```

第 7 章 ItemReader

如你所见,我们要做的就是创建自己的 HibernateTransactionManager,并通过重写 getTransactionManager()方法将其返回。然后,Spring Batch 将在合适的地方使用它[1]。

在完成所有的配置以后,实际上还需要配置 org.springframework.batch.item.database.HibernateCusorItemReader。这可能是最简单的部分。代码清单 7-52 展示了如何使用 HibernateCursorItemReaderBuilder 配置读取器:接收名称(name)、SessionFactory、查询字符串以及用于查询的任何参数。

代码清单 7-52　配置 HibernateCursorItemReader

```
...
@Bean
@StepScope
public HibernateCusorItemReader<Customer> customerItemReader(
            EntityManagerFactory entityManagerFactory,
            @Value("#{jobParameters['city']}") String city) {

    return new HibernateCursorItemReaderBuilder<Customer>()
                .name("customerItemReader")
                .sessionFactory(entityManagerFactory.unwrap(SessionFactory.class))
                .queryString("from Customer where city = :city")
                .parameterValues(Collections.singletonMap("city", city))
                .build();
}
...
```

在以上示例中,我们使用 HQL 查询作为查询数据库的方法,还有其他三种方法可用来指定想要执行的查询。表 7-3 涵盖了所有的 Hibernate 查询选项。

表 7-3　Hibernate 查询选项

选项	类型	描述
queryName	String	引用 Hibernate 配置中的命名 Hibernate 查询
queryString	String	使用你在 Spring 配置中指定的 HQL 查询
queryProvider	HibernateQueryProvider	以编程方式构建 Hibernate 查询
nativeQuery	String	用于运行原生 SQL 查询,并让 Hibernate 映射结果

这就是使 Hibernate 实现等同于使用 JdbcCursorItemReader 进行处理所需的一切。执行后将得到与上一个作业相同的输出。

2. 使用 Hibernate 进行分页数据库访问

与 JDBC 一样,Hibernate 同时支持游标数据库访问和分页数据库访问。唯一需要做的更改是,在作业配置类中指定 HibernatePagingItemReader 而不是 HibernateCursorItemReader,并为条目读取器指定页的大小。代码清单 7-53 展示了更新后的使用 Hibernate 进行分页数据库访问的条目读取器。

---

[1] 在这种方式下,将会显示一条指示 DataSourceTransactionManager 被使用的日志消息。这是 Spring Batch 中的漏洞之一,已在 4.2 版本中解决。

代码清单 7-53　使用 Hibernate 进行分页数据库访问的条目读取器

```
...
@Bean
@StepScope
public HibernatePagingItemReader<Customer> customerItemReader (
    EntityManagerFactory entityManagerFactory,
        @Value("#{jobParameters['city']}") String city) {

    return new HibernatePagingItemReaderBuilder<Customer>()
            .name("customerItemReader")
            .sessionFactory(entityManagerFactory.unwrap(SessionFactory.class))
            .queryString("from Customer where city = :city")
            .parameterValues(Collections.singletonMap("city", city))
            .pageSize(10)
            .build();
}
...
```

当映射已经存在时，使用 Hibernate 可以加快批处理的开发，并简化关系数据到领域对象的映射。但 Hibernate 并不是 ORM 领域的唯一子领域。Java Persistence API（或简称 JPA）是 Java 原生的 ORM 持久化的实现。

7.4.3　JPA

JPA 为 ORM 领域带来了标准化的方法。Hibernate 是 JPA 的早期灵感，并且在当前实现了 JPA 规范。然而，Hibernate 并不是 JPA 的替代品。例如，JPA 没有提供基于游标读取条目的方式，但 Hibernate 提供了。在本例中，你将使用 JPA 提供分页的数据库访问，类似于之前使用 Hibernate 的示例。

就像 Spring Boot 涉及的大多数东西一样，在使用 Spring Boot 时，配置 JPA 实际上非常容易。实际上，我们在前面的 Hibernate 中已经做到了。任何使用 spring-boot-starter-data-jpa 的应用都拥有在 Spring Batch 中使用 JPA 所需的所有组件。实际上，当使用 Spring Boot 的启动包时，我们甚至不需要创建自定义的 BatchConfigurer 实现，因为启动包为我们做了 JpaTransactionManager 方面的配置（类似于 Hibernate 的事务管理器）。由于我们的 Hibernate 示例使用了 JPA 注解，因此从映射的角度看，实际上不需要我们做什么。

在使用 JPA 时，我们真正需要关注的是配置条目读取器。如前所述，JPA 不支持游标数据库访问，但支持分页数据库访问。条目读取器 org.springframework.batch.item.database.JpaPagingItemReader 有四个依赖项：在 ExecutionContext 中用作条目前缀的名称(name)、Spring Boot 提供的 EntityManager、要执行的查询以及为查询的参数注入的值。代码清单 7-54 展示了为进行 JPA 数据库访问而配置的 customerItemReader。

代码清单 7-54　使用 JPA 的 customerItemReader

```
...
@Bean
@StepScope
public JpaPagingItemReader<Customer> customerItemReader (
        EntityManagerFactory entityManagerFactory,
        @Value("#{jobParameters['city']}") String city) {
```

第 7 章 ItemReader

```
        return new JpaPagingItemReaderBuilder<Customer>()
                .name("customerItemReader")
                .entityManagerFactory(entityManagerFactory)
                .queryString("select c from Customer c where c.city = :city")
                .parameterValues(Collections.singletonMap("city", city))
                .build();
    }
...
```

在执行这个作业后，将输出你在命令行中指定的城市里所有客户的姓名和地址。在 JPA 中，还有另一种指定查询的方式：Query API。为了使用 JPA 的 Query API，需要实现 org.springframework.batch.item.database.orm.JpaQueryProvider 接口。这个接口由 createQuery()方法和 setEntityManager(EntityManager em)方法组成，JpaPagingItemReader 使用后面这个方法来获取想要执行的查询。为了使事情变得容易，Spring Batch 提供了如下抽象基类供你扩展：org.springframework.batch.item.database.orm.AbstractJpaQueryProvider。代码清单 7-55 展示了能够返回相同查询(参见代码清单 7-54)的实现代码。

代码清单 7-55　CustomerByCityQueryProvider

```
...
public class CustomerByCityQueryProvider extends AbstractJpaQueryProvider {

    private String cityName;

    public Query createQuery() {
        EntityManager manager = getEntityManager();

        Query query =
            manager.createQuery("select c from Customer " +
                                "c where c.city = :city");
        query.setParameter("city", cityName);

        return query;
    }

    public void afterPropertiesSet() throws Exception {
        Assert.notNull(cityName, "City name is required");
    }

    public void setCityName(String cityName) {
        this.cityName = cityName;
    }
}
```

对于 CustomerByCityQueryProvider，可以使用 AbstractJpaQueryProvider 基类来获取 EntityManager。在这里，可以创建 JPA 查询，填充查询中的所有参数，并将它们返回给 Spring Batch 以执行。为了配置 ItemReader 使用 CustomerByCityQueryProvider 而不是之前提供的查询字符串，只需要将 queryString 参数替换为 queryProvider 参数，如代码清单 7-56 所示。

代码清单 7-56　使用 JpaQueryProvider

```
...
@Bean
@StepScope
```

185

```
public JpaPagingItemReader<Customer> customerItemReader (
            EntityManagerFactory entityManagerFactory,
            @Value("#{jobParameters['city']}") String city) {

    CustomerByCityQueryProvider queryProvider =
                    new CustomerByCityQueryProvider();
    queryProvider.setCityName(city);

    return new JpaPagingItemReaderBuilder<Customer>()
            .name("customerItemReader")
            .entityManagerFactory(entityManagerFactory)
            .queryProvider(queryProvider)
            .parameterValues(Collections.singletonMap("city", city))
            .build();
}
...
```

使用 JPA 不仅能限制应用对第三方库的依赖，而且继承了 ORM 库(如 Hibernate)的许多优点。

7.4.4 存储过程

在很多企业里，关系数据库指的不仅仅是数据表，而且指代一种生态系统，里面包含了复杂的用于各种业务目的的存储过程。对于普通的 Java 开发人员来说，存储过程虽然不是使用数据库的最友好的机制，但却是这个世界上许多数据库中的成熟工具。

什么是存储过程？存储过程是特定于数据库的代码单元，它们存储在数据库中，供某些类型的客户端在未来执行。尽管大多数企业级关系数据库都提供了存储过程，但并非所有数据库都支持它们。

在查看 StoredProcedureItemReader(Spring Batch 提供的用于从存储过程中读取数据的组件)的配置之前，我们首先应该查看将要使用的存储过程。在本例中，由于我们使用 MySQL 作为数据库，因此这里使用 MySQL 语法创建一个存储过程。这个存储过程要做的事情与到目前为止我们一直执行的查询所做的事情相同——按城市查找所有客户。代码清单 7-57 展示了在 MySQL 中创建存储过程所需的代码。

代码清单 7-57　customer_list 存储过程

```
DELIMITER //

CREATE PROCEDURE customer_list(IN cityOption CHAR(16))
  BEGIN
    SELECT * FROM CUSTOMER
    WHERE city = cityOption;
  END //

DELIMITER ;
```

为了创建 customer_list 存储过程，你需要在执行作业之前执行上述代码。这个存储过程接收一个参数(cityOption)，并且在查询中使用了这个参数。默认情况下，查询将返回 ResultSet，就像常规的 SQL 查询一样。重要的是要注意，不能只将代码放到 schema.sql 文件中就期望 Spring Boot 能运行它们。你需要直接从 MySQL 命令行执行前面的代码。

创建完存储过程后，我们来看看如何配置条目读取器，这里的条目读取器类似于 JdbcCursorItemReader，它们都配置了名称(name)、数据源(DataSource)、RowMapper 和 PreparedStatementSettter。然而，我们没有为查询定义 SQL，而是使用将要调用的存储过程的名称来配置 StoredProcedureItemReader。由于存储过程可以处理更复杂的参数，因此我们需要对参数定义进行更多的映射。StoredProcedureItemReader 将接收一个 SqlParameter 类型的对象数组，用于让存储过程接收参数。在本例中，我们将定义一个名为 cityOption 且类型为 VARCHAR 类型的参数。代码清单 7-58 展示了 StoredProcedureItemReader 的配置。

代码清单 7-58　StoredProcedureItemReader 的配置

```
...
@Bean
@StepScope
public StoredProcedureItemReader<Customer> customerItemReader(DataSource dataSource,
            @Value("#{jobParameters['city']}") String city) {

    return new StoredProcedureItemReaderBuilder<Customer>()
            .name("customerItemReader")
            .dataSource(dataSource)
            .procedureName("customer_list")
            .parameters(new SqlParameter[]{
                new SqlParameter("cityOption", Types.VARCHAR)})
            .preparedStatementSetter(
                new ArgumentPreparedStatementSetter(new Object[] {city}))
            .rowMapper(new CustomerRowMapper())
            .build();
}
...
```

这个应用在构建和执行之后，结果将与前面的关系数据库示例相同。

关系数据库在数据存储方面占据最大的市场份额。但是，一种新的数据存储类别正变得日益强大，那就是 NoSQL 数据库。Spring Data 旨在简化跨数据源操作并提供一致的编程模型，详见 7.4.5 节。

7.4.5　Spring Data

根据 Spring Data 网站上的描述："Spring Data 的使命是为数据访问提供令人熟悉且一致的基于 Spring 的编程模型，同时仍保留底层数据存储的特质。" Spring Data 不是单独的项目，而是代表一组项目，这些项目具有一致的抽象集合(也就是 Repository)，同时仍然允许用户访问受支持的每个 NoSQL 和 SQL 数据存储的特性。在本小节中，我们将研究多个 NoSQL 数据存储，以及 Spring Batch 如何使用声明性 I/O 风格使用数据。下面从 MongoDB 开始。

1. MongoDB

MongoDB 源于 2007 年成立的一家名为 10gen 的公司，MongoDB 是 10gen 公司正在开发的平台产品的一部分。该公司后来将 MongoDB 作为单独的产品发布，并更名为 MongoDB Inc.。

MongoDB 是第一批流行起来的 NoSQL 数据存储之一，但是它与大多数企业开发人员习惯的数据库不同，并且它缺乏一些传统的企业级特性(例如 ACID 事务性)，因此在当时被视为玩具。但是，MongoDB 已经发展为市面上十分流行的文档数据库。

MongoDB 的关键特性在于它不使用数据表。相反，每个数据库由一个或多个集合(Collection)组成。其中的每个集合都是一组文档(Document)，通常为 JSON 或 BSON 格式。可以通过 JavaScript 或基于 JSON 的查询语言遍历和查询这些集合及文档。这种设计使 MongoDB 既快速又动态，因为数据库模式可以根据正在查看的数据变化。MongoDB 的其他功能包括：

- 高可用性和可伸缩性——MongoDB 通过原生的复制和分片功能，分别为高可用性和可伸缩性提供了强大支持。
- 地理空间支持——MongoDB 的查询语言支持一些查询，例如确定某一点是否在特定边界之内。

MongoItemReader 是基于分页的条目读取器，它接收一个查询对象，并返回来自 MongoDB 服务器的分页数据。MongoItemReader 需要如下一些依赖项。

- MongoOperations 实现：为了执行查询，需要 MongoTemplate。
- name(名称)：如果 saveState 为 true，那么就与所有其他有状态的 Spring Batch 组件一样。
- targetType(目标类型)：返回的一个 Java 类。返回的文件将被映射到这个 Java 类。
- 一个基于 JSON 的查询或一个 Query 实例：将要执行的查询。

其他的依赖项包括：排序(sort)、索引(hint)、从 MongoDB 返回的字段、要查询的 MongoDB 集合以及查询所需的任何参数。

对本例而言，我们查看一下 Twitter 的推文(Tweet)数据。Twitter 能够方便地将 JSON 用于所有的通信机制，因此以 JSON 格式获取推文数据非常容易。我们可以从 GitHub 获得一个简单的数据集，下载网址为 https://github.com/ozlerhakan/mongodb-json-files。如果下载了推文数据集，那就可以使用命令 mongorestore -d tweets -c tweets_collection <PATH_TO_YOUR_UNZIPPED_FILE>/dump\ 2/twitter/tweets.bson 将其导入 MongoDB。这将创建一个名为 tweets 的数据库，其中包含一个名为 weets_collection 的集合。JSON 数据的根是一个名为 entities 的对象，该对象有四个字段，分别是 hashtags、symbols、user_mentions 和 urls。我们关心的字段是 hashtag。我们将编写一个查询，用于在这个数组中找出那些包含特定文本值的元素。这个文本值可由我们的作业参数传入。

要做的下一件事情是把一个合适的依赖项添加到 pom.xml 中，以便使用 MongoDB。这个依赖项是 spring-boot-starter-data-mongodb，如代码清单 7-59 所示。

代码清单 7-59　用于 MongoDB 的 Spring Boot 启动包

```
...
<dependency>
    <groupId>org.springframework.boot</groupId>
    <artifactId>spring-boot-starter-data-mongodb</artifactId>
</dependency>
...
```

在把数据加载到数据库并添加依赖项之后，就可以查询特定的标签。为了查询 MongoDB，你需要配置 MongoItemReader 的如下项。

- name(名称)：用于将状态存储在 ExecutionContext 中以便进行重启。如果 storeState 为 true，则为必需项。
- targetType(目标类型)：返回的每个文档将被反序列化为目标类型。对本例而言，使用 Map 即可。
- jsonQuery(JSON 查询)：我们想要查找与通过作业参数传递的值相等的标签。
- collection(集合)：要查询的集合，在本例中是 tweets_collection。

第 7 章 ItemReader

- parameterValues(参数值)：查询中的任何参数的值，在本例中是要搜索的标签的值。
- sorts(排序)：用于排序的字段，由于 MongoItemReader 是用于分页的 ItemReader，因此排序是必需的。
- template(模板)：用于执行查询的 MongoOperations 实现。

代码清单 7-60 展示了这里使用的 ItemReader 和 Step 的配置。

代码清单 7-60　MongoItemReader

```
...
@Bean
@StepScope
public MongoItemReader<Map> tweetsItemReader(MongoOperations mongoTemplate,
            @Value("#{jobParameters['hashTag']}") String hashtag) {

        return new MongoItemReaderBuilder<Map>()
                    .name("tweetsItemReader")
                    .targetType(Map.class)
                    .jsonQuery("{ \"entities.hashtags.text\": { $eq: ?0 }}")
                    .collection("tweets_collection")
                    .parameterValues(Collections.singletonList(hashtag))
                    .pageSize(10)
                    .sorts(Collections.singletonMap("created_at", Sort.Direction.ASC))
                    .template(mongoTemplate)
                    .build();
}

@Bean
public Step copyFileStep() {
    return this.stepBuilderFactory.get("copyFileStep")
                    .<Map, Map>chunk(10)
                    .reader(tweetsItemReader(null, null))
                    .writer(itemWriter())
                    .build();
}
...
```

在执行这个作业之前，你还需要进行一项配置。在将数据导入 MongoDB 时，我们还将它们导入了 tweets 数据库，但我们还没有告诉应用这个数据库的名称是什么。在 application.yml 中，我们需要添加属性 spring.data.mongodb.database:tweets。

在构建了应用之后，就可以使用作业参数 hashtag=nodejs 运行应用。对于导入的数据集，你应该可以看到有 20 条记录已被打印到控制台，如图 7-7 所示。

图 7-7　使用 MongoDB 时的输出

MongoDB 是当今市场上最受欢迎的 NoSQL 数据存储之一，它通过面向文档的方法来达到特定的目的。然而，还有其他方法可用于 NoSQL 数据存储。我们接下来将要探索的读取器可以让你利用 Spring Data 的存储库(Repository)抽象。

2. Spring Data 存储库

前面提到过，Spring Data 的目标是提供一致的编程模型。这种一致性的重点就是存储库抽象。存储库抽象让你只需要简单地定义一个接口，并用它扩展 Spring Data 提供的接口之一，即可执行基本的创建/读取/更新/删除(CRUD)操作。对于任何数据存储，如果有与之关联的 Spring Data 项目，那么它们就可以通过 Spring Batch 来读取——但需要这些项目支持所需的存储库机制。因此，举个例子，假定需要从没有 Spring Batch 原生支持的 Apache Cassandra 中读取数据，但是由于 Spring Data for Apache Cassandra 项目确实提供了支持，那么只需要创建存储库即可使用。

Spring Data 存储库是什么？Spring Data 提供的特性是：你可以首先定义一个接口，并用它扩展 Spring Data 提供的特定接口之一(例如 PagingAndSortingRepository)，然后 Spring Data 将负责处理这个接口的实现。更有意思的是，Spring Data 为这些存储库提供了一种查询语言，这种查询语言基于你在接口上定义的方法。例如，如果想通过 city 属性查询 CUSTOMER 表，那么可以创建签名为 public List<Customer> findByCity(String city);的方法。Spring Data 将解释这个方法，并将根据使用的数据存储生成合适的查询。

Spring Batch 适应以上这种机制的方式是：Spring Batch 利用了 Spring Data 的 PagingAndSortingRepository 存储库接口，这个存储库接口以标准化的方式定义了能够对数据进行分页和排序的存储库。RepositoryItemReader 则利用它来完成分页查询，就像 JdbcPagingItemReader 或 HibernatePagingItemReader 一样。区别就在于，RepositoryItemReader 可用于查询有关 Spring Data 项目和存储库支持的数据存储。

在前面的 JPA 示例中，我们按城市查询了 CUSTOMER 表。为此，我们需要定义一个字符串类型的查询并将其传递给 JpaPagingItemReader。这一次，我们将创建一个存储库接口，并让 Spring Data 完成以上工作。我们将使用与代码清单 7-50 相同的领域对象。

在定义条目读取器之前，我们需要创建一个扩展了 PagingAndSortingRepository 的存储库。在本例中，这个存储库将包含一个按城市查询数据库的方法。前面提到的那个签名方法无法满足需求，因为它没有利用 Spring Data 提供的分页机制。我们需要修改那个方法签名的两个地方。首先，我们需要添加如下附加参数：org.springframework.data.domain.Pageable。Pageable 接口的实现封装了与请求分页数据相关的参数，具体来说，包括页的大小、偏移量、页码、排序选项以及其他构建分页查询所需的机制。其次，我们需要让那个签名方法返回 org.springframework.data.domain.Page<Customer>而不是 List<Customer>。Page 对象包含我们传入的 Pageable 接口参数的值以及特定页的实际数据，根据运行的查询，Page 对象还包含与数据集有关的元数据，例如数据集中的元素总数(并不仅仅包含当前页面中的元素)、表明当前页面是第一页还是最后一页的指示符，等等。代码清单 7-61 展示了我们将要使用的存储库接口。

代码清单 7-61　CustomerRepository

```
...
public interface CustomerRepository extends PagingAndSortingRepository<Customer, Long> {
    Page<Customer> findByCity(String city, Pageable pageRequest);
}
```

■ 注意　如果项目中包含多个 Spring Data 启动包，则可能需要使用特定于数据存储的接口(例如 JpaRepository)，从而指示存储库属于哪个数据存储。

第 7 章 ItemReader

定义好存储库之后，现在可以定义 RepositoryItemReader，以便从中读取数据。为此，我们将使用 RepositoryItemReaderBuilder，为之传递名称(以确保能够重启)、要调用方法的除了 Pageable 参数之外的任何参数、要调用方法的名称、存储库实现本身以及定义所需的任何排序(在本例中，我们按客户的姓氏进行排序)。代码清单 7-62 展示了以上配置。

代码清单7-62　RepositoryItemReader

```
...
@Bean
@StepScope
public RepositoryItemReader<Customer> customerItemReader(CustomerRepository repository,
            @Value("#{jobParameters['city']}") String city) {

    return new RepositoryItemReaderBuilder<Customer>()
                .name("customerItemReader")
                .arguments(Collections.singletonList(city))
                .methodName("findByCity")
                .repository(repository)
                .sorts(Collections.singletonMap("lastName", Sort.Direction.ASC))
                .build();
}
...
```

有了上面这段代码，就可以执行这个作业并查看结果，如代码清单 7-63 所示。

代码清单7-63　RepositoryItemReader 作业的执行结果

```
...
2019-02-04 17:17:07.333 INFO 8219 --- [main] o.s.batch.core.job.
SimpleStepHandler      : Executing step: [copyFileStep]
2019-02-04 17:17:07.448 INFO 8219 --- [main] o.h.h.i.
QueryTranslatorFactoryInitiator: HHH000397: Using ASTQueryTranslatorFactory
Customer{id=831, firstName='Oren', middleInitial='Y', lastName='Benson', address=
'P.O. Box 201, 1204 Sed St.', city='Chicago', state='IL', zipCode='91416'}
Customer{id=297, firstName='Hermione', middleInitial='K', lastName='Kirby',
address='599-9125 Et St.', city='Chicago', state='IL', zipCode='95546'}
2019-02-04 17:17:07.657 INFO 8219 --- [main] o.s.b.c.l.support.
SimpleJobLauncher      : Job: [SimpleJob: [name=job]] completed with the following
parameters: [{city=Chicago}] and the following status: [COMPLETED]
```

到目前为止，我们已经介绍了各种文件和数据库(它们均可以作为输入源)，以及从中获取输入数据的各种方法。但在微服务和 API 中，我们并不总能保证实现对数据的直接访问。在 7.5 节，你将了解如何从现有的 Java 服务中获取数据。

7.5　现有的服务

许多公司的生产环境中都包括 Java 应用(Web 应用或其他应用)。对于这些应用，开发人员已经做了大量的分析、开发、测试和漏洞修复工作。包含这些应用的代码已经经过实战测试，并被证明可以正常工作。

那么为什么不能在批处理过程中使用这些代码呢？举个例子，你的批处理过程要求读取客户对象。然而，客户对象并没有像现在这样被映射到单个表或文件中；相反，客户数据分布在多个数据库的多个表中。另外，客户数据绝不会被物理删除；相反，而是被标记为已删除。基于 Web 的应用中已经存在用于检索客户对象的服务。如何在批处理过程中使用服务呢？接下来你将研究如何调用现有的 Spring 服务，以便为条目读取器提供数据。

在第 4 章，你已经了解了 Spring Batch 提供的一些适配器，它们能让 Tasklet 执行不同的操作，特别是 org.springframework.batch.core.step.tasklet.CallableTaskletAdapter、org.springframework.batch.core.step.tasklet.MethodInvokingTaskletAdapter 和 org.springframework.batch.core.step.tasklet.SystemCommandTasklet。这三个适配器都用于封装其他元素，以使 Spring Batch 可以与之交互。为了在 Spring Batch 中使用现有的服务，我们可使用相同的模式。

在读取输入时，可以使用 org.springframework.batch.item.adapter.ItemReaderAdapter。与 RepositoryItemReader 获取存储库的引用并调用引用中的方法类似，ItemReaderAdapter 也有两个依赖项：要调用服务的引用以及要调用方法的名称。使用 ItemReaderAdapter 时，你需要记住以下两点：

1) 每次调用的返回对象也是 ItemReader 将要返回的对象。如果服务返回了单个 Customer 对象，那么这个 Customer 对象将是传递给 ItemProcessor 并最终传递给 ItemWriter 的对象。如果服务返回了 Customer 对象的集合，那么这个集合将作为单个条目传递给 ItemProcessor 和 ItemWriter，而你还要负责遍历这个集合。

2) 一旦完成输入，服务就必须返回 null，从而向 Spring Batch 表明步骤的输入已经结束。

本例将使用一个硬编码的服务，让每次调用返回一个 Customer 对象，直至到达列表的结尾。一旦到达列表的结尾，之后的每个调用将返回 null。代码清单 7-64 中的 CustomerService 可生成一个随机的 Customer 对象列表供使用。

代码清单 7-64　CustomerService

```
...
@Component
public class CustomerService {

    private List<Customer> customers;
    private int curIndex;

    private String [] firstNames = {"Michael", "Warren", "Ann", "Terrence",
                                    "Erica", "Laura", "Steve", "Larry"};
    private String middleInitial = "ABCDEFGHIJKLMNOPQRSTUVWXYZ";
    private String [] lastNames = {"Gates", "Darrow", "Donnelly", "Jobs",
                                   "Buffett", "Ellison", "Obama"};
    private String [] streets = {"4th Street", "Wall Street", "Fifth Avenue",
                                 "Mt. Lee Drive", "Jeopardy Lane",
                                 "Infinite Loop Drive", "Farnam Street",
                                 "Isabella Ave", "S. Greenwood Ave"};
    private String [] cities = {"Chicago", "New York", "Hollywood", "Aurora",
                                "Omaha", "Atherton"};
    private String [] states = {"IL", "NY", "CA", "NE"};

    private Random generator = new Random();

    public CustomerService() {
        curIndex = 0;
```

```
    customers = new ArrayList<>();

    for(int i = 0; i < 100; i++) {
        customers.add(buildCustomer());
    }
}

private Customer buildCustomer() {
    Customer customer = new Customer();

    customer.setId((long) generator.nextInt(Integer.MAX_VALUE));
    customer.setFirstName(
        firstNames[generator.nextInt(firstNames.length - 1)]);
    customer.setMiddleInitial(
        String.valueOf(middleInitial.charAt(
            generator.nextInt(middleInitial.length() - 1))));
    customer.setLastName(
        lastNames[generator.nextInt(lastNames.length - 1)]);
    customer.setAddress(generator.nextInt(9999) + " " +
                        streets[generator.nextInt(streets.length - 1)]);
    customer.setCity(cities[generator.nextInt(cities.length - 1)]);
    customer.setState(states[generator.nextInt(states.length - 1)]);
    customer.setZip(String.valueOf(generator.nextInt(99999)));

    return customer;
}

public Customer getCustomer() {
    Customer cust = null;

    if(curIndex < customers.size()) {
        cust = customers.get(curIndex);
        curIndex++;
    }

    return cust;
}
```

最后，为了使用代码清单 7-64 中的服务，需要使用 ItemReaderAdapter。配置 customerItemReader，以便为每个条目调用 getCustomer 方法，如代码清单 7-65 所示。

代码清单 7-65　配置 ItemReaderAdapter 以使用 CustomerService

```
...
@Bean
public ItemReaderAdapter<Customer> itemReader(CustomerService customerService) {
    ItemReaderAdapter<Customer> adapter = new ItemReaderAdapter<>();

    adapter.setTargetObject(customerService);
    adapter.setTargetMethod("getCustomer");

    return adapter;
}
...
```

Spring Batch 权威指南

这就是使用现有服务作为批处理作业的数据源所需的一切。使用现有服务可以让你复用经过测试和验证的代码，而不必承担通过重写现有流程而引入新漏洞的风险。

Spring Batch 提供了各种条目读取器的实现，对此你已经了解了许多。然而，Spring Batch 框架的开发人员无法规划每一种可能的场景。因此，他们为你提供了自定义条目读取器的设施。7.6 节将介绍如何实现自定义的 ItemReader。

7.6　自定义输入

Spring Batch 为 Java 应用中常见的几乎每种输入类型提供了读取器。但是，如果 Spring Batch 没有为你使用的输入形式提供条目读取器，那你就需要自己创建一个。ItemReader 接口的 read 方法很容易实现。然而，当需要重启读取器时会发生什么？如何在执行过程中保持状态？本节将研究如何实现能够处理跨执行的状态的 ItemReader。

就像之前提到的那样，实现 Spring Batch 的 ItemReader 接口实际上非常简单。实际上，只需要稍作调整，就可以将代码清单 7-61 中的 CustomerService 转换为 ItemReader。你需要做的就是实现接口，并将 getCustomer 方法重命名为 read。代码清单 7-66 展示了更新后的代码。

代码清单 7-66　CustomerItemReader

```
...
public class CustomerItemReader implements ItemReader<Customer> {

    private List<Customer> customers;
    private int curIndex;

    private String [] firstNames = {"Michael", "Warren", "Ann", "Terrence",
                                    "Erica", "Laura", "Steve", "Larry"};
    private String middleInitial = "ABCDEFGHIJKLMNOPQRSTUVWXYZ";
    private String [] lastNames = {"Gates", "Darrow", "Donnelly", "Jobs",
                                   "Buffett", "Ellison", "Obama"};
    private String [] streets = {"4th Street", "Wall Street", "Fifth Avenue",
                                 "Mt. Lee Drive", "Jeopardy Lane",
                                 "Infinite Loop Drive", "Farnam Street",
                                 "Isabella Ave", "S. Greenwood Ave"};
    private String[] cities = {"Chicago", "New York", "Hollywood", "Aurora",
                               "Omaha", "Atherton"};
    private String[] states = {"IL", "NY", "CA", "NE"};
    private Random generator = new Random();

    public CustomerItemReader () {
        curIndex = 0;

        customers = new ArrayList<Customer>();

        for(int i = 0; i < 100; i++) {
            customers.add(buildCustomer());
        }
    }

    private Customer buildCustomer() {
```

第 7 章　ItemReader

```
    Customer customer = new Customer();

    customer.setFirstName(
        firstNames[generator.nextInt(firstNames.length - 1)]);
    customer.setMiddleInitial(
        String.valueOf(middleInitial.charAt(
            generator.nextInt(middleInitial.length() - 1))));
    customer.setLastName(
        lastNames[generator.nextInt(lastNames.length - 1)]);
    customer.setAddress(generator.nextInt(9999) + " " +
                        streets[generator.nextInt(streets.length - 1)]);
    customer.setCity(cities[generator.nextInt(cities.length - 1)]);
    customer.setState(states[generator.nextInt(states.length - 1)]);
    customer.setZip(String.valueOf(generator.nextInt(99999)));

    return customer;
}

@Override
public Customer read() {
    Customer cust = null;

    if(curIndex < customers.size()) {
        cust = customers.get(curIndex);
        curIndex++;
    }

    return cust;
}
}
```

即使你忽略了 CustomerItemReader 在每次执行作业时都会生成一个新列表的事实，你在代码清单 7-66 中编写的 CustomerItemReader 也会在每次执行作业时从列表的开头重新处理。尽管在许多情况下这就是你期望的行为，然而情况并不总是如此。相反，如果在处理了一百万条记录的一半记录后出现错误，那就需要从发生错误的块重新开始。

为了使 Spring Batch 能够在 JobRepository 中维护读取器的状态，并且从上次中断的地方重启读取器，你还需要实现 ItemStream 接口。如代码清单 7-67 所示，ItemStream 接口包含三个方法：open、update 和 close 方法。

代码清单 7-67　ItemStream 接口

```
package org.springframework.batch.item;

public interface ItemStream {

  void open(ExecutionContext executionContext) throws ItemStreamException;
  void update(ExecutionContext executionContext) throws ItemStreamException;
  void close() throws ItemStreamException;
}
```

在执行步骤时，Spring Batch 会调用 ItemStream 接口的每一个方法。open 方法用于初始化 ItemReader 中任何必要的状态，包括打开任何文件或数据库连接，以及在重启作业时恢复状态。举例来说，open 方法可用于重新加载已处理的记录数，以便在第二次执行期间跳过已经处理的记录。Spring

Batch 使用 update 方法在处理时更新状态。update 方法的一种用法是追踪已处理多少记录或块。最后，close 方法用于关闭所有必需的资源(如关闭文件等)。

你将注意到，open 和 update 方法提供了对 ExecutionContext 的访问，但 ItemReader 的实现中并没有这个句柄。这是因为 Spring Batch 在 open 方法中使用了 ExecutionContext，用于在重新启动作业时提供读取器的先前状态。在处理每个条目时，Spring Batch 还使用 update 方法来了解读取器的当前状态(当前正在处理的记录)。最后，close 方法用于清理 ItemStream 中使用的所有资源。

现在你可能想知道，如果没有 read 方法，如何将 ItemStream 接口用于 ItemReader。答案就是扩展工具类 org.springframework.batch.item.ItemStreamSuport。ItemStreamSupport 实现了 ItemStream，并且提供了工具方法 getExecutionContextKey(String key)，该方法使得基于组件的名称键是唯一的。代码清单 7-68 展示了更新后的 CustomerItemReader，以便扩展 ItemStreamSupport 基类[1]。

代码清单 7-68　扩展了 ItemStreamSupport 基类的 CustomerItemReader

```
...
public class CustomerItemReader extends ItemStreamSupport implements ItemReader<Customer> {

    private List<Customer> customers;
    private int curIndex;
    private String INDEX_KEY = "current.index.customers";
    private String [] firstNames = {"Michael", "Warren", "Ann", "Terrence",
                                    "Erica", "Laura", "Steve", "Larry"};
    private String middleInitial = "ABCDEFGHIJKLMNOPQRSTUVWXYZ";
    private String [] lastNames = {"Gates", "Darrow", "Donnelly", "Jobs",
                                   "Buffett", "Ellison", "Obama"};
    private String [] streets = {"4th Street", "Wall Street", "Fifth Avenue",
                                 "Mt. Lee Drive", "Jeopardy Lane",
                                 "Infinite Loop Drive", "Farnam Street",
                                 "Isabella Ave", "S. Greenwood Ave"};
    private String [] cities = {"Chicago", "New York", "Hollywood", "Aurora",
                                "Omaha", "Atherton"};
    private String [] states = {"IL", "NY", "CA", "NE"};

    private Random generator = new Random();

    public CustomerItemReader() {
        customers = new ArrayList<>();

        for(int i = 0; i < 100; i++) {
            customers.add(buildCustomer());
        }
    }

    private Customer buildCustomer() {
        Customer customer = new Customer();

        customer.setFirstName(
            firstNames[generator.nextInt(firstNames.length - 1)]);
        customer.setMiddleInitial(
```

---

[1] 在本例中，扩展 AbstractItemCountingItemStreamItemReader 会更高效一些，但本例演示了 ItemStreamReader 接口的用法。

```java
            String.valueOf(middleInitial.charAt(
                generator.nextInt(middleInitial.length() - 1))));
        customer.setLastName(
            lastNames[generator.nextInt(lastNames.length - 1)]);
        customer.setAddress(generator.nextInt(9999) + " " +
                streets[generator.nextInt(streets.length - 1)]);
        customer.setCity(cities[generator.nextInt(cities.length - 1)]);
        customer.setState(states[generator.nextInt(states.length - 1)]);
        customer.setZip(String.valueOf(generator.nextInt(99999)));

        return customer;
    }

    public Customer read() {
        Customer cust = null;

        if(curIndex == 50) {
            throw new RuntimeException("This will end your execution");
        }

        if(curIndex < customers.size()) {
            cust = customers.get(curIndex);
            curIndex++;
        }

        return cust;
    }

    public void close() throws ItemStreamException {
    }

    public void open(ExecutionContext executionContext) throws ItemStreamException {
        if(executionContext.containsKey(getExecutionContextKey(INDEX_KEY))) {
            int index = executionContext.getInt(getExecutionContextKey(INDEX_KEY));
            if(index == 50) {
                curIndex = 51;
            } else {
                curIndex = index;
            }
        } else {
            curIndex = 0;
        }
    }

    public void update(ExecutionContext executionContext) throws ItemStreamException {
        **executionContext.putInt(getExecutionContextKey(INDEX_KEY), curIndex);**
    }
}
```

代码清单 7-68 中的粗体部分展示了对 CustomerItemReader 所做的更新。首先，这个类扩展了 ItemStreamSupport 类，然后添加了 close、open(ExecutionContext executeContext)和 update (ExecutionContext executeContext)方法。update 方法将一个键值对添加到了 ExecutionContext 中，以指示当前正在处理的记录。open 方法则检查是否已经设置了值。如果已经设置，则意味着作业重启。在 read 方法中，为了强制结束作业，我们添加了一段代码，使得在处理了第 50 个客户之后抛出 RuntimeException 异常。在 open 方法中，如果要恢复的索引是 50，而你知道这是由于先前的代码引起的，因此将跳过对应的记录，否则将重试。注意，所有对 ExecutionContext 中使用的键的引用也都是通过 ItemStreamSupport 类提供的 getExecutionContextKey 方法进行传递的。

你需要配置的另一部分是新的 ItemReader 实现。在本例中，ItemReader 没有依赖项，因此你要做的就是使用正确的名称定义 Bean(以便在现有的 copyJob 中引用)。代码清单 7-69 展示了 CustomerItemReader 的配置。

代码清单 7-69　CustomerItemReader 的配置

```
...
@Bean
public CustomerItemReader customerItemReader() {
    CustomerItemReader customerItemReader = new CustomerItemReader();

    customerItemReader.setName("customerItemReader");

    return customerItemReader;
}
...
```

大功告成。如果执行作业，那么在处理 50 条记录后，CustomerItemReader 将抛出异常，导致作业失败。然而，如果查看作业存储库中的 BATCH_STEP_EXECUTION_CONTEXT 表，你将高兴地看到代码清单 7-70 中列出的内容。

代码清单 7-70　步骤执行上下文

```
mysql> select * from BATCH_STEP_EXECUTION_CONTEXT where STEP_EXECUTION_ID = 8495;
+-------------------+-----------------------------------------------------------+
| STEP_EXECUTION_ID | SHORT_CONTEXT
| SERIALIZED_CONTEXT |
+-------------------+-----------------------------------------------------------+
|              8495 | {"customerItemReader.current.index.customers":50,"batch.
taskletType":"org.springframework.batch.core.step.item.ChunkOrientedTasklet","batch.
stepType":"org.springframework.batch.core.step.tasklet.TaskletStep"} |
NULL               |
```

尽管有点难以理解，但你会注意到 Spring Batch 已经保存了提交计数，并将其存储在作业存储库中。由于这个原因，以及在第二次跳过第 50 位客户时的逻辑，你可以重新执行作业，因为 Spring Batch 将从中断的地方重新开始，而你的条目读取器将跳过那些导致错误的条目。

文件、数据库、服务甚至自定义的条目读取器——Spring Batch 为你提供了各种各样的输入数据源，即便完全掌握它们也只是触及表面而已。遗憾的是，我们在现实世界中使用的所有数据并不都像这里使用的数据那样干净。然而，并非所有错误都需要停止作业。在 7.7 节，你将研究 Spring Batch 允许的用来处理输入错误的一些方式。

7.7　错误处理

Spring Batch 应用的任何部分都可能出错，比如启动、读取输入、处理输入或写入输出时。在本节中，你将研究处理批处理期间可能发生的各种错误的方法。

7.7.1 跳过记录

如果在从输入读取记录时发生错误，那么有两种不同的选择。首先，可以抛出导致作业停止处理的异常。根据需要处理的记录总数以及不处理错误导致的影响，这可能是一种过激的解决方案。相反，Spring Batch 为你提供了在抛出指定的异常时跳过记录的能力。下面将研究如何使用这种技术，基于特定的异常跳过记录。

选择何时跳过记录涉及两个方面。首先是在什么条件下跳过记录，特别是要忽略哪些异常。在读取过程中发生任何错误时，Spring Batch 都会抛出异常。为了确定要跳过的记录，就需要确定要跳过的异常。其次是在让步骤执行失败之前，允许步骤跳过多少条记录。如果要从一百万条记录中跳过一两条记录，那没什么问题；但是，如果要从一百万条记录中跳过一半的记录，那么可能发生错误。

要真正地跳过记录，你需要告诉 Spring Batch 想要跳过的异常以及可以跳过多少次。假定要跳过引发任何 org.springframework.batch.item.ParseException 异常的前 10 条记录，代码清单 7-71 展示了对应的配置。

代码清单 7-71　配置跳过 10 次 ParseException 异常

```
@Bean
public Step copyFileStep() {

    return this.stepBuilderFactory.get("copyFileStep")
                .<Customer, Customer>chunk(10)
                .reader(itemReader())
                .writer(outputWriter(null))
                .faultTolerant()
                .skip(ParseException.class)
                .skipLimit(10)
                .build();
}
```

在这种场景下，要跳过的异常只有一个。然而，有时要跳过的也可能是一个详尽的异常列表。代码清单 7-71 中的配置允许跳过特定的异常，但是配置那些不想跳过的异常可能会更容易些。因此，我们可以结合使用代码清单 7-71 中的 skip(Class exception) 和 noSkipMethod(Class exception) 方法。代码清单 7-72 展示了如何跳过除了 ParseException 之外的所有异常。

代码清单 7-72　配置跳过除了 ParseException 之外的所有异常

```
@Bean
public Step copyFileStep() {
    return this.stepBuilderFactory.get("copyFileStep")
                .<Customer, Customer>chunk(10)
                .reader(itemReader())
                .writer(outputWriter(null))
                .faultTolerant()
                .skip(Exception.class)
                .noSkip(ParseException.class)
                .skipLimit(10)
                .build();
}
```

代码清单 7-72 中的配置指定了除 org.springframework.batch.item.ParseException 外，所有扩展了 java.lang.Exception 的异常都将被跳过，并且最多跳过 10 次。

还有第三种方法可用来指定要跳过的异常和次数。Spring Batch 提供了一个名为 org.springframework.batch.core.step.skip.SkipPolicy 的接口。该接口及其唯一的方法 shouldSkip 会接收抛出的异常和跳过的次数。在此基础上，这个接口的任何实现都可以决定应该跳过哪些异常以及跳过多少次。代码清单 7-73 展示了 SkipPolicy 接口的一个实现，该实现不允许跳过 java.io.FileNotFoundException 异常，但允许跳过 ParseException 异常 10 次。

代码清单 7-73　FileVerificationSkipper

```
...
public class FileVerificationSkipper implements SkipPolicy {

    public boolean shouldSkip(Throwable exception, int skipCount)
        throws SkipLimitExceededException {

        if(exception instanceof FileNotFoundException) {
            return false;
        } else if(exception instanceof ParseException && skipCount <= 10) {
            return true;
        } else {
            return false;
        }
    }
}
```

跳过记录在批处理中十分常见，这种操作将允许比单条记录大得多的处理流程继续进行，将错误产生的影响降到最小。一旦可以跳过有错误的记录，那就可能需要执行其他操作，例如将它们记录下来以备将来进行评估，7.7.2 节将讨论如何实现。

7.7.2　把无效的记录记入日志

跳过有问题的记录尽管主意不错，但这种操作本身也会引发问题。在一些场景下，跳过记录是没有问题的。假设正在做数据挖掘，并且遇到无法解决的问题，那么跳过记录是没有问题的。但是，当涉及金钱时，例如正在处理交易，那么仅仅跳过记录可能不够健壮。在这种情况下，能够记录导致错误的日志会很有帮助。在本小节中，你将研究如何使用 ItemListener 把无效的记录记入日志。

ItemReadListener 接口包含三个方法：beforeRead()、afterRead(T item) 和 onReadError(Exception ex)。为了把读取到的无效记录记入日志，可以使用 ItemListenerSupport 类并重写 onReadError 方法以记录发生的情况，也可使用包含了带有 @OnReadError 注解的方法的 POJO。需要重点指出的是，Spring Batch 能够很好地构建文件解析中的异常，从而告诉你发生的情况以及产生原因。由于大多数实际的数据库工作是由其他框架（Spring 本身、Hibernate 等）完成的，因此 Spring Batch 在数据库异常方面做得比较少。更重要的是，在开发自己的处理组件（自定义的 ItemReader、RowMapper 等）时，必须包含足够的详细信息，以便通过异常本身诊断问题。

在本例中，我们将从本章开头的客户文件中读取数据。在读取过程中，当引发异常时，就把导致异常的记录以及异常本身写入日志。为此，如果是异常是 FlatFileParseException，那么

CustomerItemListener 将接收到抛出的异常，同时你将能够访问引起问题的记录以及错误信息。代码清单 7-74 展示了 CustomerItemListener 监听器。

代码清单 7-74　CustomerItemListener 监听器

```
...
public class CustomerItemListener {

    private static final Log logger = LogFactory.getLog(CustomerItemListener.class);

    @OnReadError
    public void onReadError(Exception e) {
        if(e instanceof FlatFileParseException) {
            FlatFileParseException ffpe = (FlatFileParseException) e;

            StringBuilder errorMessage = new StringBuilder();
            errorMessage.append("An error occured while processing the " +
                    ffpe.getLineNumber() +
                    " line of the file. Below was the faulty " +
                    "input.\n");
            errorMessage.append(ffpe.getInput() + "\n");
            logger.error(errorMessage.toString(), ffpe);
        } else {
            logger.error("An error has occurred", e);
        }
    }
}
```

为了配置这个监听器，你需要更新读取文件的步骤。对于本例而言，只需要在 copyJob 作业中更新一个步骤，代码清单 7-75 展示了用于这个监听器的配置。

代码清单 7-75　配置 CustomerItemListener 监听器

```
...
@Bean
public CustomerItemListener customerListener() {
    return new CustomerItemListener ();
}

@Bean
public Step copyFileStep() {
    return this.stepBuilderFactory.get("copyFileStep")
            .<Customer, Customer>chunk(10)
            .reader(customerFileReader(null))
            .writer(outputWriter(null))
            .faultTolerant()
            .skipLimit(100)
            .skip(Exception.class)
            .listener(customerListener())
            .build();
}
...
```

如果以读取固定宽度记录的作业为例，并使用包含长度超过 63 个字符的输入记录的文件执行作业，就会引发异常。但是，由于已经把作业配置为跳过所有扩展了 Exception 的异常，因此引发的异常不会影响作业的执行结果，但 customerItemLogger 会被调用并根据需要记录条目。当执行这个作业时，你会看到两样东西。首先是因无效记录导致的 FlatFileParseException，其次是记录的日志消息。代码清单 7-76 展示了当作业出错时生成的日志消息。

代码清单 7-76　CustomerItemLogger 的输出

```
2011-05-03 23:49:22,148 ERROR main [com.apress.springbatch.chapter7.CustomerItemListener] -
<An error occured while processing the 1 line of the file. Below was the faulty input.
         Michael    TMinella 123      4th Street           Chicago      IL60606ABCDE>
```

通过使用良好的日志记录框架，就可以从 FlatFileParseException 获取解析失败的输入，并将它们记录到日志文件中。然而，这本身无法实现将错误记录写到文件中并继续执行的目标。在这种情况下，作业将写入导致问题发生的记录，然后失败。接下来你将了解在作业运行时，如何处理没有输入的情况。

7.7.3　处理没有输入的情况

SQL 查询不返回任何行并不是什么特殊情况。在许多情况下都存在空的文件。但是它们对于批处理过程有意义吗？在本小节中，你将研究 Spring Batch 如何应对输入源没有数据的情况。

默认情况下，当读取器在第一次尝试读取输入源就得到 null 时，处理方法与读取器在任何其他时间收到 null 时是一样的：读取器将认为步骤已完成。尽管这种方法在大多数情况下是可行的，但是你可能需要知道给定查询在什么时候返回零行或空文件。

如果要在未读取到任何输入的情况下引发步骤的失败或者希望采取任何其他措施(比如发送电子邮件等)，那么可以使用 StepExecutionListner。第 4 章曾使用 StepExecutionListner 记录步骤的开始和结束。在这里，可以使用 StepExecutionListner 的@AfterStep 方法查看已读取多少记录并做出响应。代码清单 7-77 展示了当未读取到任何记录时，如何将步骤标记为失败。

代码清单 7-77　EmptyInputStepFailer

```
...
public class EmptyInputStepFailer {

   @AfterStep
   public ExitStatus afterStep(StepExecution execution) {
      if(execution.getReadCount() > 0) {
         return execution.getExitStatus();
      } else {
         return ExitStatus.FAILED;
      }
   }
}
```

EmptyInputStepFailer 监听器的配置与其他任何 StepExecutionListener 监听器一样。代码清单 7-78 展示了 EmptyInputStepFailer 的配置。

代码清单 7-78　配置 EmptyInputStepFailer

```
...
@Bean
public EmptyInputStepFailer emptyFileFailer() {
```

```
            return new EmptyInputStepFailer();
    }

    @Bean
    public Step copyFileStep() {
        return this.stepBuilderFactory.get("copyFileStep")
                    .<Customer, Customer>chunk(10)
                    .reader(customerFileReader(null))
                    .writer(outputWriter(null))
                    .listener(emptyFileFailer())
                    .build();
    }
    ...
```

在配置了这个步骤之后，运行作业。当没有输入时，作业不会以 COMPLETED 状态结束，而是会失败，从而让你可以使用期望的输入重新运行这个作业。

7.8　本章小结

读取和写入占据批处理过程的绝大部分时间，因此，读取和写入也是 Spring Batch 框架中最重要的部分。在本章中，你对 Spring Batch 框架内的条目读取器的各个选项有了透彻但不全面的理解。你现在已经可以读取一个条目，接下来要做的就是对条目进行处理。第 8 章将介绍用于对条目进行处理的条目处理器。

第 8 章

ItemProcessor

第 7 章介绍了如何使用 Spring Batch 的组件读取多种类型的输入。尽管让软件获取输入是所有项目的重要方面之一，但如果不使用输入数据完成一些工作，就没有什么意义了。ItemProcessor(条目处理器)是 Spring Batch 中的组件，用来对输入进行一些操作。本章将介绍 ItemProcessor 接口，以及如何用它处理批处理条目。

- 在 8.1 节中，我们将首先概述 ItemProcessor，让你快速了解 ItemProcessor 在步骤的流程中扮演的角色。
- Spring Batch 提供了实用的 ItemProcessor 实现，例如 ItemProcessorAdapter。在 8.2 节中，我们将深入研究 Spring Batch 框架提供的每一种条目处理器。
- 在许多情况下，你可能想要开发自己的 ItemProcessor 实现。在 8.3 节中，你将编写自己的条目处理器。
- ItemProcessor 的常见用法是过滤掉 ItemReader 读入的部分条目，使它们不被步骤的 ItemWriter 写入，详见 8.3 节的"过滤条目"部分。

8.1 ItemProcessor 概述

第 7 章介绍了 ItemReader，它们是 Spring Batch 中的输入工具。在接收到输入后，你有两种选择。第一种是仅仅将它们写出(就像第 7 章那样)。在很多时候，我们是可以这样做的，比如把数据从一个系统迁移到另一个系统，或者将数据加载到数据库中以进行初始化。在这种情况下，在读取数据之后，将它们直接写出而不进行任何额外的处理是有意义的。

然而，在大多数情况下，需要对读入的数据进行处理。Spring Batch 已经对步骤进行了拆分，很好地分离了数据的读入、处理和写入之间的关注点。这样你就有机会完成一些特殊的事情，例如：

- 输入校验。在 Spring Batch 的原始版本中，校验发生在 ItemReader 中，并通过继承 ValidatingItemReader 类来实现。这种做法存在的问题是，Spring Batch 框架并没有提供继承 ValidatingItemReader 类的读取器，所以如果要进行校验，就无法使用 Spring Batch 框架包含的任何读取器。将校验步骤移到 ItemProcessor 中，就可以在处理之前进行校验，并且不考虑输入方法。从关注点分离的角度看，这更有意义。

- 复用现有的服务。就像第 7 章的 ItemReaderAdapter 用于复用输入的服务一样，Spring Batch 出于同样的原因提供了 ItemProcessorAdapter。
- 执行脚本。ItemProcessor 可以为其他开发人员或团队提供插入处理逻辑的机会。不过，其他团队可能不会使用基于 Spring 的工具集。ScriptItemProcessor 允许将某些类型的脚本的执行作为 ItemProcessor，将条目作为输入提供给脚本，并将脚本返回的内容作为输出。
- 链式 ItemProcessor。在一些情况下，需要在同一事务内，在单个条目上执行多个操作。尽管可以自行编写 ItemProcessor，并在一个类中完成所有的逻辑，但这种做法将逻辑与 Spring Batch 框架耦合了起来，而这是应该避免的。相反，Spring Batch 允许创建一个 ItemProcessor 列表，以按照顺序执行每个条目。

为此，org.springframework.batch.item.ItemProcessor 接口提供了唯一的方法 process，如代码清单 8-1 所示，用于接收一个来自 ItemReader 的条目，并返回另一个条目。

代码清单 8-1　ItemProcessor 接口

```
package org.springframework.batch.item;

public interface ItemProcessor<I, O> {

    O process(I item) throws Exception;
}
```

需要特别注意的是，ItemProcessor 接收到的输入类型并不需要与返回类型相同。Spring Batch 框架允许你读入某种类型的对象，把它们传给 ItemProcessor，并让 ItemProcessor 返回用于写入的其他类型的对象。有了这个特性，你应该会注意到，最后一个 ItemProcessor 返回的类型必须是 ItemWriter 的输入类型。你还应该意识到，如果某个 ItemProcessor 返回 null，那么对这个条目的所有处理都会停止。然而，与从 ItemReader 返回 null 不同(用于指示 Spring Batch 已经处理完所有的输入)，当 ItemProcessor 返回 null 时，对其他条目的处理仍会继续进行。

> **注意**　ItemProcessor 必须是幂等的。在容错场景下，一个条目可能被多次传给 ItemProcessor。

下面介绍如何在作业中使用 ItemProcessor。首先，我们研究一下 Spring Batch 框架提供的条目处理器。

8.2　使用 Spring Batch 提供的 ItemProcessor

前面在介绍 ItemReader 时，介绍了很多有关 Spring Batch 的基础知识。在大部分情况下，从文件读入的方式是相同的。对于大部分数据库来说，写入数据库的方式也是相同的。然而，基于具体的业务需求，对每个条目的处理是不同的。也正因为如此，每个作业才与众不同。因此，Spring Batch 框架仅仅提供了一些基础工具，以便你编写自己的逻辑或者封装现有逻辑。本节将介绍 Spring Batch 框架中包含的 ItemProcessor 实现。

8.2.1　ValidatingItemProcessor

下面继续第 7 章末尾的话题，开始讨论 Spring Batch 中的 ItemProcessor 的各个实现。之前，你已经获取到作业的输入；但是仅仅能够读取输入并不意味着输入就是有效的。数据类型和格式相关的校验可以通过 ItemReader 进行；然而，基于业务规则的校验最好在构建了条目之后进行。这就是为什么 Spring Batch 提供用来校验输入的 ItemProcessor 实现的原因。在 8.2.2 节中，你将了解如何使用 ValidatingItemProcessor 来校验输入。

8.2.2　输入校验

org.springframework.batch.item.validator.ValidatingItemProcessor(校验条目处理器)是 ItemProcessor 接口的一个实现[1]，它允许设置 Spring Batch 的 Validator 接口的一个实现，用于在处理之前校验输入。如果输入通过了校验，就进行处理；否则将抛出 org.springframework.batch.item.validator.ValidationException 异常，引发 Spring Batch 的标准错误处理。

JSR 303 是用于 Bean 校验的 Java 规范。这个规范提供了一种已在 Java 生态中被广泛采纳的校验方法。校验可通过 javax.validation.*中的代码来完成，并通过注解进行配置。这里提供了一组开箱即用的预定义了校验函数的注解；此外，你还可以创建自己的校验函数。下面首先看看如何校验类似于代码清单 8-2 所示的 Customer 类。

代码清单 8-2　Customer 类

```
...
public class Customer {
    private String firstName;
    private String middleInitial;
    private String lastName;
    private String address;
    private String city;
    private String state;
    private String zip;

    // Getters & setters go here
...
}
```

如果查看代码清单 8-2 所示的 Customer 类，那么可以很快地确定一些基本的输入校验规则。

- 非空(Not Null)：firstName、lastName、address、city、state 和 zip。
- 按字母排序(Alphabetic)：firstName、middleInitial、lastName、city 和 state。
- 大小(Size)：middleInitial 不应该超过一个字符，state 不应该超过两个字符，zip 不应该超过五个字符。

对于数据中的 zip 字段，可以进一步校验对应的是城市(city)还是州(state)的邮政编码。不过，这已经提供了良好的开端。你现在已经识别了需要校验的内容，接下来可以通过注解来描述用于

---

[1] 尽管 Spring 确实提供了自己的 Validator 接口，但是 ValidatingItemProcessor 却选择使用 Spring Batch 提供的 Validator 接口。

Customer 对象的校验器。特别是，你需要使用@NotNull、@Size 和@Pattern 等注解来应用这些规则。为了使用它们，你需要在项目中使用如下新的启动包：spring-boot-starter-validation，从而引入 JSR-303 校验工具的 Hibernate 实现。

下面创建一个新的项目来完成校验工作，在 Spring Initializr 中选择 Batch、JDBC、MySQL 和 Validation 等依赖项。创建好新的项目后，添加代码清单 8-2 中的代码作为领域对象。接下来，使用 JSR-303 注解，应用前面提到的校验规则，代码清单 8-3 展示了它们在 Customer 对象上的用法。

代码清单 8-3　使用校验注解的 Customer 对象

```
...
public class Customer {

    @NotNull(message="First name is required")
    @Pattern(regexp="[a-zA-Z]+", message="First name must be alphabetical")
    private String firstName;

    @Size(min=1, max=1)
    @Pattern(regexp="[a-zA-Z]", message="Middle initial must be alphabetical")
    private String middleInitial;

    @NotNull(message="Last name is required")
    @Pattern(regexp="[a-zA-Z]+", message="Last name must be alphabetical")
    private String lastName;

    @NotNull(message="Address is required")
    @Pattern(regexp="[0-9a-zA-Z\\. ]+")
    private String address;

    @NotNull(message="City is required")
    @Pattern(regexp="[a-zA-Z\\. ]+")
    private String city;

    @NotNull(message="State is required")
    @Size(min=2,max=2)
    @Pattern(regexp="[A-Z]{2}")
    private String state;

    @NotNull(message="Zip is required")
    @Size(min=5,max=5)
    @Pattern(regexp="\\d{5}")
    private String zip;

    // Accessors go here
...
}
```

快速查看一下代码清单 8-3 中定义的规则，你可能会问，既然@Pattern 中定义的正则表达式可能已经满足了需求，为何还要同时使用@Size 和@Pattern 注解呢？你的想法没错。然而，每一个注解只允许指定唯一的错误消息(可选)；再者，能够识别字段是长度错误还是格式错误，在将来可能很有用。

现在，你已经定义了用于 Customer 条目的校验规则。为了应用这个功能，你需要提供一种机制，让 Spring Batch 校验每一项。org.springframework.batch.item.validator.BeanValidatingItemProcessor 能帮

第 8 章 ItemProcessor

你完成所有的事情，这个条目处理器是对ValidatingItemProcessor的扩展，特点是可以利用JSR-303来提供校验。

ValidatingItemProcessor的校验能力来源于org.springframework.batch.item.validator.Validator接口的实现。这个接口只有一个方法 void validate(T value)。如果条目是有效的，那么这个方法不做任何事情；如果校验失败，那么抛出 org.springframework.batch.item.validator.ValidationException 异常。BeanValidatingItemProcessor 是 ValidatingItemProcessor 的一个特殊版本，用于创建符合JSR-303规范的校验器。

> **注意** Spring Batch 框架包含的 Validator 接口与核心 Spring 框架的 Validator 接口不同。Spring Batch 提供了适配器类 SpringValidator 来处理这种区别。

下面创建一个作业以综合使用所有这些技术，看看它们如何工作。将这个作业以逗号分隔的文件读入 Customer 对象，并在 BeanValidatingItemProcessor 中对其进行校验，然后写入一个 CSV 文件(与第 7 章类似)。代码清单 8-4 展示了将要处理的输入文件。

代码清单 8-4　customer.csv

```
Richard,N,Darrow,5570 Isabella Ave,St. Louis,IL,58540
Barack,G,Donnelly,7844 S. Greenwood Ave,Houston,CA,38635
Ann,Z,Benes,2447 S. Greenwood Ave,Las Vegas,NY,55366
Laura,9S,Minella,8177 4th Street,Dallas,FL,04119
Erica,Z,Gates,3141 Farnam Street,Omaha,CA,57640
Warren,L,Darrow,4686 Mt. Lee Drive,St. Louis,NY,94935
Warren,M,Williams,6670 S. Greenwood Ave,Hollywood,FL,37288
Harry,T,Smith,3273 Isabella Ave,Houston,FL,97261
Steve,O,James,8407 Infinite Loop Drive,Las Vegas,WA,90520
Erica,Z,Neuberger,513 S. Greenwood Ave,Miami,IL,12778
Aimee,C,Hoover,7341 Vel Avenue,Mobile,AL,35928
Jonas,U,Gilbert,8852 In St.,Saint Paul,MN,57321
Regan,M,Darrow,4851 Nec Av.,Gulfport,MS,33193
Stuart,K,Mckenzie,5529 Orci Av.,Nampa,ID,18562
Sydnee,N,Robinson,894 Ornare. Ave,Olathe,KS,25606
```

注意，在以上输入文件的第 4 行中，中间名缩写 9S 是无效的，这会导致校验在此处失败。在定义了输入文件后，接下来配置作业。我们将要运行的作业包含一个唯一的步骤：读取输入，将它们传给 ValidatingItemProcessor 的一个实例，然后写到标准输出。代码清单 8-5 展示了作业的配置。

代码清单 8-5　ValidationJob

```
...
@EnableBatchProcessing
@SpringBootApplication
public class ValidationJob {

    @Autowired
    public JobBuilderFactory jobBuilderFactory;

    @Autowired
    public StepBuilderFactory stepBuilderFactory;
```

```java
    @Bean
    @StepScope
    public FlatFileItemReader<Customer> customerItemReader(
            @Value("#{jobParameters['customerFile']}")Resource inputFile) {

        return new FlatFileItemReaderBuilder<Customer>()
                    .name("customerItemReader")
                    .delimited()
                    .names(new String[] {"firstName",
                                         "middleInitial",
                                         "lastName",
                                         "address",
                                         "city",
                                         "state",
                                         "zip"})
                    .targetType(Customer.class)
                    .resource(inputFile)
                    .build();
    }

    @Bean
    public ItemWriter<Customer> itemWriter() {
        return (items) -> items.forEach(System.out::println);
    }

    @Bean
    public BeanValidatingItemProcessor<Customer> customerValidatingItemProcessor() {
        return new BeanValidatingItemProcessor<>();
    }

    @Bean
    public Step copyFileStep() {
        return this.stepBuilderFactory.get("copyFileStep")
                    .<Customer, Customer>chunk(5)
                    .reader(customerItemReader(null))
                    .processor(customerValidatingItemProcessor())
                    .writer(itemWriter())
                    .build();
    }

    @Bean
    public Job job() throws Exception {
        return this.jobBuilderFactory.get("job")
                    .start(copyFileStep())
                    .build();
    }

    public static void main(String[] args) {
        SpringApplication.run(ValidationJob.class, "customerFile=/input/customer.csv");
    }
}
```

下面解释代码清单 8-5 中的 ValidationJob 类，我们首先从输入文件和读取器的定义开始。这是一个简单的分隔文件读取器，用于将文件中的字段映射到 Customer 对象。接下来是输出配置，用于为 ItemWriter 定义一个 lambda，将内容写到标准输出。定义了输入输出后，customerValidatingItemProcessor

Bean 开始承担 ItemProcessor 的角色。默认情况下，BeanValidatingItemProcessor 仅仅把条目从 ItemReader 传给 ItemWriter，对于本例来说这是可行的。

在定义了所有的 Bean 之后，就可以构建步骤，步骤是这个文件的另一部分。对于步骤来说，需要定义读取器、处理器和写入器。在有了步骤的定义之后，剩下要做的就是完成作业本身的配置。

为了运行作业，下面在项目的 target 目录中执行代码清单 8-6 所示的命令。

代码清单 8-6　运行 copyJob

```
java -jar itemProcessors-0.0.1-SNAPSHOT.jar customerFile=/input/customer.csv
```

前面提到过，因为输入中包含错误数据，所以无法通过校验。作业在运行时会因为抛出的 ValidationException 异常而失败。为了让作业能够成功完成，我们必须修复输入以通过校验。代码清单 8-7 展示了作业在输入校验失败时的输出。

代码清单 8-7　copyJob 的输出

```
2019-02-05 17:19:35.287 INFO 39336 --- [   main] o.s.batch.core.job.
SimpleStepHandler: Executing step: [copyFileStep]
2019-02-05 17:19:35.462 ERROR 39336 --- [   main] o.s.batch.core.step.
AbstractStep: Encountered an error executing step copyFileStep in job
org.springframework.batch.item.validator.ValidationException: Validation failed
for Customer{firstName='Laura', middleInitial='9S', lastName='Minella', address='8177
4th Street',city='Dallas', state='FL', zip='04119'}:
Field error in object 'item' on field 'middleInitial': rejected value [9S]; codes [Size.
item.middleInitial,Size.middleInitial,Size.java.lang.String,Size]; arguments [org.
springframework.context.support.DefaultMessageSourceResolvable: codes [item.
middleInitial,middleInitial]; arguments[]; default message [middleInitial],1,1]; default
message [size must be between 1 and 1]
Field error in object 'item' on field 'middleInitial': rejected value [9S]; codes [Pattern.
item.middleInitial,Pattern.middleInitial,Pattern.java.lang.String,Pattern]; arguments
[org.springframework.context.support.DefaultMessageSourceResolvable: codes [item.
middleInitial,middleInitial]; arguments[]; default message [middleInitial],[Ljavax.
validation.constraints.Pattern$Flag;@3fd05b3e,org.springframework.validation.
beanvalidation.
SpringValidatorAdapter$ResolvableAttribute@4eb9f2af]; default message [Middle
initial must be alphabetical]at org.springframework.batch.item.validator.SpringValidator.
validate(SpringValidator.java:54) ~[spring-batch-infrastructure-4.1.1.RELEASE.ja
r:4.1.1.RELEASE]
...
```

这就是在 Spring Batch 中使用 JSR-303 规范添加条目校验所需的一切。不过，如果想实现自己的校验，该如何做呢？为此，我们将 ItemProcessor 从 BeanValidatingItemProcessor 修改为 ValidatingItemProcessor，并注入 Validator 接口的实现。

假定 lastName(姓氏)字段在整个数据集中必须是唯一的。为了校验记录是否符合上述要求，我们将实现一个有状态的检验器以记录已经处理过的姓氏。如果想在作业重启之间持久化状态，那么这个检验器也需要通过扩展 ItemStreamSupport 来实现 ItemStream 接口，并且在每次提交时都把这些姓氏保存在 ExecutionContext 中。代码清单 8-8 展示的检验器实现了以上逻辑。

代码清单 8-8　校验姓氏在数据集中的唯一性

```
public class UniqueLastNameValidator extends ItemStreamSupport
```

```
    implements Validator<Customer> {

    private Set<String> lastNames = new HashSet<>();

    @Override
    public void validate(Customer value) throws ValidationException {
        if(lastNames.contains(value.getLastName())) {
            throw new ValidationException("Duplicate last name was found: "
                + value.getLastName());
        }

        this.lastNames.add(value.getLastName());
    }

    @Override
    public void open(ExecutionContext executionContext) {
        String lastNames = getExecutionContextKey("lastNames");

        if(executionContext.containsKey(lastNames)) {
            this.lastNames = (Set<String>) executionContext.get(lastNames);
        }
    }

    @Override
    public void update(ExecutionContext executionContext) {
        executionContext.put(getExecutionContextKey("lastNames"), this.lastNames);
    }
}
```

在上述代码的开头，UniqueLastNameValidator 类扩展了 ItemStreamSupport，以便在作业执行之间持久化状态，同时还实现了 Validator 接口。我们把姓氏保存在接下来定义的 lastNames 集合中。validate(Customer customer)方法用于完成如下校验：如果姓氏不在 lastNames 集合中，那就保存它们；如果在处理前面的记录时已经把姓氏添加到 lastNames 集合中，那么抛出 ValidationExecption 异常，以表明遇到错误数据。UniqueLastNameValidator 类的最后两个方法是为了在作业执行之间持久化状态。open 方法先确定是否在前面的执行中保存了 lastNames 字段，如果是，就在步骤处理开始前恢复。update 方法（事务提交后，为每个块调用一次）把当前状态保存在 ExecutionContext 中，以防止在处理下一个块时出现故障。

在创建了 Validator 接口的实现后，我们还需要进行配置。这种校验机制的配置包含三部分：首先定义一个 UniqueLastNameValidator 类型的 Bean，然后把它注入 ValidatingItemProcessor，最后把这个 UniqueLastNameValidator 类型的 Bean 注册为步骤中的流(Stream)，这样 Spring Batch 就知道要为步骤调用与 ItemStream 相关的方法。代码清单 8-9 展示了以上配置。

代码清单 8-9 UniqueLastNameValidator 的配置

```
...
@Bean
public UniqueLastNameValidator validator() {
    UniqueLastNameValidator uniqueLastNameValidator = new UniqueLastNameValidator();

    uniqueLastNameValidator.setName("validator");
```

第 8 章 ItemProcessor

```
        return uniqueLastNameValidator;
}

@Bean
public ValidatingItemProcessor<Customer> customerValidatingItemProcessor() {
    return new ValidatingItemProcessor<>(validator());
}

@Bean
public Step copyFileStep() {

return this.stepBuilderFactory.get("copyFileStep")
            .<Customer, Customer>chunk(5)
            .reader(customerItemReader(null))
            .processor(customerValidatingItemProcessor())
            .writer(itemWriter)
            .stream(validator())
            .build();
}
...
```

如果使用代码清单 8-4 中定义的数据运行更新后的配置，那么需要尝试三次才能完成作业：

1) 在第一次尝试中提交第一个块(5 条记录)，然后在第二个块中失败。

2) 移除第 6 行的输入文件，然后再次运行。这一次尝试会跳过第一个块中的记录，提交第二个块，然后在处理第三个块时因为找到相同的姓氏而失败。

3) 移除第 13 行(第二次尝试后输入文件中的第 12 行)，最后运行一次以查看作业是否已完成。

在处理条目的过程中，使用 ValidatingItemProcessor 及其子类 BeanValidatingItemProcessor 进行校验是有用的。然而，这些只是 Spring Batch 提供的 ItemProcessor 接口实现的三个主要部分之一。在 8.2.2 节中，你将看到如何使用 ItemProcessorAdapter 把现有的服务作为 ItemProcessor 来使用。

8.2.3 ItemProcessorAdapter

第 7 章曾经使用 ItemReaderAdapter 把现有的服务作为作业的输入。Spring Batch 还允许通过使用 org.springframework.batch.item.adapter.ItemProcessorAdapter，把已经开发的多种服务作为条目处理器。

下面创建一个服务，目的是把客户的姓名(包括名字、中间名以及姓氏)转换成大写。我们把这个服务命名为 UpperCaseNameService，其中只有一个方法，用于将 Customer 输入对象复制到另一个新的 Customer 输出对象(以使之幂等)，然后在新的实例中把姓名转换成大写并返回。代码清单 8-10 展示了 UpperCaseNameService。

代码清单 8-10　UpperCaseNameService

```
...
@Service
public class UpperCaseNameService {

    public Customer upperCase(Customer customer) {
        Customer newCustomer = new Customer(customer);

        newCustomer.setFirstName(newCustomer.getFirstName().toUpperCase());
        newCustomer.setMiddleInitial(newCustomer.getMiddleInitial().toUpperCase());
```

```
            newCustomer.setLastName(newCustomer.getLastName().toUpperCase());
            return newCustomer;
        }
    }
```

在定义了这个服务之后,就可以把前面作业中的校验功能替换为这个服务提供的用于将姓名转换成大写的功能。代码清单 8-11 展示了用于配置新的 ItemProcessor 的代码以及更新后的步骤。

代码清单 8-11　ItemProcessorAdapter 的配置

```
...
@Bean
public ItemProcessorAdapter<Customer, Customer> itemProcessor(UpperCaseNameService service)
{
    ItemProcessorAdapter<Customer, Customer> adapter = new ItemProcessorAdapter<>();

    adapter.setTargetObject(service);
    adapter.setTargetMethod("upperCase");

    return adapter;
}

@Bean
public Step copyFileStep() {
    return this.stepBuilderFactory.get("copyFileStep")
                .<Customer, Customer>chunk(5)
                .reader(customerItemReader(null))
                .processor(itemProcessor(null))
                .writer(itemWriter())
                .build();
}
...
```

代码清单 8-11 为 ItemProcessorAdapter 定义了 Bean,并注入了代码清单 8-10 中定义的 UpperCaseNameService。这个适配器需要设置两个必选值和一个可选值。两个必选值分别是目标对象(将要调用的实例)和目标方法(将要在实例上调用的方法)。还有一个配置选项,用于提供一个数组类型的参数,但是,你在 ItemProcessorAdapter 上传给它的任何值都会被忽略。

在配置了适配器之后,你需要更新引用它的步骤,如代码清单 8-11 所示。在完成了配置之后,当运行作业时,输出会以大写形式展示客户的姓名,如代码清单 8-12 所示。

代码清单 8-12　ItemProcessorAdapterJob 的输出

```
...
2019-02-05 22:23:19.185 INFO 45123 --- [main] o.s.batch.core.job.SimpleStepHandler:
Executing step: [copyFileStep]
Customer{firstName='RICHARD', middleInitial='N', lastName='DARROW', address='5570
Isabella Ave', city='St. Louis', state='IL', zip='58540'}
Customer{firstName='BARACK', middleInitial='G', lastName='DONNELLY', address='7844 S.
greenwood Ave', city='Houston', state='CA', zip='38635'}
Customer{firstName='ANN', middleInitial='Z', lastName='BENES', address='2447 S.
Greenwood Ave', city='Las Vegas', state='NY', zip='55366'}
Customer{firstName='LAURA', middleInitial='9S', lastName='MINELLA', address='8177 4th
Street', city='Dallas', state='FL', zip='04119'}
```

```
Customer{firstName='ERICA', middleInitial='Z', lastName='GATES', address='3141 Farnam
Street', city='Omaha', state='CA', zip='57640'}
...
```

自从 Groovy 语言出现以来，脚本语言和 JVM 对它们的支持有了很大的提升。现在，你可以在 JVM 上运行 Ruby、JavaScript、Groovy 等其他脚本语言。8.2.3 节将介绍如何使用脚本实现条目处理器。

8.2.4　ScriptItemProcessor

脚本语言提供了很多特别的机会。脚本通常更易于创建和修改，所以对于需要频繁修改的组件，脚本能够提供很大的灵活性。在构建原型时，如果使用脚本语言，就不必再受制于静态类型语言(例如 Java)中所有的"繁文缛节"。通过在 ItemProcessor 中执行脚本，Spring Batch 把这种灵活性注入了批处理作业。org.springframework.batch.item.support.ScriptItemProcessor 允许设定一段脚本，从而接收 ItemProcessor 的输入，返回的对象是 ItemProcessor 的输出。本节将实现与 ItemProcessorAdapter 相同的功能。不过，我们并不会使用 Java 服务来执行逻辑，而是将其中的逻辑放在 JavaScript 脚本中。

首先，你需要勇于"走出舒适区"，编写一些 JavaScript 脚本。这些脚本实际上非常简单。默认情况下，ScriptItemProcessor 会把来自 ItemProcessor 的输入绑定到变量条目(这个值是可配置的，如果有需要，你可以修改它)。接下来，就可以在它上面执行 JavaScript 功能，并把它返回给 ItemProcessor。在这里，我们把名字、中间名和姓氏转换成大写，参见代码清单 8-13。

代码清单 8-13　upperCase.js

```
item.setFirstName(item.getFirstName().toUpperCase());
item.setMiddleInitial(item.getMiddleInitial().toUpperCase());
item.setLastName(item.getLastName().toUpperCase());
item;
```

在编写了脚本之后，就可以让作业使用 ScriptItemProcessor。我们只需要为这个条目处理器提供一个 Resource 作为依赖项，用它指向将要使用的脚本文件(也可以通过内联的方式使用字符串定义脚本)。代码清单 8-14 展示了 ScriptItemProcessor 的配置。

代码清单 8-14　ScriptItemProcessor 的配置

```
...
@Bean
@StepScope
public ScriptItemProcessor<Customer, Customer> itemProcessor(
    @Value("#{jobParameters['script']}") Resource script) {

    ScriptItemProcessor<Customer, Customer> itemProcessor =
            new ScriptItemProcessor<>();

    itemProcessor.setScript(script);

    return itemProcessor;
}
...
```

对于以上配置，我们需要给作业添加一个额外的作业参数，也就是脚本文件的位置，并使用如下命令执行作业：java -jar copyJob.jar customerFile=/input/customer.csv script=/upperCase.js。执行以上命令后，输出将与代码清单 8-12 中使用 ItemProcessorAdapter 的作业的输出相同。

在某些情况下，在一次事务中，对条目只应用单个动作可能是一种限制。例如，如果需要在部分条目上完成一组计算，那么可能想要过滤掉那些不需要处理的条目。在 8.2.4 节中，你将了解如何对步骤中的每个条目进行配置以执行一组条目处理器。

8.2.5 CompositeItemProcessor

我们可以把步骤拆分成三个阶段(读取、处理和写入)，以划分组件之间的职责。然而，对于给定的条目，如果把业务逻辑与某个 ItemProcessor 耦合起来，那么意义可能不大。Spring Batch 允许在一个步骤中链接多个 ItemProcessor，以维护与业务逻辑相同的职责区域划分方式。在本小节中，你将了解如何使用组合(Composition)，从而在步骤的条目处理阶段实现更复杂的编排。

下面首先从 org.springframework.batch.item.support.CompositeItemProcessor(组合条目处理器)开始，作为 ItemProcessor 接口的一个实现，它按顺序把条目的处理委托给了一个列表，这个列表包含 ItemProcessor 接口的多个实现。每个条目处理器都会返回自己的结果，这些结果又被传给下一个条目处理器，直到所有条目处理器都被调用。无论返回什么类型，这种模式都会发生。所以，如果第一个 ItemProcessor 接收字符串作为输入，那么只要下一个 ItemProcessor 接收 Product 作为输入，就可以返回一个 Product 对象。最后的结果会被传给你为当前步骤配置的 ItemWriter。需要重点注意的是，CompositeItemProcessor 与其他的 ItemProcessor 类似，只要委托的任何条目处理器返回 null，这个条目就不再被处理。图 8-1 展示了 CompositeItemProcessor 中的处理是如何进行的。

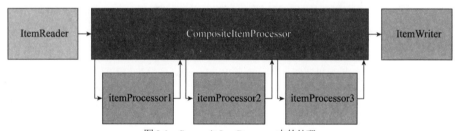

图 8-1　CompositeItemProcessor 中的处理

如图 8-1 所示，CompositeItemProcessor 是多个 ItemProcessor 的封装器，并且会按顺序调用它们。当一个条目处理器完之后，就会使用前一个条目处理器返回的条目调用下一个条目处理器。

在下面的示例中，我们将采用本章前面的示例并一次性应用它们。我们将创建链式的条目处理器，按顺序执行下面的操作：

1) 校验输入，过滤掉错误记录。

2) 使用 UpperCaseNameService 服务把姓名转换成大写。

3) 使用 JavaScript 脚本把 address、city 和 state 字段转换成小写。

下面首先配置用于校验接收的 Customer 对象的 ItemProcessor。为此，我们使用与代码清单 8-9 相同的配置，并做一点点修改。当条目非法时，ValidatingItemProcessor 默认会抛出 ValidationException 异常。然而，在某些情况下，这可能是一种过于激烈的度量方法(例如对于这里的示例)。正因为如此，我们还可以配置 ItemProcessor 以过滤掉那些不能通过校验的条目。这正是此时应用逻辑的方法。代码清单 8-15 展示了更新后的配置。

代码清单 8-15　配置 ValidatingItemProcessor 以过滤条目

```
...
@Bean
public UniqueLastNameValidator validator() {
    UniqueLastNameValidator uniqueLastNameValidator = new UniqueLastNameValidator();

    uniqueLastNameValidator.setName("validator");

    return uniqueLastNameValidator;
}

@Bean
public ValidatingItemProcessor<Customer> customerValidatingItemProcessor() {
    ValidatingItemProcessor<Customer> itemProcessor =
        new ValidatingItemProcessor<>(validator());

    itemProcessor.setFilter(true);

    return itemProcessor;
}
...
```

执行完第一个条目处理器之后，还有三个条目处理器。第二个条目处理器是前面配置的用于将客户姓名转换成小写的 ItemProcessorAdapter。具体配置与前面使用过的 ItemProcessAdapter 的配置相同，参见代码清单 8-16。

代码清单 8-16　upperCaseItemProcessor

```
...
@Bean
public ItemProcessorAdapter<Customer, Customer> upperCaseItemProcessor(
    UpperCaseNameService service) {

    ItemProcessorAdapter<Customer, Customer> adapter = new ItemProcessorAdapter<>();

    adapter.setTargetObject(service);
    adapter.setTargetMethod("upperCase");

    return adapter;
}
...
```

对于第三个条目处理器，首先编写一个新的脚本，用于将所有的与客户地址相关的字段转换成小写，而不是把客户姓名相关的字段转换成大写，参见代码清单 8-17。

代码清单 8-17　lowerCase.js

```
item.setAddress(item.getAddress().toLowerCase());
item.setCity(item.getCity().toLowerCase());
```

```
item.setState(item.getState().toLowerCase());
item;
```

为了使用这个脚本，我们还要配置 ScriptItemProcessor，参见代码清单 8-18。

代码清单 8-18　lowerCaseItemProcessor

```
...
@Bean
@StepScope
public ScriptItemProcessor<Customer, Customer> lowerCaseItemProcessor(
        @Value("#{jobParameters['script']}") Resource script) {

    ScriptItemProcessor<Customer, Customer> itemProcessor =
            new ScriptItemProcessor<>();

    itemProcessor.setScript(script);

    return itemProcessor;
}
...
```

最后，为了把它们结合起来，我们还需要配置 CompositeItemProcessor。这个条目处理器将接收一个 ItemProcessor 列表，然后依次执行，所以顺序是很重要的。代码清单 8-19 展示了 CompositeItemProcessor 的配置及其委托链。

代码清单 8-19　CompositeItemProcessor 的配置及其委托链

```
...
@Bean
public CompositeItemProcessor<Customer, Customer> itemProcessor() {
    CompositeItemProcessor<Customer, Customer> itemProcessor =
            new CompositeItemProcessor<>();

    itemProcessor.setDelegates(Arrays.asList(
                customerValidatingItemProcessor(),
                upperCaseItemProcessor(null),
                lowerCaseItemProcessor(null)));

    return itemProcessor;
}
...
```

在应用了这些更改之后，运行作业。输出中的所有姓名都被转换成了大写，所有地址相关的字段也都被转换成了大写，并且过滤掉了两条记录(Warren M. Darrow 的记录和 Regan M. Darrow 的记录)。代码清单 8-20 展示了这个作业的输出。

代码清单 8-20　CompositeItemProcessor 作业的输出

```
...
2019-02-05 23:46:49.884 INFO 46774 --- [main] o.s.batch.core.job.SimpleStepHandler:
Executing step: [copyFileStep]
Customer{firstName='RICHARD', middleInitial='N', lastName='DARROW', address='5570
isabella ave', city='st. louis', state='il', zip='58540'}
```

```
Customer{firstName='BARACK', middleInitial='G', lastName='DONNELLY', address=
'7844 s.greenwood ave', city='houston', state='ca', zip='38635'}
Customer{firstName='ANN', middleInitial='Z', lastName='BENES', address='2447 s.
greenwood ave', city='las vegas', state='ny', zip='55366'}
...
```

CompositeIemProcessor 允许把其他组件组合成单个条目处理器,从而实现良好的软件设计。然而,如果不想把每个条目都发送给列表中的每个条目处理器,该怎么办呢?如果想让一些条目进入 ItemProcessorA,而让另一些条目进入 ItemProcessorB,又该怎么办?此时就需要使用 ClassifierCompositeItemProcessor 组件。

ClassifierCompositeItemProcessor(分类组合条目处理器)使用 org.springframework.classify.Classifier 类的实现来选择使用哪个条目处理器。这个分类器的 classify(C classifiable)方法的实现必须接收一个条目作为输入,并返回合适的将要使用的条目处理器。为了演示这一点,我们将创建一个分类器:如果邮政编码是奇数,则返回把名称转换成大写的条目处理器;如果邮政编码是偶数,则返回把地址转换成小写的条目处理器。

我们需要编写的关键部分是分类器的实现。首先判断邮政编码是偶数还是奇数,然后返回正确的委托。代码清单 8-21 列出了 ZipCodeClassifier 分类器的代码。

代码清单 8-21　ZipCodeClassifier

```
...
public class ZipCodeClassifier implements Classifier<Customer, ItemProcessor
<Customer, Customer>> {

    private ItemProcessor<Customer, Customer> oddItemProcessor;
    private ItemProcessor<Customer, Customer> evenItemProcessor;

    public ZipCodeClassifier(ItemProcessor<Customer, Customer> oddItemProcessor,
            ItemProcessor<Customer, Customer> evenItemProcessor) {

        this.oddItemProcessor = oddItemProcessor;
        this.evenItemProcessor = evenItemProcessor;
    }

    @Override
    public ItemProcessor<Customer, Customer> classify(Customer classifiable) {
        if(Integer.parseInt(classifiable.getZip()) % 2 == 0) {
            return evenItemProcessor;
        }
        else {
            return oddItemProcessor;
        }
    }
}
```

在创建了分类器的实现之后,就可以在作业中配置分类器和 ClassifierCompositeItemProcessor。首先配置 ZipCodeClassifier,分别把 upperCaseItemProcessor 和 lowerCaseItemProcessor 注入构造函数。然后配置 ClassifierCompositeItemProcessor,把分类器设置成唯一的依赖项。代码清单 8-22 展示了配置代码。

代码清单 8-22　配置 ClassifierCompositeItemProcessor

```
...
@Bean
public Classifier classifier() {
    return new ZipCodeClassifier(upperCaseItemProcessor(null),
                lowerCaseItemProcessor(null));
}

@Bean
public ClassifierCompositeItemProcessor<Customer, Customer> itemProcessor() {
    ClassifierCompositeItemProcessor<Customer, Customer> itemProcessor =
            new ClassifierCompositeItemProcessor<>();

    itemProcessor.setClassifier(classifier());

    return itemProcessor;
}
...
```

构建和执行把以上配置作为条目处理器的作业，输入如代码清单 8-23 所示。

代码清单 8-23　ClassifierCompositeItemProcessor 作业的输出

```
...
2019-02-06 00:17:11.833 INFO 47362 --- [main] o.s.b.c.l.support.SimpleJobLauncher:
Job: [SimpleJob: [name=job]] launched with the following parameters:
[{customerFile=/input/customer.csv, script=/lowerCase.js}]
2019-02-06 00:17:11.882 INFO 47362 --- [ main] o.s.batch.core.job.
SimpleStepHandler: Executing step: [copyFileStep]
Customer{firstName='Richard', middleInitial='N', lastName='Darrow', address='5570
isabella ave', city='st. louis', state='il', zip='58540'}
Customer{firstName='BARACK', middleInitial='G', lastName='DONNELLY', address='7844 S.
Greenwood Ave', city='Houston', state='CA', zip='38635'}
Customer{firstName='Ann', middleInitial='Z', lastName='Benes', address='2447 s.
greenwood ave', city='las vegas', state='ny', zip='55366'}
...
```

ClassifierCompositeItemProcessor 是在条目处理器中构建复杂流程的另一种方式。在结合使用 CompositeItemProcessor 之后，你将能够在步骤中的条目处理阶段构建任何复杂程度的条目处理器链。

尽管你在前面的章节中已经过滤了条目，但是还没有实现自己的条目处理器。在 8.3 节中，你将编写自己的条目处理器，用于从 ItemWriter 中过滤条目。

8.3　编写自己的条目处理器

条目处理器实际上是 Spring Batch 框架中最容易自行实现的部分，这是设计使然。在各个环境和业务用例之间，输入输出都是标准的。无论文件中包含的是财务数据还是科学数据，读入文件的方式都是相同的。无论对象是什么样子，写入数据库的方式也是相同的。然而，条目处理器是处理过程中的业务逻辑所在之处。因此，在实践中总是需要创建它们的自定义实现。在本节中，你将了解如何创建自定义的条目处理器，用以过滤掉读入的特定条目，使它们不被写出。

第 8 章 ItemProcessor

过滤条目

在 8.2.4 节中，基于邮政编码的奇偶性，我们执行了不同的逻辑。我们下面将编写一个条目处理器，用于过滤掉偶数的邮政编码，只保留奇数的邮政编码，以便写入。

所以，当这些记录经过条目处理器时，如何过滤掉这些记录呢？Spring Batch 的做法很简单，就是确保把任何导致条目处理器返回 null 的条目都过滤掉。Spring Batch 不仅仅会过滤掉它们对下游的影响(其他的 ItemProcessor 或任何与步骤相关的 ItemWriter)，而且会记录过滤掉的记录数量并保存在作业存储库中。

为了实现这个条目处理器，我们需要创建一个实现了 ItemProcessor 接口的类。我们将把逻辑放在 process 方法中：如果邮政编码是偶数，那么返回 null；如果是奇数，那么返回未经修改的参数。代码清单 8-24 展示了相应的代码。

代码清单 8-24　EvenFilteringItemProcessor

```
...
public class EvenFilteringItemProcessor implements ItemProcessor<Customer, Customer> {

    @Override
    public Customer process(Customer item) {
        return Integer.parseInt(item.getZip()) % 2 == 0 ? null: item;
    }
}
```

剩下要做的事情是配置作业以使用这个条目处理器，参见代码清单 8-25。

代码清单 8-25　配置自定义的条目处理器

```
...
@Bean
public EvenFilteringItemProcessor itemProcessor() {
    return new EvenFilteringItemProcessor();
}
...
```

在配置了新的条目处理器之后，运行作业，你会看到 EvenFilteringItemProcessor 过滤掉了 9 条记录，留下了 6 条记录并写入标准输出，如代码清单 8-26 所示。

代码清单 8-26　自定义的条目处理器作业的输出

```
2019-02-06 00:31:30.808 INFO 47626 --- [main] o.s.batch.core.job.SimpleStepHandler:
Executing step: [copyFileStep]
Customer{firstName='Barack', middleInitial='G', lastName='Donnelly', address='7844 S.
Greenwood Ave', city='Houston', state='CA', zip='38635'}
Customer{firstName='Laura', middleInitial='9S', lastName='Minella', address='8177 4th
Street', city='Dallas', state='FL', zip='04119'}
Customer{firstName='Warren', middleInitial='L', lastName='Darrow', address='4686 Mt. Lee
Drive', city='St. Louis', state='NY', zip='94635'}
Customer{firstName='Harry', middleInitial='T', lastName='Smith', address='3273 Isabella
Ave', city='Houston', state='FL', zip='97261'}
Customer{firstName='Jonas', middleInitial='U', lastName='Gilbert', address='8852 In St.',
city='Saint Paul', state='MN', zip='57321'}
```

```
Customer{firstName='Regan', middleInitial='M', lastName='Darrow', address='4851 Nec Av.',
city='Gulfport', state='MS', zip='33193'}
2019-02-06 00:31:30.949 INFO 47626 --- [ main] o.s.b.c.l.support.
SimpleJobLauncher : Job: [SimpleJob: [name=job]] completed with the following
parameters: [{customerFile=/input/customer.csv}] and the following status: [COMPLETED]
```

如果要过滤记录，那么最好能以一种简单的方式知道到底将要过滤多少记录。Spring Batch 在作业存储库中记录了过滤掉的条目数量。使用代码清单 8-4 中的数据运行作业，可以看到读入了 15 条记录，过滤掉了 9 条记录，写入了 6 条记录。代码清单 8-27 展示了作业存储库中的数据。

代码清单 8-27　作业存储库中记录的过滤掉的条目数量

```
mysql> select step_execution_id as id, step_name, status, commit_count, read_count, filter_
count, write_count from SPRING_BATCH.BATCH_STEP_EXECUTION;
+----+-------------+-----------+--------------+------------+--------------+-------------+
| id | step_name   | status    | commit_count | read_count | filter_count | write_count |
+----+-------------+-----------+--------------+------------+--------------+-------------+
|  1 |copyFileStep|COMPLETED  |            4 |         15 |            9 |           6 |
+----+-------------+-----------+--------------+------------+--------------+-------------+
1 row in set (0.01 sec)
```

你在第 4 章中了解过跳过条目的方法：使用异常来标识那些不进行处理的记录。这两种方法之间的不同之处在于，本节介绍的方法用于在技术上仍有效的记录。在这里，客户是有邮政编码数据的。但是，业务规则要求不处理这条记录，所以，你决定在步骤的执行结果中过滤掉这条记录。

条目处理器尽管是一个十分简单的概念，但是在 Spring Batch 框架中，它会占用任何批处理开发人员的大量时间。条目处理器是业务逻辑所在之处，并被应用于正在处理的条目。

8.4　本章小结

在条目处理器中，可以将业务逻辑应用于作业中正在处理的条目。Spring Batch 并不会试图帮助你，而是只做框架本身应该做的事情，从而放手让你决定如何根据需要应用业务逻辑。第 9 章将深入研究 ItemWriter，它是 Spring Batch 核心组件的最后一部分。

第 9 章

ItemWriter

计算机能做的事情太神奇了：它们所能处理的数字和图像都非常令人惊讶。但是，计算机如果不能通过输出来传达自己要做的事情，就没有任何作用。ItemWriter(条目写入器)是 Spring Batch 的输出设施。当需要使用某种格式输出 Spring Batch 的处理结果时，Spring Batch 就会用它来实现交付。本章将介绍 Spring Batch 提供的不同类型的 ItemWriter，以及如何开发特定需求场景下的 ItemWriter。本章将要讨论的话题如下。

- ItemWriter 概述：与步骤执行中另一端的 ItemReader 类似，不同的 ItemWriter 有细微的差别。9.1 节将从较高层次讨论 ItemWriter 的工作方式。
- 基于文件的 ItemWriter：基于文件的输出是最容易设置的，也是批处理中最常用的形式之一。因此，在探索 ItemWriter 的实现时，应从写入平面文件和 XML 文件入手。
- 基于数据库的 ItemWriter：当考虑数据存储时，关系数据库是企业的首选。但是，当需要处理大量的数据时，挑战随之而来。9.2 节将介绍 Spring Batch 如何使用自己独特的架构来处理这些挑战。
- NoSQL ItemWriter：尽管关系数据库可能是首选，但是很多开发人员在使用 NoSQL(如 MongoDB、Apache Geode 和 Noe4j 等)来满足存储需求。9.4 节将介绍支持多种 NoSQL 存储的 ItemWriter 的实现。
- 输出到其他目标的 ItemWriter：文件和数据库并不是企业软件输出的唯一媒介。例如，系统会发送 e-mail，写入 Java 消息服务(Java Messaging Service，JMS)，并且将数据保存到其他系统中。9.5 节将介绍 Spring Batch 支持的一些不常见但仍然非常有用的输出方法。
- 复合的 ItemWriter：读入的数据通常来自单个源，然后通过 ItemWriter 进行充实。写入数据则与之不同，将输出发送到多个目标是很常见的。9.6 节将介绍用于处理多个资源或多种输出格式的 ItemWriter。

为了介绍 ItemWriter，我们首先看看它们是如何工作的，以及它们在步骤中扮演的角色。

9.1 ItemWriter 概述

ItemWriter 是 Spring Batch 的输出设施。当 Spring Batch 刚刚出现时，ItemWriter 在本质上与 ItemReader 相同。ItemWriter 可写入处理过的每一个条目。然而，由于 Spring Batch 2 引入了基于块的处理，ItemWriter 的角色发生了变化，写入处理后的每一个条目已经没有意义。

有了基于块的处理之后，ItemWriter 不再写入单个条目，而是写入整块的数据。因此，org.springframework.batch.item.ItemWriter 接口与 ItemReader 接口略微不同。在代码清单 9-1 中，ItemWriter 的 write(List<T> items) 方法将接收一个条目列表，而第 7 章的 ItemReader 接口仅仅返回来自 read 方法的单个条目。

代码清单 9-1　ItemWriter

```
package org.springframework.batch.item;

import java.util.List;

public interface ItemWriter<T> {

    void write(List<? extends T> items) throws Exception;
}
```

为了说明 ItemWriter 如何适配到步骤中，图 9-1 所示的序列图逐一介绍了步骤中的各个过程。步骤通过 ItemReader 读入每个单独的条目，将其传给 ItemProcessor 进行处理。这样的交互一直持续，直到处理完块中的所有条目。当完成一个块的处理之后，条目被传给 ItemWriter，以进行写入。

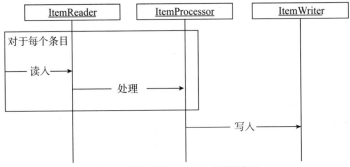

图 9-1　使用了 ItemWriter 的步骤交互

由于引入了基于块的处理，相比以前，对 ItemWriter 的调用已大大减少，不过需要以不同的方式进行处理。以处理非事务资源(如文件)为例。当写入文件失败时，无法回滚已经写入的内容。因此，为了在写入时减少错误，必须有额外的保障措施。

Spring Batch 提供了很多用于处理绝大多数输出场景的条目写入器。与介绍条目读取器时一样，我们首先从 FlatFileItemWriter 开始。

9.2 基于文件的 ItemWriter

在企业级的批处理中,大量的数据是以文件的形式移动的。这么做的原因如下:文件简单且可靠。另外,文件很容易备份。如果需要重新开始,那么恢复起来也很容易。本节将介绍如何生成多种格式的平面文件,包括格式化的记录(如固定宽度的记录)和以逗号分隔的文件,以及 Spring Batch 如何处理创建文件的问题。

9.2.1 FlatFileItemWriter

org.springframework.batch.item.file.FlatFileItemWriter 是 ItemWriter 的实现,用于生成文本文件的输出。FlatFileItemWriter 在很多方面与 FlatFileItemReader 相似,前者解决了 Java 中基于文件的输出问题,提供了干净且一致的接口。图 9-2 展示了 FlatFileItemWriter 的各个部分。

如图 9-2 所示,FlatFileItemWriter 包含了用于写入的 Resource 和 LineAggregator 接口的实现。org.springframework.batch.item.file.transform.LineAggregator 接口可用来代替第 7 章中 FlatFileItemReader 里的 LineMapper。LineMapper 负责将字符串转换成对象,而这里的 LineAggregator 负责生成对象的字符串输出。

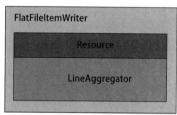

图 9-2 FlatFileItemWriter 的各个部分

FlatFileItemWriter 有许多有意思的配置选项,如表 9-1 所示。

表 9-1 FlatFileItemWriter 的配置选项

选项	类型	默认值	描述
encoding	String	UTF-8	用于文件的字符编码
footerCallback	FlatFileFooterCallback	null	在写入文件的最后一项之后执行
headerCallback	FlatFileHeaderCallback	null	在写入文件的第一项之前执行
lineAggregator	LineAggregator	null(必选)	将单独的条目转换成用于输出的字符串
lineSeparator	String	系统的 line.separator	用于生成文件的换行符
resource	Resource	null(必选)	写入的文件或流
saveState	boolean	true	决定条目写入器的状态是否应该在处理发生时保存在 ExecutionContext 中
shouldDeleteIfEmpty	boolean	false	如果为 true 并且没有写入数据(不包含头/尾记录),那么在条目写入器关闭时会删除文件
appendAllowed	boolean	false	如果为 true 并且要写入的文件已经存在,那么输出会追加到这个文件,而不是替换这个文件。如果为 true,那么 shouldDeleteIfExists 会自动被设置为 false
shouldDeleteIfExists	boolean	true	如果为 true 并且在运行作业之前将要写入的文件已经存在,那么会删除该文件并创建新的文件
transactional	boolean	true	如果为 true 并且事务在当前是活跃的,那么把数据写入文件的操作会延迟,直到事务提交

与 FlatFileItemReader 的 LineMapper 不同，这里的 LineAggregator 没有任何的强依赖项，但你需要注意一个与之相关的接口 org.springframework.batch.item.file.transform.FieldExtractor，这个接口作为访问给定条目中必要字段的一种方式，被用于大部分的 LineAggregator 实现。Spring Batch 提供了 FieldExtractor 接口的两个实现：org.springframework.batch.item.file.transform.BeanWrapperFieldExtractor 使用类的 getter 方法来访问传统 Java Bean 的属性，org.springframework.batch.item.file.transform.PassThroughFieldExtractor 返回条目(对于单个纯字符串的条目来说很有用)。

在本小节中，你会了解到 LineAggregator 的一些实现。但在讨论如何格式化文件之前，我们先花点时间讨论一下事务。Spring Batch 的事务模型已被嵌入基于块的处理中，尽管通常不会对步骤的读取端产生影响(事务性的队列除外)，但对写入端有很大的影响。那么事务是如何与平面文件一起工作的呢？

■ **注意** FlatFileItemWriter 会将输出数据的持久化延迟到提交之前的最后一刻，以减少写入数据后又需要回滚的次数。

FlatFileItemWriter 被设计为在事务循环内，尽可能推迟真正的写入操作。它使用 TransactionSynchronizationAdapter 的 beforeCommit(boolean readOnly)方法来完成真正的写入，这意味着其他的处理已经完成，剩下要做的唯一一事情是让 PlatformTransactionManager 在数据实际被推到磁盘之前提交事务。这样就可以让任何可能导致问题的交互在数据被写入磁盘之前失败，因为在将数据刷新到磁盘后，并没有实用的方法可用来回滚数据。所有的非事务数据存储都使用与之类似的机制在事务中持久化数据。

我们接下来看看 FlatFileItemWriter 的实际工作方式。为此，我们首先介绍如何生成格式化的文件。

格式化的文本文件

当从输入端查看文本文件时，有三种不同的类型：固定宽度的文件、带分隔符的文件以及 XML 文件。从输出端看，也有带分隔符的文件和 XML 文件。但是固定宽度的文件并不仅仅意味着宽度固定，在这种情况下，文件中实际上是格式化的记录。下面将介绍如何把批处理的输出构造为格式化的文本文件。

固定宽度的文件和格式化的输出文件为什么不同呢？从技术上讲，二者并没有区别。它们都包含固定格式的某种类型记录，所以这种格式无论是定义了列的宽度还是定义了其他规则，都不重要。然而，通常来说，输入文件仅仅包含了数据，并且数据是通过列定义的，而输出文件可以是定宽的，也可以很自由(就像本章后面的格式化作业中的那样)。

下面这个示例将生成客户的列表和他们居住的地方。首先，从 Spring Initializr 中创建一个新的项目，并使用常见的依赖项：Batch、JDBC 和 MySQL。有了项目之后，我们看看输入文件。代码清单 9-2 展示了 customer.csv 文件的内容。

代码清单 9-2 customer.csv

```
Richard,N,Darrow,5570 Isabella Ave,St. Louis,IL,58540
Warren,L,Darrow,4686 Mt. Lee Drive,St. Louis,NY,94935
Barack,G,Donnelly,7844 S. Greenwood Ave,Houston,CA,38635
Ann,Z,Benes,2447 S. Greenwood Ave,Las Vegas,NY,55766
Erica,Z,Gates,3141 Farnam Street,Omaha,CA,57640
```

第 9 章　ItemWriter

```
Warren,M,Williams,6670 S. Greenwood Ave,Hollywood,FL,37288
Harry,T,Darrow,3273 Isabella Ave,Houston,FL,97261
Steve,O,Darrow,8407 Infinite Loop Drive,Las Vegas,WA,90520
```

代码清单 9-2 使用的文件与到目前为止本书一直使用的客户文件十分类似。然而，这个作业的输出略有不同。在这个示例中，对于每个客户，将输出一条完整的句子：Richard Darrow lives at 5570 Isabella Ave in St. Louis, IL.。代码清单 9-3 展示了输出文件的内容。

代码清单 9-3　格式化的客户输出

```
Richard N Darrow lives at 5570 Isabella Ave in St. Louis, IL.
Warren L Darrow lives at 4686 Mt. Lee Drive in St. Louis, NY.
Barack G Donnelly lives at 7844 S. Greenwood Ave in Houston, CA.
Ann Z Benes lives at 2447 S. Greenwood Ave in Las Vegas, NY.
Laura 9S Minella lives at 8177 4th Street in Dallas, FL.
Erica Z Gates lives at 3141 Farnam Street in Omaha, CA.
Warren M Williams lives at 6670 S. Greenwood Ave in Hollywood, FL.
Harry T Darrow lives at 3273 Isabella Ave in Houston, FL.
Steve O Darrow lives at 8407 Infinite Loop Drive in Las Vegas, WA.
Erica Z Minella lives at 513 S. Greenwood Ave in Miami, IL.
```

这如何实现呢？对于这个示例，可使用单一的步骤读取输入文件并写进输出文件。这个示例不会使用 ItemProcessor，只需要编写 Customer 类即可，参见代码清单 9-4。

代码清单 9-4　Customer.java

```
...
public class Customer {

    private static final long serialVersionUID = 1L;

    private long id;
    private String firstName;
    private String middleInitial;
    private String lastName;
    private String address;
    private String city;
    private String state;
    private String zip;

    // Accessors go here
    ...
}
```

在代码清单 9-4 中，Customer 对象的字段已映射到 customer.csv 文件中的字段。定义了条目之后，开始配置作业。输入端应该与第 7 章中的类似。代码清单 9-5 展示了作为资源的输入文件的配置(资源的值可通过作业参数传入)、FlatFileItemReader 的配置以及必不可少的对 Customer 对象的引用[1]。

代码清单 9-5　配置格式化作业的输入

```
...
@Bean
```

---

[1] 稍后使用 Customer 对象时，将有一个 id 属性，该属性目前不在数据文件中。

```
@StepScope
public FlatFileItemReader<Customer> customerFileReader(
        @Value("#{jobParameters['customerFile']}")Resource inputFile) {

    return new FlatFileItemReaderBuilder<Customer>()
            .name("customerFileReader")
            .resource(inputFile)
            .delimited()
            .names(new String[] {"firstName",
                    "middleInitial",
                    "lastName",
                    "address",
                    "city",
                    "state",
                    "zip"})
            .targetType(Customer.class)
            .build();
}
...
```

代码清单9-5中的配置应该不会让人觉得意外。我们首先定义了一个步骤作用域(StepScope)Bean，以便能够通过作业参数注入输入文件的位置。然后，我们使用 FlatFileItemReaderBuilder 构建了 FlatFileItemReader。其中定义了名称(name)，注入了资源(Resource)，指明了文件是以逗号分隔的文件，并且提供了每一列的名称和返回的对象类型。Spring Batch 将使用这些信息创建一个 DefaultLineMapper，这个 DefaultLineMapper 使用了 DelimitedLineTokenizer 和 BeanWrapperFieldSetMapper，它们二者分别用于转换文件行和填充领域对象。

对于输出端，我们将使用 FlatFileItemWriter 和 LineAggregator。本例使用 Spring Batch 提供的 org.springframework.batch.item.file.transform.FormatterLineAggregator。手动配置 FieldExtractor 和 LineAggregator 需要很大的工作量，所以 Spring Batch 框架将这些组件的配置合并在了 FlatFileItemWriterBuilder 中。代码清单9-6展示了作业的输出配置。

代码清单9-6　格式化作业的输出配置

```
...
@Bean
@StepScope
public FlatFileItemWriter<Customer> customerItemWriter(
        @Value("#{jobParameters['outputFile']}") Resource outputFile) {

    return new FlatFileItemWriterBuilder<Customer>()
                .name("customerItemWriter")
                .resource(outputFile)
                .formatted()
                .format("%s %s lives at %s %s in %s, %s.")
                .names(new String[] {"firstName",
                        "lastName",
                        "address",
                        "city",
                        "state",
                        "zip"})
                .build();
}
...
```

第 9 章 ItemWriter

代码清单 9-6 首先配置了步骤作用域 Bean，以便通过作业参数注入输出文件的位置。然后使用 FlatFileItemWriterBuilder 构建 ItemWriter 本身，我们指定了名称(name)和将要写入的资源(Resource)。然后，让 Spring Batch 生成一个格式化的文件，返回的 FormattedBuilder 可用于配置输出格式和要抽取的字段名，字段名的顺序与输出格式中的顺序相同。最后调用 build 方法以真正地创建 ItemWriter 实例。

在配置了所有的输入输出之后，剩下要做的就是进行步骤和作业的配置。代码清单 9-7 展示了格式化作业的完整配置，其中包括前面的输入输出。

代码清单 9-7　FormattedTextFileJob.java

```java
...
@EnableBatchProcessing
@Configuration
public class FormattedTextFileJob {

    private JobBuilderFactory jobBuilderFactory;

    private StepBuilderFactory stepBuilderFactory;

    public FormattedTextFileJob(JobBuilderFactory jobBuilderFactory,
            StepBuilderFactory stepBuilderFactory) {

        this.jobBuilderFactory = jobBuilderFactory;
        this.stepBuilderFactory = stepBuilderFactory;
    }

    @Bean
    @StepScope
    public FlatFileItemReader<Customer> customerFileReader(
            @Value("#{jobParameters['customerFile']}")Resource inputFile) {
        return new FlatFileItemReaderBuilder<Customer>()
                .name("customerFileReader")
                .resource(inputFile)
                .delimited()
                .names(new String[] {"firstName",
                        "middleInitial",
                        "lastName",
                        "address",
                        "city",
                        "state",
                        "zip"})
                .targetType(Customer.class)
                .build();
    }

    @Bean
    @StepScope
    public FlatFileItemWriter<Customer> customerItemWriter(
            @Value("#{jobParameters['outputFile']}") Resource outputFile) {
        return new FlatFileItemWriterBuilder<Customer>()
                .name("customerItemWriter")
                .resource(outputFile)
                .formatted()
```

```
                            .format("%s %s lives at %s %s in %s, %s.")
                            .names(new String[] {"firstName",
                                    "lastName",
                                    "address",
                                    "city",
                                    "state",
                                    "zip"})
                            .build();
    }

    @Bean
    public Step formatStep() {
        return this.stepBuilderFactory.get("formatStep")
                .<Customer, Customer>chunk(10)
                .reader(customerFileReader(null))
                .writer(customerItemWriter(null))
                .build();
    }

    @Bean
    public Job formatJob() {
        return this.jobBuilderFactory.get("formatJob")
                .start(formatStep())
                .incrementer(new RunIdIncrementer())
                .build();
    }
}
```

使用 Maven 的 ./mvnw clean install 命令构建项目后，就可以使用代码清单 9-8 中的命令和 Spring Boot 执行格式化作业了。

代码清单 9-8　从命令行执行格式化作业

```
java -jar itemWriters-0.0.1-SNAPSHOT.jar
customerFile=/data/customer.csv outputFile=file:/output/formattedCustomers.txt
```

当使用代码清单 9-2 所示的输入运行作业时，结果将得到一个新的文件 formattedCustomers.txt，里面的内容如代码清单 9-9 所示。

代码清单 9-9　formattedCustomers.txt

```
Richard Darrow lives at 5570 Isabella Ave St. Louis in IL, 58540.
Warren Darrow lives at 4686 Mt. Lee Drive St. Louis in NY, 94935.
Barack Donnelly lives at 7844 S. Greenwood Ave Houston in CA, 38635.
Ann Benes lives at 2447 S. Greenwood Ave Las Vegas in NY, 55366.
Erica Gates lives at 3141 Farnam Street Omaha in CA, 57640.
Warren Williams lives at 6670 S. Greenwood Ave Hollywood in FL, 37288.
Harry Darrow lives at 3273 Isabella Ave Houston in FL, 97261.
Steve Darrow lives at 8407 Infinite Loop Drive Las Vegas in WA, 90520.
```

格式化输出的方法可用于很多不同的需求。无论是把条目格式化成人类可读的输出(就像上面这个示例)，还是格式化成固定宽度的文件，都只需要配置 LineAggregator 使用的格式化字符串即可。

另一种常见的平面文件类型是带分隔符的文件。例如，customer.csv 就是带逗号分隔符的文件。接下来介绍如何输出带分隔符的文件。

第 9 章 ItemWriter

带分隔符的文件

与前面介绍的格式化文件不同，带分隔符的文件并没有预定义的格式。相反，带分隔符的文件包含了以预定义的分隔符分隔的值列表。下面介绍如何使用 Spring Batch 生成带分隔符的文件。

为了查看如何生成带分隔符的文件，下面这个作业会使用相同的输入。对于输出，重构 ItemWriter 以生成新的带分隔符的文件。在此，修改字段的顺序，并把分隔符从逗号(,)修改为分号(;)。代码清单 9-10 展示了更新后的格式化作业的输出样例。

代码清单 9-10　格式化作业的带分隔符的输出

```
58540;IL;St. Louis;5570 Isabella Ave;Darrow;Richard
94935;NY;St. Louis;4686 Mt. Lee Drive;Darrow;Warren
38635;CA;Houston;7844 S. Greenwood Ave;Donnelly;Barack
55366;NY;Las Vegas;2447 S. Greenwood Ave;Benes;Ann
57640;CA;Omaha;3141 Farnam Street;Gates;Erica
37288;FL;Hollywood;6670 S. Greenwood Ave;Williams;Warren
97261;FL;Houston;3273 Isabella Ave;Darrow;Harry
90520;WA;Las Vegas;8407 Infinite Loop Drive;Darrow;Steve
```

为了生成代码清单 9-10 所示的输出，只需要更新 LineAggregator 的配置即可。这里并不使用 FormatterLineAggregator，而是使用 Spring Batch 的 org.springframework.batch.item.file.transform. DelimitedLineAggregator 实现。先使用相同的 BeanWrapperFieldExtractor 抽取出包含 Object 类型的数组，再使用 DelimitedLineAggregator 将数组中的元素用配置的分隔符连接起来。再一次，Spring Batch 的 FlatFileItemWriterBuilder 提供了另一个构建器，用于配置生成带分隔符的文件所需的 LineAggregator 资源，方法是调用 delimited()。代码清单 9-11 展示了用于这个条目写入器的最新配置。

代码清单 9-11　flatFileOutputWriter 的配置

```
...
@Bean
@StepScope
public FlatFileItemWriter<Customer> customerItemWriter(
        @Value("#{jobParameters['outputFile']}") Resource outputFile) {

    return new FlatFileItemWriterBuilder<Customer>()
                    .name("customerItemWriter")
                    .resource(outputFile)
                    .delimited()
                    .delimiter(";")
                    .names(new String[] {"zip",
                                "state",
                                "city",
                                "address",
                                "lastName",
                                "firstName"})
                    .build();
}
...
```

将 FormatterLineAggregator 的配置修改为 Spring Batch 的 DelimitedLineAggregator 之后，剩下要做的修改就是移除对格式化的依赖，并且定义分隔字符。在使用相同的命令 ./mvnw clean install 构建项目后，就可以使用代码清单 9-12 所示的命令运行作业。

代码清单 9-12　运行作业以生成带分隔符的输出

```
java -jar itemWriters-0.0.1-SNAPSHOT.jar
customerFile=/input/customer.csv outputFile=file:/output/delimitedCustomers.txt
```

更新后的格式化作业的运行结果如代码清单 9-13 所示。

代码清单 9-13　写入带分隔符的文件后的格式化作业的运行结果

```
58540;IL;St. Louis;5570 Isabella Ave;Darrow;Richard
94935;NY;St. Louis;4686 Mt. Lee Drive;Darrow;Warren
38635;CA;Houston;7844 S. Greenwood Ave;Donnelly;Barack
55366;NY;Las Vegas;2447 S. Greenwood Ave;Benes;Ann
57640;CA;Omaha;3141 Farnam Street;Gates;Erica
37288;FL;Hollywood;6670 S. Greenwood Ave;Williams;Warren
97261;FL;Houston;3273 Isabella Ave;Darrow;Harry
90520;WA;Las Vegas;8407 Infinite Loop Drive;Darrow;Steve
```

在 Spring Batch 中创建平面文件是很简单的。在领域对象之外，无须编写任何代码就可以读入文件并转换格式——无论是格式化的文件还是带分隔符的文件。下面将介绍 Spring Batch 提供的一些用于处理写入文件的高级选项。

文件管理选项

读取文件时文件必须存在，否则通常被视为错误。输出文件与之不同，在处理过程中，输出文件可能存在，也可能不存在，这可能是正常的，也可能是因为发生了错误。Spring Batch 为你提供了基于具体需求处理这些不同场景的能力。下面介绍如何配置 FlatFileItemWriter 以处理不同的文件创建场景。

在前面的表 9-1 中，FlatFileItemWriter 有两个可用于创建文件的选项：shouldDeleteIfEmpty 和 shouldDeleteIfExists。shouldDeleteIfEmpty 实际上处理的是步骤完成时做什么，默认为 false。如果作业在执行时没有写入任何条目(可能写入了头记录或尾记录，但没有写入条目)，并且 shouldDeleteIfEmpty 为 true，那么在步骤完成时这个文件会被删除。默认情况下，这个文件在创建后是空的。如代码清单 9-14 所示，更新 flatFileOutputWriter 的配置，将 shouldDeleteIfEmpty 设置为 true，让格式化作业处理一个空的文件，因而运行之后并没有留下输出文件。

代码清单 9-14　配置格式化作业，使其在没有写入条目时删除输出文件

```
...
@Bean
@StepScope
public FlatFileItemWriter<Customer> delimitedCustomerItemWriter(
        @Value("#{jobParameters['outputFile']}") Resource outputFile) {

    return new FlatFileItemWriterBuilder<Customer>()
            .name("customerItemWriter")
            .resource(outputFile)
            .delimited()
            .delimiter(";")
            .names(new String[] {"zip",
                    "state",
                    "city",
                    "address",
```

```
                    "lastName",
                    "firstName"})
                .shouldDeleteIfEmpty(true)
                .build();
}
...
```

如果使用空的 customer.csv 文件作为输入,那么在运行更新后的格式化作业之后,就不会生成输出。注意在这种情况下,我们依然创建、打开并关闭了这个文件。事实上,如果将步骤配置为写入头记录或尾记录,那么也会写入它们。如果写入文件的条目数量为 0,那么在步骤结束时还会删除这个文件。

下一个与文件创建/删除相关的配置参数是 shouldDeleteIfExists 标识。这个标识的默认值为 true,因此会删除与步骤要写入的文件同名的文件。例如,如果作业的写入文件是/output/jobRun,那么启动作业后,如果这个文件已存在,那么 Spring Batch 会删除这个文件并创建一个新的文件。如果文件/output/jobRun 存在,并且标识 shouldDeleteIfExists 被设置为 false,那么当步骤尝试创建新文件时,就会抛出 org.springframework.batch.item.ItemStreamException 异常。代码清单 9-15 展示了如何配置格式化作业以保留已经存在的输出文件。

代码清单 9-15　配置格式化作业以保留已经存在的输出文件

```
...
@Bean
@StepScope
public FlatFileItemWriter<Customer> delimitedCustomerItemWriter(

            @Value("#{jobParameters['outputFile']}") Resource outputFile) {

    return new FlatFileItemWriterBuilder<Customer>()
                .name("customerItemWriter")
                .resource(outputFile)
                .delimited()
                .delimiter(";")
                .names(new String[] {"zip",
                    "state",
                    "city",
                    "address",
                    "lastName",
                    "firstName"})
                .shouldDeleteIfExists(false)
                .build();
}
...
```

运行代码清单 9-15 中配置好的作业之后,就会看到前面提到的 ItemStreamException 异常,如代码清单 9-16 所示。

代码清单 9-16　写入不应该存在但却存在的文件时的输出结果

```
2018-04-16 15:38:55.269 INFO 76152 --- [main] o.s.batch.core.job.SimpleStepHandler:
Executing step: [delimitedStep]
2018-04-16 15:38:55.316 ERROR 76152 --- [main] o.s.batch.core.step.AbstractStep:
Encountered an error executing step delimitedStep in job delimitedJob
```

```
org.springframework.batch.item.ItemStreamException: File already exists: [/Users/mminella/
Documents/IntelliJWorkspace/def-guide-spring-batch/Chapter9/target/formattedCustomers.txt]
    at org.springframework.batch.item.util.FileUtils.setUpOutputFile(FileUtils.java:56)
~[spring-batch-infrastructure-4.0.1.RELEASE.jar:4.0.1.RELEASE]
    at org.springframework.batch.item.file.FlatFileItemWriter$OutputState.
initializeBufferedWriter(FlatFileItemWriter.java:572)~[spring-batch-infrastructure-
4.0.1.RELEASE.jar:4.0.1.RELEASE]
    at org.springframework.batch.item.file.FlatFileItemWriter$OutputState.
access$000(FlatFileItemWriter.java:414) ~[spring-batch-infrastructure-4.0.1.RELEASE.
jar:4.0.1.RELEASE]
    at org.springframework.batch.item.file.FlatFileItemWriter.doOpen(FlatFileItemWriter.
java:348)~[spring-batch-infrastructure-4.0.1.RELEASE.jar:4.0.1.RELEASE]
    at org.springframework.batch.item.file.FlatFileItemWriter.open(FlatFileItemWriter.
java:338)~[spring-batch-infrastructure-4.0.1.RELEASE.jar:4.0.1.RELEASE]
```

当期望保存每次运行的输出时,使用 shouldDeleteIfExists 参数是个好主意,这样可以防止意外覆盖旧的文件。

最后一个与文件创建相关的选项是 appendAllowed。当调用 .append(boolean value) 方法将这个选项(默认为 false)设置为 true 时,Spring Batch 会自动将 shouldDeleteIfExists 标识设置为 false。当输出文件不存在时,则创建新的文件,否则追加数据。当需要从多个步骤写入输出文件时,这个选项会很有用。代码清单 9-17 展示了当输出文件存在时追加数据的格式化作业的配置。

代码清单 9-17　当输出文件存在时追加数据

```
...
@Bean
@StepScope
public FlatFileItemWriter<Customer> delimitedCustomerItemWriter(
            @Value("#{jobParameters['outputFile']}") Resource outputFile) {

    return new FlatFileItemWriterBuilder<Customer>()
            .name("customerItemWriter")
            .resource(outputFile)
            .delimited()
            .delimiter(";")
            .names(new String[] {"zip",
                    "state",
                    "city",
                    "address",
                    "lastName",
                    "firstName"})
            .append(true)
            .build();
}
...
```

有了以上配置,就可以使用相同的输出文件(和不同的输入文件)多次运行作业,Spring Batch 会将当前作业的输出追加到已经存在的输出文件的末尾。

如你所见,在处理基于平面文件的输出时有很多可用选项,从将记录转换为任何格式,生成包含分隔符的文件,甚至到使用不同的选项让 Spring Batch 处理已经存在的输出文件。然而,平面文件并不是唯一的输出文件类型。XML 是 Spring Batch 支持的另一种文件类型,9.2.2 节将介绍这种文件类型。

9.2.2 StaxEventItemWriter

第 7 章在读取 XML 文件时，探索了 Spring Batch 如何以片段的方式查看 XML 文档。每一个片段都是用于处理的每个条目的 XML 呈现形式。在 ItemWriter 端，也存在相同的概念。Spring Batch 把 ItemWriter 收到的每一个条目生成为一个 XML 片段，然后把这个 XML 片段写入文件。本小节将介绍如何让 Spring Batch 将 XML 作为输出媒介。

为了使用 Spring Batch 写入 XML，需要使用 org.springframework.batch.item.xml.StaxEventItemWriter。与 ItemReader 类似，XML 流式 API(Streaming API for XML，StAX)的实现允许 Spring Batch 在处理完每个块之后，写入 XML 片段。与 FlatFileItemWriter 类似，StaxEventItemWriter 每次生成一个 XML 块，并在本地事务提交之前将其写入文件，从而防止在写入文件时出错的回滚问题。

StaxEventItemReader 的配置包括 Resource(要读取的资源)、根元素的名称(每个 XML 片段的根标签)以及用来将 XML 输入转换成对象的 Unmarshaller。StaxEventItemWriter 的配置几乎与之相同，也包括用于写入的 Resource(要写入的资源)、根元素的名称(生成的每个 XML 片段的根标签)以及用来把每个条目转换成 XML 片段的 Marshaller。

表 9-2 列出了 StaxEventItemWriter 的可用选项。

表 9-2 StaxEventItemWriter 的可用选项

选项	类型	默认值	描述
encoding	String	UTF-8	用于文件的字符编码
footerCallback	StaxWriterCallback	null	在写入文件的最后一项时执行
headerCallback	StaxWriterCallback	null	在写入文件的第一项之前执行
marshaller	Marshaller	null (必选)	将单独的条目转换成用于输出的 XML 片段
overwriteOutput	boolean	true	默认情况下，如果输出文件已经存在，那么将输出文件替换成新的。如果设置这个选项为 false 并且文件存在，那么抛出 ItemStreamException 异常
resource	Resource	null(必选)	写入的文件或流
rootElementAttributes	Map<String, String>	null	这个键值对会被追加到每个 XML 片段的根标签，其中键是属性的名称，值是属性的值
rootTagName	String	null(必选)	用于定义 XML 文档的根标签
saveState	boolean	true	决定是否让 Spring Batch 跟踪 ItemWriter 的状态(比如写入的数量，等等)
transactional	boolean	true	如果为 true，那么输出的写入操作会被延迟，直到事务提交，以防止回滚问题
version	String	"1.0"	要写入的文件的 XML 版本

为了查看 StaxEventItemWriter 的修改方式，下面修改格式化作业以输出 XML 格式的客户信息。使用与前面示例中相同的输入，代码清单 9-18 展示了更新后的格式化作业的输出。

代码清单 9-18　customer.xml

```
<?xml version="1.0" encoding="UTF-8"?>
<customers>
  <customer>
    <id>0</id>
    <firstName>Richard</firstName>
    <middleInitial>N</middleInitial>
```

```
        <lastName>Darrow</lastName>
        <address>5570 Isabella Ave</address>
        <city>St. Louis</city>
        <state>IL</state>
        <zip>58540</zip>
    </customer>
    ...
</customers>
```

为了生成代码清单 9-18 所示的输出，可以复用格式化作业的配置，但是需要把 flatFileOutputWriter 替换为新的 xmlOutputWriter，从而使用 ItemWriter 的实现 StaxEventItemWriter。为了使用新的 ItemWriter，需要提供的配置如代码清单 9-19 所示：要写入的资源(Resource)、对 org.springframework.oxm.Marshaller 实现的引用以及根标签的名称(这个示例中是 customer)。

代码清单 9-19　使用了 StaxEventItemWriter 的格式化作业的配置

```
...
@Configuration
public class XmlFileJob {

    private JobBuilderFactory jobBuilderFactory;

    private StepBuilderFactory stepBuilderFactory;
    public XmlFileJob(JobBuilderFactory jobBuilderFactory,
                StepBuilderFactory stepBuilderFactory) {

        this.jobBuilderFactory = jobBuilderFactory;
        this.stepBuilderFactory = stepBuilderFactory;
    }

    @Bean
    @StepScope
    public FlatFileItemReader<Customer> customerFileReader(
                @Value("#{jobParameters['customerFile']}")Resource inputFile) {

            return new FlatFileItemReaderBuilder<Customer>()
                        .name("customerFileReader")
                        .resource(inputFile)
                        .delimited()
                        .names(new String[] {"firstName",
                                "middleInitial",
                                "lastName",
                                "address",
                                "city",
                                "state",
                                "zip"})
                        .targetType(Customer.class)
                        .build();
    }

    @Bean
    @StepScope
    public StaxEventItemWriter<Customer> xmlCustomerWriter(
        @Value("#{jobParameters['outputFile']}") Resource outputFile) {
```

```
    Map<String, Class> aliases = new HashMap<>();
    aliases.put("customer", Customer.class);

XStreamMarshaller marshaller = new XStreamMarshaller();

marshaller.setAliases(aliases);

marshaller.afterPropertiesSet();

return new StaxEventItemWriterBuilder<Customer>()
            .name("customerItemWriter")
            .resource(outputFile)
            .marshaller(marshaller)
            .rootTagName("customers")
            .build();
}

@Bean
public Step xmlFormatStep() throws Exception {
    return this.stepBuilderFactory.get("xmlFormatStep")
            .<Customer, Customer>chunk(10)
            .reader(customerFileReader(null))
            .writer(xmlCustomerWriter(null))
            .build();
}

@Bean
public Job xmlFormatJob() throws Exception {
    return this.jobBuilderFactory.get("xmlFormatJob")
            .start(xmlFormatStep())
            .build();
}
}
```

前面的代码清单 9-7 配置了原始的格式化作业，代码清单 9-19 则仅仅修改了格式化作业的 delimitedCustomerItemWriter(Resource file)方法。xmlOutputWriter 是对 StaxEventItemWriter 的引用，并且定义了三个依赖项：要写入的资源、Marshaller 的实现以及 Marshaller 为生成每个 XML 片段而使用的根标签的名称。

我们使用内联方式配置了 Marshaller，这个对象用于为作业处理过的每一个条目生成一个 XML 片段。在使用了 Spring 的 org.springframework.oxm.xstream.XStreamMarshaller 类之后，你需要提供的唯一配置是别名映射(也就是代码中的 aliases)，以便 Marshaller 处理遇到的每一种类型。默认情况下，Marshaller 使用属性的名称作为标签名，但这里已经为 Customer 类提供了别名。这是因为 XStreamMarshaller 默认使用类的完全限定名称作为每个 XML 片段的根标签(也就是使用 com.apress. springbatch.chatper8.Customer 而不是 customer)。

为了编译和运行作业，我们还需要再做一些修改。POM 文件需要添加用于处理 XML 的依赖项，也就是 Spring 的"对象/XML 映射"(Object/XML Mapping, OXM)库以及用来处理 XML 的 XStream 库。代码清单 9-20 展示了对 POM 所做的必要更新。

代码清单 9-20 Spring 的 OXM 库的 Maven 依赖项

```
...
<dependency>
        <groupId>org.springframework</groupId>
        <artifactId>spring-oxm</artifactId>
</dependency>
<dependency>
        <groupId>com.thoughtworks.xstream</groupId>
        <artifactId>xstream</artifactId>
        <version>1.4.10</version>
</dependency>
...
```

有了更新后的 POM 和配置好的作业之后，就可以构建和运行格式化作业，生成 XML 输出。从命令行运行 ./mvnw clean install 后，就可以使用代码清单 9-21 所示的命令执行格式化作业。

代码清单 9-21 执行格式化作业以生成 XML

```
java -jar itemWriters-0.0.1-SNAPSHOT.jar
customerFile=/input/customer.csv outputFile=file:/output/xmlCustomer.xml
```

当查看 XML 结果时，你会注意到结果显然是由库生成的，并且没有调整格式。然而，通过使用 IDE 或其他的文本编辑器，就可以清晰地看到输出符合预期。代码清单 9-22 展示了生成的 XML 输出结果。

代码清单 9-22 格式化作业的 XML 输出结果

```xml
<?xml version="1.0" encoding="UTF-8"?>
<customers>
        <customer>
                <id>0</id>
                <firstName>Richard</firstName>
                <middleInitial>N</middleInitial>
                <lastName>Darrow</lastName>
                <address>5570 Isabella Ave</address>
                <city>St. Louis</city>
                <state>IL</state>
                <zip>58540</zip>
        </customer>
   ...
</customers>
```

只需要几行 Java 代码，就可以使用 Spring 支持的 XML Marshaller 的强大功能轻松地生成 XML 输出。

在企业环境中，支持 XML 格式的输入输出是非常重要的，就像处理平面文件一样。然而，尽管文件在批处理中扮演重要角色，但是它们在企业的其他处理中并不常见。相反，关系数据库已经取代它们的位置。因此，批处理过程不仅需要能够从数据库中读取数据，还需要能够将数据写入数据库。9.3 节将介绍一些更常见的使用 Spring Batch 写入数据库的方式。

9.3 基于数据库的 ItemWriter

与基于文件的输出相比，写入数据库也受到一些不同的约束。首先，与文件不同，数据库是事务性资源。因此，可以在事务中包含物理写入，而不必像基于文件的处理那样将它们分开。另外，对数据库的访问有很多不同选项。JDBC、Java 持久化 API(Java Persistence API，JPA)和 Hibernate 都提供了独特但引人注目的模型来处理数据库的写入。本节介绍如何使用 JDBC、Hibernate 和 JPA 将批处理的输出写入数据库。

9.3.1 JdbcBatchItemWriter

写入数据库的第一种方式，就是大部分人已经学过的使用 Spring 访问数据库——使用 JDBC。Spring Batch 的 JdbcBatchItemWriter 使用了 JdbcTemplate 及其批量执行 SQL 的功能，可以一次性地执行一个块中的所有 SQL。本小节介绍如何使用 JdbcBatchItemWriter 将步骤的输出写入数据库。

org.springframework.batch.item.database.JdbcBatchItemWriter 只不过是对 Spring 的 org.springframework.jdbc.support.JdbcTemplate 的简单封装，根据使用的命名参数，前者使用 JdbcTemplate.batchUpdate()或 JdbcTemplate.execute()方法执行大量的数据库插入和更新操作。需要注意的是，对于单个块，Spring 可以使用 PreparedStatement 的批量更新功能一次性地执行所有的 SQL 语句，而不必进行多次调用。这极大改善了性能，同时仍然允许在当前事务中完成所有的执行。

为了查看 JdbcBatchItemWriter 如何工作，下面再次使用与基于文件的写入器相同的输入，但这里填充 CUSTOMER 数据表而不是写入文件。图 9-3 展示了 CUSTOMER 表的设计，客户信息将被填充到这个表中。

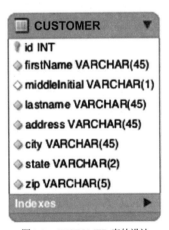

图 9-3　CUSTOMER 表的设计

如图 9-3 所示，CUSTOMER 表中的字段几乎与 customer.csv 中的元素一一对应。唯一的区别是 id 字段，可通过数据库来填充。为了在 CUSTOMER 表中插入值，可使用以下两种方式之一构建 SQL：使用问号(?)作为值的占位符，或者使用命名参数(例如，name)作为值的占位符。以这两种方式填充值时需要使用略微不同的方法。我们先看看使用问号的方式，如代码清单 9-23 所示。

代码清单 9-23　用于将值插入 CUSTOMER 表的 SQL 预备语句

```
insert into CUSTOMER (firstName, middleInitial, lastName, address, city,
state, zip) values (?, ?, ?, ?, ?, ?, ?)
```

如你所见，SQL 预备语句并没有什么特殊之处。然而，使用 SQL 语句只是配置 JdbcBatchItemWriter 的众多选项之一。表 9-3 列出了所有的配置选项。

表 9-3　JdbcBatchItemWriter 的配置选项

选项	类型	默认值	描述
assertUpdates	boolean	true	如果为 true，那么 JdbcBatchItemWriter 会验证每一条导致插入或更新的条目。如果所有的条目都没有触发任何记录的插入和更新，就会抛出 EmptyResultDataAccessException 异常
dataSource	DataSource	null(必选)	用于访问所需的数据库
itemPreparedStatementSetter	ItemPreparedStatementSetter	null	如果提供了标准的 PreparedStatement(使用?作为参数)，那么 JdbcBatchItemWriter 将使用这个类来填充参数值
itemSqlParameterSourceProvider	ItemSqlParameterSourceProvider	null	如果在提供的 SQL 中使用了命名参数，那么 JdbcBatchItemWriter 将使用这个类来填充参数值
simpleJdbcTemplate	SimpleJdbcTemplate	null	允许注入 SimpleJdbcOperations 接口的实现
sql	String	null(必选)	对于每个条目而言将要执行的 SQL

为了在格式化作业中使用 JdbcBatchItemWriter，可以将 xmlOutputWriter 替换为新的 jdbcBatchWriterBean。由于已经在查询语句中使用了标准的 PreparedStatement 语法(即使用问号)，因此需要提供数据源(DataSource)、将要执行的 SQL 以及 org.springframework.batch.item.database.ItemPreparedStatementSetter 接口的实现。没错，你还需要编写一些代码。

ItemPreparedStatementSetter 接口抽象了从每个条目中抽取值并将它们设置到预备语句的操作，如代码清单 9-24 所示。

代码清单 9-24　ItemPreparedStatementSetter 接口

```
package org.springframework.batch.item.database;

import java.sql.PreparedStatement;
import java.sql.SQLException;

public interface ItemPreparedStatementSetter<T> {
    void setValues(T item, PreparedStatement ps) throws SQLException;
}
```

为了实现 ItemPreparedStatementSetter 接口，你需要创建自己的 CustomerItemPreparedStatementSetter 类。这个类实现了 ItemPreparedStatementSetter 接口要求的唯一的 setValues(T item, PreparedStatement ps) 方法，从而使用标准的预备语句 API，把合适的条目填充到预备语句中。代码清单 9-25 展示了 CustomerItemPreparedStatementSetter 类的代码。

第 9 章 ItemWriter

代码清单 9-25　CustomerItemPreparedStatementSetter.java

```java
...
public class CustomerItemPreparedStatementSetter implements
        ItemPreparedStatementSetter<Customer> {

    public void setValues(Customer customer, PreparedStatement ps)
            throws SQLException {

        ps.setString(1, customer.getFirstName());
        ps.setString(2, customer.getMiddleInitial());
        ps.setString(3, customer.getLastName());
        ps.setString(4, customer.getAddress());
        ps.setString(5, customer.getCity());
        ps.setString(6, customer.getState());
        ps.setString(7, customer.getZip());
    }
}
```

如代码清单 9-25 所示,在设置每一条 SQL 预备语句的值时,并没有使用什么魔法。有了这些代码后,就可以更新格式化作业的配置,让它把数据写入数据库。代码清单 9-26 展示了新的 JdbcBatchItemWriter 的配置。

代码清单 9-26　JdbcBatchItemWriter 的配置

```java
...
@Bean
@StepScope
public JdbcBatchItemWriter<Customer> jdbcCustomerWriter(DataSource dataSource) throws Exception {
        return new JdbcBatchItemWriterBuilder<Customer>()
                    .dataSource(dataSource)
                    .sql("INSERT INTO CUSTOMER (first_name, " +
                            "middle_initial, " +
                            "last_name, " +
                            "address, " +
                            "city, " +
                            "state, " +
                            "zip) VALUES (?, ?, ?, ?, ?, ?, ?)")
                    .itemPreparedStatementSetter(new
                        CustomerItemPreparedStatementSetter())
                    .build();
}
...
```

在代码清单 9-26 中可以看到,新的 JdbcBatchItemWriter 引用了来自 Spring Boot 的数据源 dataSource(CUSTOMER 表与 Spring Batch 提供的用于作业存储库的表在同一模式中)。SQL 预备语句的值与代码清单 9-23 中定义的 SQL 预备语句相同。最后一个依赖项是对 CustomerItemPreparedStatementSetter 的引用。

新的 JdbcBatchItemWriter 要做的最后一部分配置工作是更新步骤,以引用这个新的条目写入器。为此,只需要更新格式化作业的配置,将之前引用 XmlOutputWriter 的地方修改为引用 JdbcBatchItemWriter。代码清单 9-27 展示了被配置为写入数据库的 JdbcFormatJob.java 的完整代码。

代码清单 9-27　被配置为使用 JDBC 写入数据库的格式化作业

```
...
@Configuration
public class JdbcFormatJob {

    private JobBuilderFactory jobBuilderFactory;

    private StepBuilderFactory stepBuilderFactory;

    public JdbcFormatJob(JobBuilderFactory jobBuilderFactory,
                StepBuilderFactory stepBuilderFactory) {

        this.jobBuilderFactory = jobBuilderFactory;
        this.stepBuilderFactory = stepBuilderFactory;
    }

    @Bean
    @StepScope
    public FlatFileItemReader<Customer> customerFileReader(
                @Value("#{jobParameters['customerFile']}")Resource inputFile) {

        return new FlatFileItemReaderBuilder<Customer>()
                    .name("customerFileReader")
                    .resource(inputFile)
                    .delimited()
                    .names(new String[] {"firstName",
                                        "middleInitial",
                                        "lastName",
                                        "address",
                                        "city",
                                        "state",
                                        "zip"})
                    .targetType(Customer.class)
                    .build();
    }

    @Bean
    @StepScope
    public JdbcBatchItemWriter<Customer> jdbcCustomerWriter(DataSource dataSource)
            throws Exception {
        return new JdbcBatchItemWriterBuilder<Customer>()
                    .dataSource(dataSource)
                    .sql("INSERT INTO CUSTOMER(first_name, " +
                                        "middle_initial, " +
                                        "last_name, " +
                                        "address, " +
                                        "city, " +
                                        "state, " +
                                        "zip) VALUES (?, ?, ?, ?, ?, ?, ?)")
                    .itemPreparedStatementSetter(
                            new CustomerItemPreparedStatementSetter())
                    .build();
    }

    @Bean
```

第 9 章 ItemWriter

```
    public Step xmlFormatStep() throws Exception {
        return this.stepBuilderFactory.get("xmlFormatStep")
                        .<Customer, Customer>chunk(10)
                        .reader(customerFileReader(null))
                        .writer(jdbcCustomerWriter(null))
                        .build();
    }

    @Bean
    public Job xmlFormatJob() throws Exception {
        return this.jobBuilderFactory.get("xmlFormatJob")
                        .start(xmlFormatStep())
                        .build();
    }
}
```

由于已经在 POM 中配置了 JDBC 驱动，并且为作业存储库配置了数据源，因此只需要执行 ./mvnw clean install 命令进行构建，然后执行代码清单 9-28 所示的命令，就能看到更新后的格式化作业的执行结果。

代码清单 9-28　用于执行格式化作业的命令

```
java -jar itemWriters-0.0.1-SNAPSHOT.jar customerFile=/input/customer.csv
```

输出不在文件中，而在数据库中。有两种方式可以用来进行确认。第一种是进入数据库并验证输出。代码清单 9-29 展示了数据库中的结果。

代码清单 9-29　使用了 JdbcBatchItemWriter 的格式化作业的执行结果

```
mysql> select id, first_name, middle_initial as middle, last_name, address, city, state as
st, zip from SPRING_BATCH.CUSTOMER;
+----+------------+--------+----------+------------------------+-----------+----+-------+
| id | first_name | middle | last_name| address                | city      | st | zip   |
+----+------------+--------+----------+------------------------+-----------+----+-------+
|  1 | Richard    | N      | Darrow   | 5570 Isabella Ave      | St. Louis | IL | 58540 |
|  2 | Warren     | L      | Darrow   | 4686 Mt. Lee Drive     | St. Louis | NY | 94935 |
|  3 | Barack     | G      | Donnelly | 7844 S. Greenwood Ave  | Houston   | CA | 38635 |
|  4 | Ann        | Z      | Benes    | 2447 S. Greenwood Ave  | Las Vegas | NY | 55366 |
|  5 | Erica      | Z      | Gates    | 3141 Farnam Street     | Omaha     | CA | 57640 |
|  6 | Warren     | M      | Williams | 6670 S. Greenwood Ave  | Hollywood | FL | 37288 |
|  7 | Harry      | T      | Darrow   | 3273 Isabella Ave      | Houston   | FL | 97261 |
|  8 | Steve      | O      | Darrow   | 8407 Infinite Loop Drive| Las Vegas| WA | 90520 |
+----+------------+--------+----------+------------------------+-----------+----+-------+
8 rows in set (0.00 sec)
```

SQL 预备语句的符号表示法很有用，因为大部分 Java 开发人员都很熟悉它。不过，使用 Spring 的 JdbcTemplate 提供的命名参数是一种更加安全的方式，并且也是在大部分 Spring 环境中填充参数的首选方式。记住这一点后，就可以对配置做两个小小的修改以使用这个功能：

- 更新配置，移除 ItemPreparedStatementSetter 的实现，然后替换为 ItemSqlParameterSourceProvider 接口的实现。
- 更新 SQL 语句，为参数使用命名参数而不是问号。

与 ItemPreparedStatementSetter 略有不同，org.springframework.batch.item.database.ItemSqlParameter-

243

SourceProvider 接口并不会为语句设置用于执行的参数。相反，ItemSqlParameterSourceProvider 接口实现的职责是从条目中抽取参数值，然后将它们作为 org.springframework.jdbc.core.namedparam.SqlParameterSource 对象返回。

这种方式的好处在于不仅更安全(无须担心配置中的 SQL 与 ItemPreparedStatementSetter 的代码同步问题)，而且 Spring Batch 提供了 ItemSqlParameterSourceProvider 接口的实现，允许使用"惯例优于代码"的方式从条目中抽取值。这个示例使用 Spring Batch 的 BeanPropertyItemSqlParameterSourceProvider 从条目中抽取值，以用于填充 SQL。JdbcBatchItemWriterBuilder 的 beanMapped()方法能使以上操作变得容易。代码清单 9-30 展示了在应用了这些修改之后的 JdbcBatchItemWriter。

代码清单 9-30　使用了 BeanPropertyItemSqlParameterSourceProvider 的 JdbcBatchItemWriter

```
...
@Bean
public JdbcBatchItemWriter<Customer> jdbcCustomerWriter(DataSource dataSource)
        throws Exception {
        return new JdbcBatchItemWriterBuilder<Customer>()
                    .dataSource(dataSource)
                    .sql("INSERT INTO CUSTOMER (first_name, " +
                                        "middle_initial, " +
                                        "last_name, " +
                                        "address, " +
                                        "city, " +
                                        "state, " +
                                        "zip) VALUES (:firstName, " +
                                        ":middleInitial, " +
                                        ":lastName, " +
                                        ":address, " +
                                        ":city, " +
                                        ":state, " +
                                        ":zip)")
                    .beanMapped()
                    .build();
}
...
```

你很快就会注意到，代码清单 9-30 并没有引用 ItemPreparedStatementSetter 的实现。通过使用以上配置，我们不需要编写任何自定义代码。结果依然是相同的。

尽管与基于 JDBC 的其他持久性框架相比，原生的 JDBC 以速度著称，但是其他框架在企业中也十分流行。下面看看如何使用其中最流行的一种框架(Hibernate)写入数据库。

9.3.2　HibernateItemWriter

当大部分的数据表和应用已经通过 Hibernate 进行映射时，复用它们将是合理的选择。第 7 章已经介绍了 Hibernate 如何作为条目读取器的组件工作。本小节将介绍如何使用 HibernateItemWriter 将变更写入数据库。

与 JdbcBatchItemWriter 类似，org.springframework.batch.item.database.HibernateItemWriter 是对 Hibernate 的 Session API 的简单封装。在完成一个块的处理之后，条目列表会被传给 HibernateItemWriter，HibernateItemWriter 则对每一个没有与 Session 关联的条目调用 Hibernate 的 Session.saveOrUpdate(Object item)方法。当保存或更新所有条目之后，HibernateItemWriter 调用 Session#flush()方法，一次性地执行所有变更。这种方式提供的批量操作功能与 JdbcBatchItemWriter 类似，并且无须直接处理 SQL。

　　HibernateItemWriter 的配置很简单。除了配置真正的 ItemWriter 之外，其他的你都应该已经很熟悉，因为它们与同样支持 Hibernate 的 ItemReader 的配置和编码是一样的。为了让格式化作业使用 Hibernate，需要做出如下修改。

- POM：更新 POM 以合并 Hibernate 依赖项。
- application.yml：配置 CurrentSessionContext 类，设置属性 spring.jpa.properties.hibernate.current_session_context_class 为 org.springframework.orm.hibernate5.SpringSessionContext。
- Customer.java：由于需要使用注解来配置用于 Customer 对象的映射，因此也为 Customer 类添加注解。
- SessionFactory：配置 SessionFactory 和新的 TransactionManager 以支持 Hibernate。
- HibernateItemWriter：使用 HibernateItemWriter 配置新的 ItemWriter。

　　下首先更新 POM。为了让 Hibernate 与 Spring Batch 一起工作，我们将使用 Spring JPA 启动包，因为它提供了我们所需的一切。代码清单 9-31 展示了支持 Hibernate 的 POM 添加项。

代码清单 9-31　支持 Hibernate 的 POM 添加项

```
...
<dependency>
    <groupId>org.springframework.boot</groupId>
    <artifactId>spring-boot-starter-data-jpa</artifactId>
</dependency>
...
```

现在可以开始更新格式化作业了。我们从唯一需要编写的代码开始：为 Customer 类添加注解，从而映射到 CUSTOMER 表。代码清单 9-32 展示了更新后的 Customer 类。

代码清单 9-32　映射到 CUSTOMER 表的 Customer.java

```
...
@Entity
@Table(name = "CUSTOMER")
public class Customer implements Serializable {
    private static final long serialVersionUID = 1L;

    @Id
    @GeneratedValue(strategy = GenerationType.IDENTITY)
    private long id;
    private String firstName;
    private String middleInitial;
    private String lastName;
    private String address;
    private String city;
    private String state;
    private String zip;

    // Accessors go here
    ....
}
```

Spring Batch 权威指南

这里使用的注解与第 7 章为 ItemReader 示例使用的注解相同。我们为 Customer 类添加的注解非常直观，因为 CUSTOMER 表中的列名与 Customer 类的字段是一一对应的。你还需要注意的是，这里并没有使用任何特定于 Hibernate 的注解。这里使用的所有注解都是 JPA 支持的注解，从而允许在不修改任何代码的前提下，从 Hibernate 切换到任何支持 JPA 的实现。

需要添加的下一个类是自定义的 HibernateBatchConfigurer，这个类提供了 HibernateTransactionManager，用于代替通常的 DataSourceTransactionManager，如代码清单 9-33 所示。

代码清单 9-33　HibernateBatchConfigurer.java

```
...
@Component
public class HibernateBatchConfigurer implements BatchConfigurer {

    private DataSource dataSource;
    private SessionFactory sessionFactory;
    private JobRepository jobRepository;
    private PlatformTransactionManager transactionManager;
    private JobLauncher jobLauncher;
    private JobExplorer jobExplorer;
    public HibernateBatchConfigurer(DataSource dataSource,
        EntityManagerFactory entityManagerFactory) {

        this.dataSource = dataSource;
        this.sessionFactory = entityManagerFactory.unwrap(SessionFactory.class);
    }

    @Override
    public JobRepository getJobRepository() throws Exception {
        return this.jobRepository;
    }

    @Override
    public PlatformTransactionManager getTransactionManager() throws Exception {
            return this.transactionManager;
    }

    @Override
    public JobLauncher getJobLauncher() throws Exception {
        return this.jobLauncher;
    }

    @Override
    public JobExplorer getJobExplorer() throws Exception {
        return this.jobExplorer;
    }

    @PostConstruct
    public void initialize() {
        try {
                HibernateTransactionManager transactionManager =
                new HibernateTransactionManager(sessionFactory);
                transactionManager.afterPropertiesSet();
```

第 9 章 ItemWriter

```
            this.transactionManager = transactionManager;

            this.jobRepository = createJobRepository();
            this.jobExplorer = createJobExplorer();
            this.jobLauncher = createJobLauncher();
        }
        catch (Exception e) {
            throw new BatchConfigurationException(e);
        }
    }

    private JobLauncher createJobLauncher() throws Exception {
        SimpleJobLauncher jobLauncher = new SimpleJobLauncher();
        jobLauncher.setJobRepository(this.jobRepository);
        jobLauncher.afterPropertiesSet();

        return jobLauncher;
    }

    private JobExplorer createJobExplorer() throws Exception {
        JobExplorerFactoryBean jobExplorerFactoryBean = new JobExplorerFactoryBean();

        jobExplorerFactoryBean.setDataSource(this.dataSource);
        jobExplorerFactoryBean.afterPropertiesSet();

        return jobExplorerFactoryBean.getObject();
    }

    private JobRepository createJobRepository() throws Exception {
        JobRepositoryFactoryBean jobRepositoryFactoryBean = new
        JobRepositoryFactoryBean();

        jobRepositoryFactoryBean.setDataSource(this.dataSource);
        jobRepositoryFactoryBean.setTransactionManager(this.transactionManager);
        jobRepositoryFactoryBean.afterPropertiesSet();

        return jobRepositoryFactoryBean.getObject();
    }
}
```

最后要做的就是配置 HibernateItemWriter 了。这可能是最容易进行的 ItemWriter 配置了，因为其他的组件和 Hibernate 框架会替你完成所有的工作。HibernateItemWriter 需要一个必选的依赖项和一个可选的依赖项：前者是一个对 SessionFactory 的引用；后者是一个标识(本例并不会使用)，用于指明 ItemWriter 在调用 flush 方法之后是否应该为 Session 调用 clear 方法(默认为 true)。代码清单 9-34 展示了在使用新的 HibernateItemWriter 之后的作业配置。

代码清单 9-34　使用了 Hibernate 的 HibernateImportJob.java

```
...
@Configuration
public class HibernateImportJob {

    private JobBuilderFactory jobBuilderFactory;
```

```java
        private StepBuilderFactory stepBuilderFactory;

    public HibernateImportJob(JobBuilderFactory jobBuilderFactory,
                StepBuilderFactory stepBuilderFactory) {

        this.jobBuilderFactory = jobBuilderFactory;
        this.stepBuilderFactory = stepBuilderFactory;
    }

    @Bean
    @StepScope
    public FlatFileItemReader<Customer> customerFileReader(
            @Value("#{jobParameters['customerFile']}")Resource inputFile) {

        return new FlatFileItemReaderBuilder<Customer>()
                    .name("customerFileReader")
                    .resource(inputFile)
                    .delimited()
                    .names(new String[] {"firstName",
                                        "middleInitial",
                                        "lastName",
                                        "address",
                                        "city",
                                        "state",
                                        "zip"})
                    .targetType(Customer.class)
                    .build();
    }

    @Bean
    public HibernateItemWriter<Customer> hibernateItemWriter(
        EntityManagerFactory entityManager) {

        return new HibernateItemWriterBuilder<Customer>()
                    .sessionFactory(entityManager.unwrap(SessionFactory.class))
                    .build();
    }

    @Bean
    public Step hibernateFormatStep() throws Exception {
        return this.stepBuilderFactory.get("jdbcFormatStep")
                    .<Customer, Customer>chunk(10)
                    .reader(customerFileReader(null))
                    .writer(hibernateItemWriter(null))
                    .build();
    }

    @Bean
    public Job hibernateFormatJob() throws Exception {
        return this.jobBuilderFactory.get("hibernateFormatJob")
                    .start(hibernateFormatStep ())
                    .build();
    }
}
```

以上作业配置仅仅修改了 hibernateBatchItemWriter 的配置及其在 hibernateFormatStep 中的引用。

你在前面看到过，HibernateItemWriter 只需要一个对 SessionFactory 的引用，而 Spring Boot 已经为你提供。执行这个作业后，返回的结果与前面使用 JdbcBatchItemWriter 的示例相同。

当其他框架完成所有繁重的工作时，Spring Batch 的配置将变得极其简单，就像前面使用 Hibernate 的示例那样。作为 Hibernate 官方规范的一员，JPA 是另一个可用于写入数据库的数据库访问框架。

9.3.3 JpaItemWriter

JPA(Java Persistence API，Java 持久化 API)提供了与 Hibernate 非常相似的功能，并且与 Hibernate 的配置几乎完全相同。与 Hibernate 类似，JPA 能够完成数据库写入的繁重工作，所以 Spring Batch 需要完成的工作非常少。本小节介绍如何配置 JPA，使其完成数据库写入操作。

org.springframework.batch.item.writer.JpaItemWriter 其实是对 JPA 的 javax.persistence.EntityManager 的简单封装。当处理完一个块之后，这个块中的条目列表会被传给 JpaItemWriter。这个写入器循环遍历列表中的条目，为每个条目调用 EntityManager 的 merge(T entity)方法，然后在保存了所有条目之后调用 flush 方法。

为了查看 JpaItemWriter 在实践中如何运作，下面使用与前面相同的客户数据作为输入，将它们插入相同的 CUSTOMER 表中。为了将 JPA 挂接到作业中，你需要完成下面两件事情：

- 创建 BatchConfigurer 的一个实现，进而创建一个 JpaTransactionManager，作用与前面 Hibernate 版本中的相同。
- 配置 JpaItemWriter，用于保存作业中读入的条目。

剩下要做的工作由 Spring Boot 完成。我们可以从使用 JpaBatchConfigurer 编写所需的代码开始。除两个很小的改动外，这里的代码与前面的 Hibernate 版本完全相同。首先，在构造方法中保存一个 EntityManagerFactory 而不是一个 SessionFactory。然后，在 initialize 方法中创建一个 JpaTransactionManager 而不是 HibernateTransactionManager。代码清单 9-35 展示了配置的 JpaBatchConfigurer。

代码清单 9-35　JpaBatchConfigurer.java

```
...
@Component
public class JpaBatchConfigurer implements BatchConfigurer {

    private DataSource dataSource;
    private EntityManagerFactory entityManagerFactory;
    private JobRepository jobRepository;
    private PlatformTransactionManager transactionManager;
    private JobLauncher jobLauncher;
    private JobExplorer jobExplorer;

    public JpaBatchConfigurer(DataSource dataSource,
            EntityManagerFactory entityManagerFactory) {
        this.dataSource = dataSource;
        this.entityManagerFactory = entityManagerFactory;
    }

    @Override
    public JobRepository getJobRepository() throws Exception {
        return this.jobRepository;
    }
```

```java
@Override
public PlatformTransactionManager getTransactionManager() throws Exception {
    return this.transactionManager;
}

@Override
public JobLauncher getJobLauncher() throws Exception {
    return this.jobLauncher;
}

@Override
public JobExplorer getJobExplorer() throws Exception {
    return this.jobExplorer;
}

@PostConstruct
public void initialize() {

    try {
        JpaTransactionManager transactionManager =
                new JpaTransactionManager(entityManagerFactory);
        transactionManager.afterPropertiesSet();

        this.transactionManager = transactionManager;

        this.jobRepository = createJobRepository();
        this.jobExplorer = createJobExplorer();
        this.jobLauncher = createJobLauncher();
    }
    catch (Exception e) {
        throw new BatchConfigurationException(e);
    }
}

private JobLauncher createJobLauncher() throws Exception {
    SimpleJobLauncher jobLauncher = new SimpleJobLauncher();

    jobLauncher.setJobRepository(this.jobRepository);
    jobLauncher.afterPropertiesSet();

    return jobLauncher;
}

private JobExplorer createJobExplorer() throws Exception {
    JobExplorerFactoryBean jobExplorerFactoryBean =
            new JobExplorerFactoryBean();

    jobExplorerFactoryBean.setDataSource(this.dataSource);
    jobExplorerFactoryBean.afterPropertiesSet();

    return jobExplorerFactoryBean.getObject();
}

private JobRepository createJobRepository() throws Exception {
    JobRepositoryFactoryBean jobRepositoryFactoryBean =
```

第 9 章　ItemWriter

```
            new JobRepositoryFactoryBean();

    jobRepositoryFactoryBean.setDataSource(this.dataSource);
    jobRepositoryFactoryBean.setTransactionManager(this.transactionManager);
    jobRepositoryFactoryBean.afterPropertiesSet();

    return jobRepositoryFactoryBean.getObject();
    }
}
```

由于第 8 章已经使用 JPA 注解来映射 Customer 对象，因此在这个示例中并不需要做任何改动。作业配置的最后一方面是使用 JPA 配置 JpaItemWriter。JpaItemWriter 只需要一个依赖项，也就是对 EntityManagerFactory 的一个引用，就可以获取和使用一个 EntityManager 实例，如代码清单 9-26 所示。

代码清单 9-36　使用了 JpaItemWriter 的格式化作业

```
...
@Configuration
public class JpaImportJob {

    private JobBuilderFactory jobBuilderFactory;

    private StepBuilderFactory stepBuilderFactory;

    public JpaImportJob(JobBuilderFactory jobBuilderFactory,
            StepBuilderFactory stepBuilderFactory) {

        this.jobBuilderFactory = jobBuilderFactory;
        this.stepBuilderFactory = stepBuilderFactory;
    }

    @Bean
    @StepScope
    public FlatFileItemReader<Customer> customerFileReader(
            @Value("#{jobParameters['customerFile']}")Resource inputFile) {
        return new FlatFileItemReaderBuilder<Customer>()
                    .name("customerFileReader")
                    .resource(inputFile)
                    .delimited()
                    .names(new String[] {"firstName",
                                        "middleInitial",
                                        "lastName",
                                        "address",
                                        "city",
                                        "state",
                                        "zip"})
                    .targetType(Customer.class)
                    .build();
    }

    @Bean
    public JpaItemWriter<Customer> jpaItemWriter(
        EntityManagerFactory entityManagerFactory) {

        JpaItemWriter<Customer> jpaItemWriter = new JpaItemWriter<>();
```

```
            jpaItemWriter.setEntityManagerFactory(entityManagerFactory);

            return jpaItemWriter;
        }

        @Bean
        public Step jpaFormatStep() throws Exception {
            return this.stepBuilderFactory.get("jpaFormatStep")
                    .<Customer, Customer>chunk(10)
                    .reader(customerFileReader(null))
                    .writer(jpaItemWriter(null))
                    .build();
        }

        @Bean
        public Job jpaFormatJob() throws Exception {
            return this.jobBuilderFactory.get("jpaFormatJob")
                    .start(jpaFormatStep())
                    .build();
        }
    }
```

现在可以快速地使用 ./mvnw clean install 命令构建作业了，然后使用代码清单 9-37 所示的命令执行作业，返回的结果与其他的数据库示例相同。

代码清单 9-37　用于执行使用 JPA 配置的格式化作业的命令

```
java -jar itemWriters-0.0.1-SNAPSHOT.jar customerFile=/input/customer.csv
```

无论如何，关系数据库在现代企业中仍起着举足轻重的作用。可以看到，使用 Spring Batch 可以很容易地把作业的执行结果写入数据库。但是，关系数据库并不是唯一的选择，对 Spring Batch 来说是这样，对企业来说也是这样。9.4 将介绍 Spring Data 支持的其他 NoSQL 存储。

9.4　NoSQL ItemWriter

在第 7 章，你已经了解了 Spring Data 项目及其如何将公共编程模型引入许多不同的数据存储中。Spring Batch 利用 Spring Data 为你提供了写入 NoSQL 数据存储的能力，特别是 MongoDB、Noe4j、Pivotal Gemfire 和 Apache Geode，并且通过 CrudRepository 提供了对其他 Spring Data 项目的支持。本节将介绍如何将这些 NoSQL 数据存储集成到 Spring Batch 项目中。

9.4.1　MongoDB

MongoDB 具有高性能、高可伸缩性等数据存储特性，这使其成为对企业非常有吸引力的 NoSQL 数据存储。Spring Batch 支持通过 MongoItemWriter 把对象作为文档存储到 MongoDB 集合中。

为了使用 MongoDB，我们需要对 Customer 对象进行少量修改。首先，MongoDB 不支持长整型 (long) 的 id，但支持字符串类型 (String) 的 id。其次，可以移除 Customer 对象上的 JPA 注解，因为 MongoDB 中并没有数据表。代码清单 9-38 展示了用于 MongoDB 的 Customer 领域对象。

第 9 章 ItemWriter

代码清单 9-38　用于 MongoDB 的 Customer.java

```java
...
public class Customer implements Serializable {
    private static final long serialVersionUID = 1L;

    @Id
    private String id;
    private String firstName;
    private String middleInitial;
    private String lastName;
    private String address;
    private String city;
    private String state;
    private String zip;

    // Getters and setters removed

    @Override
    public String toString() {
        return "Customer{" +
                        "id=" + id +
                        ", firstName='" + firstName + '\"' +
                        ", middleInitial='" + middleInitial + '\"' +
                        ", lastName='" + lastName + '\"' +
                        ", address='" + address + '\"' +
                        ", city='" + city + '\"' +
                        ", state='" + state + '\"' +
                        ", zip='" + zip + '\"' +
                        '}';
    }
}
```

在更新了 Customer 领域对象之后，你需要在 pom.xml 中添加正确的依赖项，以引入 MongoDB 依赖项。与使用 Spring Boot 添加其他新功能时类似，我们必须引入正确的启动包，在这里也就是代码清单 9-39 所示的 spring-boot-starter-data-mongodb。

代码清单 9-39　spring-boot-starter-data-mongodb

```xml
...
<dependency>
        <groupId>org.springframework.boot</groupId>
        <artifactId>spring-boot-starter-data-mongodb</artifactId>
</dependency>
...
```

你要做的下一项配置是在 application.yml 里指向 MongoDB 数据库。默认情况下，Spring Boot 会使用标准的登录凭证查找本地数据库，所以我们并不需要配置。不过，我们必须告诉应用将要写入的数据库的名称，在我们的示例中是 customerdb，通过 spring.data.mongodb.database 属性进行设置即可。

用于 MongdDB 的 ItemWriter 与基于 Hibernate 和 JPA 的 ItemWriter 的工作方式相同。映射由用于领域对象的注解处理，所以 ItemWriter 所需的配置很少。在本例中，你需要配置数据库中用于写入的集合名称(customer)，提供一个 MongoOperations 实例并调用 build 方法。其他唯一的配置选项是 delete

Spring Batch 权威指南

标识,用于指明对于匹配的条目 ItemWriter 应该删除还是保存。默认情况是保存它们。代码清单 9-40 展示了 mongoFormatJob 作业的配置,其中包括更新后的 MongoItemWriter。

代码清单 9-40　mongoFormatJob

```
...
@Configuration
public class MongoImportJob {

    private JobBuilderFactory jobBuilderFactory;

    private StepBuilderFactory stepBuilderFactory;

    public MongoImportJob(JobBuilderFactory jobBuilderFactory,
            StepBuilderFactory stepBuilderFactory) {
        this.jobBuilderFactory = jobBuilderFactory;
        this.stepBuilderFactory = stepBuilderFactory;
    }

    @Bean
    @StepScope
    public FlatFileItemReader<Customer> customerFileReader(
            @Value("#{jobParameters['customerFile']}")Resource inputFile) {

        return new FlatFileItemReaderBuilder<Customer>()
                .name("customerFileReader")
                .resource(inputFile)
                .delimited()
                .names(new String[] {"firstName",
                                    "middleInitial",
                                    "lastName",
                                    "address",
                                    "city",
                                    "state",
                                    "zip"})
                .targetType(Customer.class)
                .build();
    }

    @Bean
    public MongoItemWriter<Customer> mongoItemWriter(MongoOperations mongoTemplate) {
        return new MongoItemWriterBuilder<Customer>()
                .collection("customers")
                .template(mongoTemplate)
                .build();
    }

    @Bean
    public Step mongoFormatStep() throws Exception {
        return this.stepBuilderFactory.get("mongoFormatStep")
                .<Customer, Customer>chunk(10)
                .reader(customerFileReader(null))
                .writer(mongoItemWriter(null))
                .build();
    }
```

```
@Bean
public Job mongoFormatJob() throws Exception {
    return this.jobBuilderFactory.get("mongoFormatJob")
                .start(mongoFormatStep())
                .build();
}
```

MongoDB 不支持事务的 ACID 特性(原子性、一致性、隔离性和持久性)。因此，Spring Batch 把 MongoDB 当作其他不支持事务的数据存储：把数据写入缓冲区，直到提交发生之前，才在最后时刻执行写操作。

在编写了作业之后，如果构建并运行作业，就会发现集合中插入了 9 个文档，你在 MongoDB 的 GUI 客户端 Robot 3T 将看到如图 9-4 所示的结果。

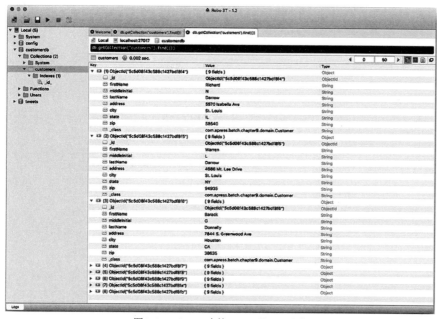

图 9-4　Robot 3T 中的 mongoFormatJob

尽管 MongoDB 可能是市场上最流行的 NoSQL 数据存储，但是以文档的格式存储数据并不适用于所有情况。这就是 NoSQL 运动背后的理念：使用适合于数据的存储。9.4.2 节将介绍另一种 NoSQL 数据存储 Neo4j，Neo4j 是一种图数据库，它使用了与众不同的模型。

9.4.2　Noe4j

在万物互联的时代，我们的生活已经变成无数巨大的图。这些图包括 Facebook 上的朋友、LinkedIn 上的社交连接，等等。这些图中的每个节点都有几乎完全不同的关系形式，导致关系数据存储在存储这类数据时变得十分低效。Noe4j 是当今市场上领先的图数据库，拥有几百万次的下载量，并且正在以每月 5 万次的速度增加。Spring Data 为 Spring Portfolio 带来了对 Noe4j 的支持。本小节将会介绍如何使用 Neo4jItemWriter 把记录写入 Noe4j 数据库。

为了上手 Noe4j，就像使用 JPA 和 MongoDB 时那样，我们使用注解将数据映射到它们在数据库中的表示形式。为此，我们将使用适当的注解更新 Customer 类，将其映射到 Noe4j 数据库。首先，在类级别使用@NodeEntity 注解，从而告诉 ItemWriter 这个类代表图中的一个节点。Noe4j 还支持使用@Relationship 注解，以允许映射图中各个节点之间的关系。在我们的示例中，节点之间没有任何关系，但是我们需要设置节点的标识符 id。使用关系数据库时，我们使用了长整型(long)的 id；MongoDB 则需要字符串类型(String)的 id；对于 Noe4j，我们将使用 UUID 类型。代码清单 9-14 展示了更新后的 Customer 领域对象，它已经被映射到 Noe4j。

代码清单 9-41　映射到 Neo4j 的 Customer 领域对象

```
...
@NodeEntity
public class Customer implements Serializable {
    private static final long serialVersionUID = 1L;

    @Id
    @GeneratedValue(strategy = UuidStrategy.class)
    private UUID id;
    private String firstName;
    private String middleInitial;
    private String lastName;
    private String address;
    private String city;
    private String state;
    private String zip;

    // Getters and setters removed
    @Override
    public String toString() {
        return "Customer{" +
                        "id=" + id +
                        ", firstName='" + firstName + '\"' +
                        ", middleInitial='" + middleInitial + '\"' +
                        ", lastName='" + lastName + '\"' +
                        ", address='" + address + '\"' +
                        ", city='" + city + '\"' +
                        ", state='" + state + '\"' +
                        ", zip='" + zip + '\"' +
                        '}';
    }
}
```

为了在应用中使用 Noe4j，我们需要在 pom.xml 中添加合适的 Spring Boot 启动包。此时此刻，引入启动包 spring-boot-starter-data-neo4j 对你来说并不意外，如代码清单 9-42 所示。

代码清单 9-42　Neo4j 依赖项

```
...
<dependency>
    <groupId>org.springframework.boot</groupId>
    <artifactId>spring-boot-starter-data-neo4j</artifactId>
```

在应用中包含正确的依赖项之后，我们还需要在 application.yml 中添加正确的配置。首先，建议使用社区版的 Noe4j 服务器。为此，需要配置用户名、密码和 URI，如代码清单 9-43 所示。

代码清单 9-43　Noe4j 依赖项

```
spring:
  data:
...
  neo4j:
    username: neo4j
    password: password
    embedded:
      enabled: false
    uri: bolt://localhost:7687
```

使用 Noe4j 作为 ItemWriter 的最后一部分工作是真正地配置 ItemWriter。Neo4jItemWriter 只需要一个依赖项——一个 org.neo4j.orgm.session.SessionFactory 实例，可由 Spring Boot 通过前面添加的启动包来提供。代码清单 9-44 展示了用于写入 Noe4j 的完整的作业配置。

代码清单 9-44　Neo4jImportJob

```java
...
@Configuration
public class Neo4jImportJob {

    private JobBuilderFactory jobBuilderFactory;

    private StepBuilderFactory stepBuilderFactory;

    public Neo4jImportJob(JobBuilderFactory jobBuilderFactory,
            StepBuilderFactory stepBuilderFactory) {
        this.jobBuilderFactory = jobBuilderFactory;
        this.stepBuilderFactory = stepBuilderFactory;
    }

    @Bean
    @StepScope
    public FlatFileItemReader<Customer> customerFileReader(
            @Value("#{jobParameters['customerFile']}")Resource inputFile) {

        return new FlatFileItemReaderBuilder<Customer>()
                    .name("customerFileReader")
                    .resource(inputFile)
                    .delimited()
                    .names(new String[] {"firstName",
                                "middleInitial",
                                "lastName",
                                "address",
                                "city",
                                "state",
                                "zip"})
                    .targetType(Customer.class)
```

```
            .build();
}

@Bean
public Neo4jItemWriter<Customer> neo4jItemWriter(SessionFactory sessionFactory) {
    return new Neo4jItemWriterBuilder<Customer>()
            .sessionFactory(sessionFactory)
            .build();
}

@Bean
public Step neo4jFormatStep() throws Exception {
    return this.stepBuilderFactory.get("neo4jFormatStep")
            .<Customer, Customer>chunk(10)
            .reader(customerFileReader(null))
            .writer(neo4jItemWriter(null))
            .build();
}

@Bean
public Job neo4jFormatJob() throws Exception {
    return this.jobBuilderFactory.get("neo4jFormatJob")
            .start(neo4jFormatStep())
            .build();
}
```

在构建和运行 Neo4jImportJob 后,就可以打开服务器提供的 Noe4j 浏览器,然后验证结果是否已经成功导入数据库。打开 Neo4j 浏览器后,就可以执行等价于 SELECT firstName, lastName FROM customer;的查询"暗语"——MATCH(c:Customer) RETURN c.firstName, c.lastName。返回的结果如图 9-5 所示。

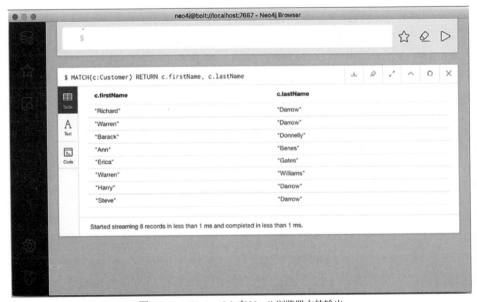

图 9-5　neo4jFormatJob 在 Noe4j 浏览器中的输出

图数据库为解决正确的问题提供了一些强大的工具。不过，Spring Batch 还提供了另一种数据存储。9.4.3 节将介绍如何在内存中使用特定的键值存储提高性能。

9.4.3 Pivotal Gemfire 和 Apache Geode

在金融世界里，一毫秒也是意义重大的。对于零售交易中的欺诈检测，没有时间等待从多跳网络或慢速磁盘中检索数据。在这些环境中，数据必须缓存在内存中以便快速访问。这种环境就是 Pivotal Gemfire 诞生的地方。本小节将介绍 Pivotal Gemfire 及其开源版本 Apache Geode，还将介绍如何使用 Spring Batch 将数据写入其中。

Pivotal Gemfire 是一张内存数据网格，核心是高性能的分布式 HashMap。HashMap 是一种键值存储，可以把所有的数据保存在内存中，并且由于采用了网络拓扑结构，读取时间极快。这与 Spring Batch 又有什么关系呢？很简单！Pivotal Gemfire 的主要用法是缓存数据。为了让缓存发挥作用，需要事先启动缓存。Spring Batch 提供了一款在应用启动时高效启动冷缓存的优秀工具。

与前面相同，首先更新领域对象的映射。对于 Pivotal Gemfire，由于将要使用键值存储，因此这里的键将来自领域对象，而领域对象本身则作为值。我们真正需要做的是指定要把值持久化到的区域（与 MongoDB 中的集合类似）。因此，下面为 Customer 类添加 @Region 注解，如代码清单 9-45 所示。

代码清单 9-45　Neo4jImportJob

```
...
@Region(value = "Customers")
public class Customer implements Serializable {

    private static final long serialVersionUID = 1L;
    private long id;
    private String firstName;
    private String middleInitial;
    private String lastName;
    private String address;
    private String city;
    private String state;
    private String zip;

    // Accessors removed for brevity

    @Override
    public String toString() {
        return "Customer{" +
                        "id=" + id +
                        ", firstName='" + firstName + '\'' +
                        ", middleInitial='" + middleInitial + '\'' +
                        ", lastName='" + lastName + '\'' +
                        ", address='" + address + '\'' +
                        ", city='" + city + '\'' +
                        ", state='" + state + '\'' +
                        ", zip='" + zip + '\'' +
                        '}';
    }
}
```

在定义了领域对象之后，你还需要更新 pom.xml 文件以引入 Pivotal Gemfire 依赖项。为此，你需要在 POM 中修改两处地方。首先，添加两个新的依赖项，一个是 Spring Data Gemfire，另一个是 Spring Shell(Spring Data Gemfire 需要它)。其次，排除 spring-boot-starter-batch 上的日志依赖项，这是因为 Spring Boot 默认使用的日志和 Pivotal Gemfire 默认使用的日志有冲突。代码清单 9-46 展示了更新后的 POM。

代码清单 9-46　用于 Pivotal Gemfire 的 pom.xml

```
...
<dependency>
        <groupId>org.springframework.boot</groupId>
        <artifactId>spring-boot-starter-batch</artifactId>
        <exclusions>
                <exclusion>
                        <groupId>org.springframework.boot</groupId>
                        <artifactId>spring-boot-starter-logging</artifactId>
                </exclusion>
        </exclusions>
</dependency>
...
<dependency>
        <groupId>org.springframework.data</groupId>
        <artifactId>spring-data-gemfire</artifactId>
</dependency>
<dependency>
        <groupId>org.springframework.shell</groupId>
        <artifactId>spring-shell</artifactId>
</dependency>
...
```

在添加了这些依赖项之后，就可以配置 Pivotal Gemfire 了。到目前为止，我们使用过的其他数据存储都运行在应用之外，而 Pivotal Gemfire 则与它们不同，Pivotal Gemfire 运行在应用之内，这么做是因为能够限制网络跳数。

为了进行配置，首先为配置类添加@PeerCacheApplication 注解。使用注解后，就会在应用中启动 Pivotal Gemfire 服务。我们将提供两个选项：Pivotal Gemfire 应用的名称以及日志的级别。接下来需要配置区域。前面提到过，Pivotal Gemfire 中的区域与 MongoDB 中的集合相似。Pivotal Gemfire 允许通过 Spring 直接设定这些配置，而不是使用外部机制。一旦配置好了区域，就需要为 ItemWriter 创建模板 GemfireTemplate。定义好这个模板后，就可以创建 ItemWriter 了。GemfireItemWriter 需要配置两个地方：第一个地方是刚刚创建的模板，第二个地方是 org.springframework.core.convert.converter.Converter。Converter 实例用于把写入 Povotal Gemfire 的条目转换成使用的键。在本例中，可通过使用 Spring Batch 提供的一个实现(即 SpELItemKeyMapper)来达到这个目的。这个实现将采用一个 SqEL 表达式，从当前的条目中创建键。与前面几个作业的配置相比，这里多了一个 CommandLineRunner。我们将用它来验证作业是否成功运行，这是因为我们没有使用 GUI 来查看结果，而且 Pivatal Gemfire 会随着应用的关闭而关闭。这个 CommandLineRunner 会查询一个 Pivotal Gemfire 实例，并列出其中的条目。代码清单 9-47 配置了这个 Pivotal Gemfire 实例以及用于加载它的 Spring Batch 作业。

第 9 章 ItemWriter

代码清单 9-47　GemfireImportJob

```java
...
@Configuration
@PeerCacheApplication(name = "AccessingDataGemFireApplication",
                     logLevel = "info")
public class GemfireImportJob {

    private JobBuilderFactory jobBuilderFactory;

    private StepBuilderFactory stepBuilderFactory;

    public GemfireImportJob(JobBuilderFactory jobBuilderFactory,
            StepBuilderFactory stepBuilderFactory) {

        this.jobBuilderFactory = jobBuilderFactory;
        this.stepBuilderFactory = stepBuilderFactory;
    }

    @Bean
    @StepScope
    public FlatFileItemReader<Customer> customerFileReader(
            @Value("#{jobParameters['customerFile']}")Resource inputFile) {

        return new FlatFileItemReaderBuilder<Customer>()
                    .name("customerFileReader")
                    .resource(inputFile)
                    .delimited()
                    .names(new String[] {"firstName",
                                         "middleInitial",
                                         "lastName",
                                         "address",
                                         "city",
                                         "state",
                                         "zip"})
                    .targetType(Customer.class)
                    .build();
    }

    @Bean
    public GemfireItemWriter<Long, Customer> gemfireItemWriter(
        GemfireTemplate gemfireTemplate) {

        return new GemfireItemWriterBuilder<Long, Customer>()
                    .template(gemfireTemplate)
                    .itemKeyMapper(new SpELItemKeyMapper<>(
                            "firstName + middleInitial + lastName"))
                    .build();
    }

    @Bean
    public Step gemfireFormatStep() throws Exception {
        return this.stepBuilderFactory.get("gemfireFormatStep")
                    .<Customer, Customer>chunk(10)
                    .reader(customerFileReader(null))
                    .writer(gemfireItemWriter(null))
```

```java
                .build();
    }

    @Bean
    public Job gemfireFormatJob() throws Exception {
        return this.jobBuilderFactory.get("gemfireFormatJob")
                    .start(gemfireFormatStep())
                    .build();
    }

    @Bean(name="customer")
    public Region<Long, Customer> getCustomer(final GemFireCache cache) throws Exception {
        LocalRegionFactoryBean<Long, Customer> customerRegion =
                new LocalRegionFactoryBean<>();
        customerRegion.setCache(cache);
        customerRegion.setName("customer");
        customerRegion.afterPropertiesSet();
        Region<Long, Customer> object = customerRegion.getRegion();
        return object;
    }

    @Bean
    public GemfireTemplate gemfireTemplate() throws Exception {
        return new GemfireTemplate(getCustomer(null));
    }

    @Bean
    public CommandLineRunner validator(final GemfireTemplate gemfireTemplate) {
        return args -> {
            List<Object> customers =
                gemfireTemplate.find("select * from /customer").asList();
            for (Object customer : customers) {
                System.out.println(">> object: " + customer);
            }
        };
    }
}
```

有了以上配置，就可以构建和运行作业。结果就是把所有加载到这个 Pivotal Gemfire 区域中的条目写入标准输出，如代码清单 9-48 所示。

代码清单 9-48　GemfireImportJob 的输出

```
...
[info 2019/02/09 12:59:40.617 CST <main> tid=0x1] Job: [SimpleJob: [name=gemfireFormatJob]]
completed with the following parameters: [{customerFile=/data/customer.csv}] and the
following status: [COMPLETED]

>> object: Customer{id=0, firstName='Ann', middleInitial='Z', lastName='Benes',
address='2447 S. Greenwood Ave', city='Las Vegas', state='NY', zip='55366'}
>> object: Customer{id=0, firstName='Warren', middleInitial='M', lastName='Williams',
address='6670 S. Greenwood Ave', city='Hollywood', state='FL', zip='37288'}
>> object: Customer{id=0, firstName='Erica', middleInitial='Z', lastName='Gates',
address='3141 Farnam Street', city='Omaha', state='CA', zip='57640'}
>> object: Customer{id=0, firstName='Warren', middleInitial='L', lastName='Darrow',
address='4686 Mt. Lee Drive', city='St. Louis', state='NY', zip='94935'}
```

```
>> object: Customer{id=0, firstName='Richard', middleInitial='N', lastName='Darrow',
address='5570 Isabella Ave', city='St. Louis', state='IL', zip='58540'}
>> object: Customer{id=0, firstName='Steve', middleInitial='O', lastName='Darrow',
address='8407 Infinite Loop Drive', city='Las Vegas', state='WA', zip='90520'}
>> object: Customer{id=0, firstName='Harry', middleInitial='T', lastName='Darrow',
address='3273 Isabella Ave', city='Houston', state='FL', zip='97261'}
>> object: Customer{id=0, firstName='Barack', middleInitial='G', lastName='Donnelly',
address='7844 S. Greenwood Ave', city='Houston', state='CA', zip='38635'}
[info 2019/02/09 12:59:40.660 CST <Distributed system shutdown hook> tid=0x21] VM is exiting
- shutting down distributed system
...
```

Pivotal Gemfire 和 Apache Geode 都可以为各种工作负载提供最佳的类性能。Spring Batch 提供了有效的机制来更新它们。与 Spring Data 相关的最后一个 ItemWriter 是 RepositoryItemWriter。就像第 7 章讨论的 RepositoryItemReader，RepositoryItemWriter 利用了 Spring Data 的 Repository(存储库)抽象，从而把记录写到 Spring Data 支持的任何数据存储中。

9.4.4 Repository 抽象

Spring Data 的 Repository 抽象提供了非常有用的方式用于构建 ItemWriter。在第 7 章，你了解了如何使用 Spring Data 的 PagingAndSortingRepository 为 Spring Data 支持的任何数据存储创建 ItemReader。ItemReader 和 ItemWriter 对存储库的使用区别在于使用了哪个类型的存储库。在读取时，使用 PagingAndSortingRepository。然而在写入时，由于不需要关心分页和排序，因此我们使用 PagingAndSortingRepository 的父接口，也就是 org.springframework.data.repository.CrudRepository。在本小节中，我们将使用 RepositoryItemWriter 把数据持久化到 Spring Data 支持的数据存储中。

为了查看 RepositoryItemWriter 如何工作，我们将把数据写入 CUSTOMER 表中(就像本章前面的 JPA 示例那样)。不过，我们并不使用 JpaItemWriter，而是使用 RepositoryItemWriter。我们首先从领域对象开始。由于这个示例在底层使用了 JPA，因此实际上可以使用与 JPA 示例相同的领域对象配置。为了方便起见，代码清单 9-49 列出了 Customer 领域对象的 JPA 映射。

代码清单 9-49　Customer 领域对象的 JPA 映射

```
...
@Entity
@Table(name = "CUSTOMER")
public class Customer implements Serializable {
    private static final long serialVersionUID = 1L;

    @Id
    @GeneratedValue(strategy = GenerationType.IDENTITY)
    private long id;
    private String firstName;
    private String middleInitial;
    private String lastName;
    private String address;
    private String city;
    private String state;
    private String zip;

    // Getters and setters removed for brevity
```

```
    @Override
    public String toString() {
        return "Customer{" +
                "id=" + id +
                ", firstName='" + firstName + '\"' +
                ", middleInitial='" + middleInitial + '\"' +
                ", lastName='" + lastName + '\"' +
                ", address='" + address + '\"' +
                ", city='" + city + '\"' +
                ", state='" + state + '\"' +
                ", zip='" + zip + '\"' +
                '}';
    }
}
```

在定义好了领域对象之后，就可以创建存储库的定义。由于我们仅仅把 Customer 领域对象保存到数据库中，因此我们只需要创建一个扩展了 CrudRepository 的接口。Spring Data 会完成其余的事情。代码清单 9-50 展示了 CustomerRepository 的定义。

代码清单 9-50　CustomerRepository

```
...
public interface CustomerRepository extends CrudRepository<Customer, Long> {
}
```

由于前面已经引入了 JPA 相关的依赖项，因此这里并不需要再次添加，只需要配置作业即可。RepositoryItemWriter 有两个依赖项：将要使用的存储库和将要调用的方法。其余要做的事情是告诉 Spring 启用存储库功能以及在哪里查找存储库，这需要使用@EnableJpaRepositories 注解，并指定存储库所在包中的类。代码清单 9-51 展示了作业的配置。

代码清单 9-51　RepositoryImportJob

```
...
@Configuration
@EnableJpaRepositories(basePackageClasses = Customer.class)
public class RepositoryImportJob {

    private JobBuilderFactory jobBuilderFactory;

    private StepBuilderFactory stepBuilderFactory;

    public RepositoryImportJob(JobBuilderFactory jobBuilderFactory,
            StepBuilderFactory stepBuilderFactory) {
        this.jobBuilderFactory = jobBuilderFactory;
        this.stepBuilderFactory = stepBuilderFactory;
    }

    @Bean
    @StepScope
    public FlatFileItemReader<Customer> customerFileReader(
            @Value("#{jobParameters['customerFile']}")Resource inputFile) {
```

```java
        return new FlatFileItemReaderBuilder<Customer>()
                .name("customerFileReader")
                .resource(inputFile)
                .delimited()
                .names(new String[] {"firstName",
                                     "middleInitial",
                                     "lastName",
                                     "address",
                                     "city",
                                     "state",
                                     "zip"})
                .targetType(Customer.class)
                .build();
}

@Bean
public RepositoryItemWriter<Customer> repositoryItemWriter(
    CustomerRepository repository) {

    return new RepositoryItemWriterBuilder<Customer>()
                .repository(repository)
                .methodName("save")
                .build();
}

@Bean
public Step repositoryFormatStep() throws Exception {
    return this.stepBuilderFactory.get("repositoryFormatStep")
                .<Customer, Customer>chunk(10)
                .reader(customerFileReader(null))
                .writer(repositoryItemWriter(null))
                .build();
}

@Bean
public Job repositoryFormatJob() throws Exception {
    return this.jobBuilderFactory.get("repositoryFormatJob")
                .start(repositoryFormatStep())
                .build();
}
}
```

在构建和运行这个作业之后，就可以在 MySQL 中查看 CUSTOMER 表以验证结果，如代码清单 9-52 展示所示。

代码清单 9-52　RepositoryImportJob 作业的执行结果

```
mysql> select id, first_name, middle_initial as middle, last_name, address, city, state as
st, zip from SPRING_BATCH.CUSTOMER;
+----+------------+--------+----------+-------------------------+-----------+----+-------+
| id | first_name | middle | last_name| address                 | city      | st | zip   |
+----+------------+--------+----------+-------------------------+-----------+----+-------+
| 1  | Richard    | N      | Darrow   | 5570 Isabella Ave       | St. Louis | IL | 58540 |
| 2  | Warren     | L      | Darrow   | 4686 Mt. Lee Drive      | St. Louis | NY | 94935 |
| 3  | Barack     | G      | Donnelly | 7844 S. Greenwood Ave   | Houston   | CA | 38635 |
| 4  | Ann        | Z      | Benes    | 2447 S. Greenwood Ave   | Las Vegas | NY | 55366 |
```

```
| 5 | Erica   | Z | Gates    | 3141 Farnam Street      | Omaha     | CA | 57640 |
| 6 | Warren  | M | Williams | 6670 S. Greenwood Ave   | Hollywood | FL | 37288 |
| 7 | Harry   | T | Darrow   | 3273 Isabella Ave       | Houston   | FL | 97261 |
| 8 | Steve   | O | Darrow   | 8407 Infinite Loop Drive| Las Vegas | WA | 90520 |
+---+---------+---+----------+-------------------------+-----------+----+-------+
8 rows in set (0.01 sec)
```

我们已经看到，Spring Batch 能够与很多数据库进行交互。然而，Spring Batch 并非只能写入数据库和文件。

9.5 输出到其他目标的 ItemWriter

在与条目的最终处理结果进行通信时，并不是只能使用文件或数据库。企业往往使用很多其他工具来保存处理后的条目。第 7 章介绍了如何使用 Spring Batch 调用现有的 Spring 服务来获取数据。毫无意外地，Spring Batch 在写入端提供了类似的功能。Spring Batch 还暴露了 Spring 强大的用于与 JMS 交互的 JmsItemWriter。最后，如果有在批处理过程中发送 e-mail 的需求，那么 Spring Batch 也能够处理。本节将介绍如何使用 Spring Batch 提供的 ItemWriter 调用现有的 Java 服务、写入 JMS 目标以及发送邮件。

9.5.1 ItemWriterAdapter

在使用 Spring 的大部分企业中，已经存在大量的久经考验的服务，没有理由不在批处理中复用它们。你在第 7 章已经了解了如何把它们当成作业的输入源。下面将介绍如何通过 ItemWriterAdapter 把现有的 Spring 服务作为 ItemWriter。

org.springframework.batch.item.adapter.ItemWriterAdapter 仅仅是对所配置服务的简单包装。与其他 ItemWriter 一样，write 方法接收用于写入的类型为 List<T> 的条目列表。ItemWriterAdapter 循环遍历这个条目列表，对其中的每个条目调用配置过的服务方法。需要注意的是，ItemWriterAdapter 调用的方法仅接收正在处理的条目的类型参数。例如，如果步骤正在处理 Car 类型的对象，那么调用的方法必须接收类型为 Car 的参数。

为了配置 ItemWriterAdapter，需要如下两个依赖项。

- targetObject：用于包含要调用的方法的 Spring Bean。
- targetMethod：用于对每个条目进行调用的方法。

> **注意** ItemWriterAdapter 正在调用的方法所接收的参数类型必须是当前步骤正在处理的条目类型。

下面看看实践中的 ItemWriterAdapter 示例。代码清单 9-53 展示的服务用于把客户条目写入标准输出(System.out)。

代码清单 9-53　CustomerService.java

```
package com.apress.springbatch.chapter9;

@Service
public class CustomerService {
```

第 9 章 ItemWriter

```
    public void logCustomer(Customer cust) {
        System.out.println("I just saved " + cust);
    }
}
```

如代码清单 9-53 所示，CustomService 短小精干、恰到好处。为了在作业中使用这个服务，需要将其配置为新的 ItemWriterAdapter 的目标。我们选择使用与本章其他作业相同的配置，代码清单 9-54 展示了 ItemWriterAdapter 的配置，其中使用了 CustomerService 的 logCustomer(Customer cust)方法。

代码清单 9-54　ItemWriterAdapter 的配置

```
...
@Bean
public ItemWriterAdapter<Customer> itemWriter(CustomerService customerService) {
    ItemWriterAdapter<Customer> customerItemWriterAdapter = new ItemWriterAdapter<>();

    customerItemWriterAdapter.setTargetObject(customerService);
    customerItemWriterAdapter.setTargetMethod("logCustomer");

    return customerItemWriterAdapter;
}

@Bean
public Step formatStep() throws Exception {
    return this.stepBuilderFactory.get("jpaFormatStep")
                .<Customer, Customer>chunk(10)
                .reader(customerFileReader(null))
                .writer(itemWriter(null))
                .build();
}
@Bean
public Job itemWriterAdapterFormatJob() throws Exception {
    return this.jobBuilderFactory.get("itemWriterAdapterFormatJob")
                .start(formatStep())
                .build();
}
...
```

代码清单 9-54 首先配置 ItemWriter 为 itemWriterAdapter。itemWriterAdapter 有两个依赖项，分别是指向 customerService 的引用和 logCustomer 方法的名称。然后，在步骤中引用 itemWriterAdapter，以便在作业中使用。

为了执行这个作业，与其他作业一样，执行./mvnw clean install 命令以构建作业。在构建了作业之后，就可以像以前一样执行 jar 文件。作业的输出如代码清单 9-55 所示。

代码清单 9-55　ItemWriterAdapter 的输出

```
2018-05-03 21:55:01.287 INFO 61906 --- [main] o.s.b.c.l.support.SimpleJobLauncher:
Job: [SimpleJob: [name=itemWriterAdapterFormatJob]] launched with the following parameters:
[{customerFile=/data/customer.csv, outputFile=file:/Users/mminella/Documents/
IntelliJWorkspace/def-guide-spring-batch/Chapter9/target/formattedCustomers.xml}]
2018-05-03 21:55:01.299 INFO 61906 --- [main] o.s.batch.core.job.SimpleStepHandler:
Executing step: [jpaFormatStep]
Customer{id=0, firstName='Richard', middleInitial='N', lastName='Darrow', address='5570
Isabella Ave', city='St. Louis', state='IL', zip='58540'}
```

267

```
Customer{id=0, firstName='Warren', middleInitial='L', lastName='Darrow', address='4686 Mt.
Lee Drive', city='St. Louis', state='NY', zip='94935'}
Customer{id=0, firstName='Barack', middleInitial='G', lastName='Donnelly', address='7844 S.
Greenwood Ave', city='Houston', state='CA', zip='38635'}
Customer{id=0, firstName='Ann', middleInitial='Z', lastName='Benes', address='2447 S.
Greenwood Ave', city='Las Vegas', state='NY', zip='55366'}
Customer{id=0, firstName='Erica', middleInitial='Z', lastName='Gates', address='3141 Farnam
Street', city='Omaha', state='CA', zip='57640'}
Customer{id=0, firstName='Warren', middleInitial='M', lastName='Williams', address='6670 S.
Greenwood Ave', city='Hollywood', state='FL', zip='37288'}
Customer{id=0, firstName='Harry', middleInitial='T', lastName='Darrow', address='3273
Isabella Ave', city='Houston', state='FL', zip='97261'}
Customer{id=0, firstName='Steve', middleInitial='O', lastName='Darrow', address='8407
Infinite Loop Drive', city='Las Vegas', state='WA', zip='90520'}
2018-05-03 21:55:01.373 INFO 61906 --- [main] o.s.b.c.l.support.SimpleJobLauncher:
Job: [SimpleJob: [name=itemWriterAdapterFormatJob]] completedwith the following
parameters: [{customerFile=/data/customer.csv, outputFile=file:/Users/
mminella/Documents/IntelliJWorkspace/def-guide-spring-batch/Chapter9/target/
formattedCustomers.xml}] and the following status: [COMPLETED]
```

如你所料，利用 Spring Batch 可以轻松地在步骤中使用处理过的条目调用现有的服务。不过，如果服务接收的对象与处理的条目不同，该怎么办呢？如果想从条目中抽取值，然后传给服务，那么 Spring Batch 对此已经提供了支持。

9.5.2　PropertyExtractingDelegatingItemWriter

　　ItemWriterAdapter 的用法非常简单：把正在处理的条目传给已经存在的 Spring 服务即可。然而，很少有软件会那么简单直接。因此，Spring Batch 提供了从条目中抽取值并将它们作为参数传给服务的机制。下面介绍 PropertyExtractingDelegatingItemWriter（"属性抽取委托条目写入器"）以及如何将其应用到已存在的服务。

　　尽管名称很长，但是 org.springframework.batch.item.adapter.PropertyExtractingDelegatingItemWriter 与 ItemWriterAdapter 很类似。就像 ItemWriterAdapter，PropertyExtractingDelegatingItemWriter 也会调用引用的 Spring 服务上的特定方法。不同之处在于，ItemWriterAdapter 会不加处理地传递步骤正在处理的条目，而 PropertyExtractingDelegatingItemWriter 仅仅传递条目指定的属性。例如，如果有一个类型为 Product 的条目，里面包含了数据库中的 id、name、price 和 SKU 字段，那么在使用 ItemWriterAdapter 时必须把整个 Product 对象传给服务。但在使用 PropertyExtractingDelegatingItemWriter 时，可以指定只把数据库中的 id 和 price 字段作为参数传给服务。

　　下面使用我们熟悉的客户输入，在 CustomerService 中添加一个方法，以允许打印正在处理的客户条目的 address(地址)，并使用 PropertyExtractingDelegatingItemWriter 调用这个新的方法。代码清单 9-56 展示了更新后的 CustomerService。

代码清单 9-56　添加了 logCustomerAddress 方法的 CustomerService

```
...
@Service
public class CustomerService {

    public void logCustomer(Customer customer) {
```

第 9 章 ItemWriter

```
            System.out.println(customer);
    }

    public void logCustomerAddress(String address,
            String city,
            String state,
            String zip) {
        System.out.println(
                String.format("I just saved the address:\n%s\n%s, %s\n%s",
                        address,
                        city,
                        state,
                        zip));
    }
}
```

如代码清单 9-56 所示，我们添加了新的方法 logCustomerAddress(String address, String city, String state, String zip)。不过，这个新的方法并不接收 Customer 条目，而是接收条目指定的值。为了调用这个方法，可以使用 PropertyExtractingDelegatingItemWriter 从每个 Customer 条目中抽取与地址相关的字段(address、city、state 和 zip 字段)，并使用接收到的值调用服务。为了配置这个条目写入器，可以传入一个有序的属性列表(用于从条目中抽取属性)以及用于调用的目标对象和方法。传入的这个属性列表的顺序与属性所需参数的顺序相同。Spring 支持使用点分记号法(例如 address.city)和索引属性(例如 e-mail[5])。就像 ItemWriterAdapter 一样，这个条目写入器的实现也暴露了属性 arguments。由于条目写入器会动态地抽取参数，因此这里并没有使用 arguments 属性。代码清单 9-57 展示了更新后的作业，我们调用了 logCustomerAddress(String address, String city, String state, String zip)方法而不是处理整个 Customer 条目。

代码清单 9-57 将作业配置为调用 CustomerService 上的 logCustomerAddress 方法

```
...
@Bean
public PropertyExtractingDelegatingItemWriter<Customer> itemWriter(CustomerService
customerService) {
    PropertyExtractingDelegatingItemWriter<Customer> itemWriter =
            new PropertyExtractingDelegatingItemWriter<>();

    itemWriter.setTargetObject(customerService);
    itemWriter.setTargetMethod("logCustomerAddress");
    itemWriter.setFieldsUsedAsTargetMethodArguments(
            new String[] {"address", "city", "state", "zip"});
    return itemWriter;
}

@Bean
public Step formatStep() throws Exception {
    return this.stepBuilderFactory.get("formatStep")
            .<Customer, Customer>chunk(10)
            .reader(customerFileReader(null))
            .writer(itemWriter(null))
            .build();
```

```
}

@Bean
public Job propertiesFormatJob() throws Exception {
    return this.jobBuilderFactory.get("propertiesFormatJob")
                .start(formatStep())
                .build();
}
...
```

当运行整个作业时，将向 System.out 写入一行格式化的地址，如代码清单 9-58 所示。

代码清单 9-58　使用了 PropertyExtractingDelegatingItemWriter 之后的输出

```
2018-05-03 22:15:06.509 INFO 62192 --- [main] o.s.b.c.l.support.SimpleJobLauncher:
Job: [SimpleJob: [name=propertiesFormatJob]] launched with the following parameters:
[{customerFile=/data/customer.csv, outputFile=file:/Users/mminella/Documents/
IntelliJWorkspace/def-guide-spring-batch/Chapter9/target/formattedCustomers.xml}]
2018-05-03 22:15:06.523 INFO 62192 --- [main] o.s.batch.core.job.SimpleStepHandler:
Executing step: [formatStep]
I just saved the address:
5570 Isabella Ave
St. Louis, IL
58540
I just saved the address:
4686 Mt. Lee Drive
St. Louis, NY
94935
I just saved the address:
7844 S. Greenwood Ave
Houston, CA
38635
I just saved the address:
2447 S. Greenwood Ave
Las Vegas, NY
55366
I just saved the address:
3141 Farnam Street
Omaha, CA
57640
I just saved the address:
6670 S. Greenwood Ave
Hollywood, FL
37288
I just saved the address:
3273 Isabella Ave
Houston, FL
97261
I just saved the address:
8407 Infinite Loop Drive
Las Vegas, WA
90520
2018-05-03 22:15:06.598 INFO 62192 --- [main] o.s.b.c.l.support.SimpleJobLauncher:
Job: [SimpleJob: [name=propertiesFormatJob]] completed with the following parameters:
[{customerFile=/data/customer.csv, outputFile=file:/Users/mminella/Documents/
IntelliJWorkspace/def-guide-spring-batch/Chapter9/target/formattedCustomers.xml}]
```

```
and the following status: [COMPLETED]
2018-05-03 22:15:06.599 INFO 62192 --- [Thread-7] s.c.a.AnnotationConfigApplication-
Context : Closing org.springframework.context.annotation.AnnotationConfigApplicationContext
@22635ba0: startup date [Thu May 03 22:15:04 CDT 2018]; root of context hierarchy
```

Spring Batch 为你提供了把任何 Spring 服务作为 ItemWriter 进行复用的能力。这些代码在企业的生产环境中久经考验,复用它们不太可能引入新的漏洞,并且有益于加快开发速度。9.5.3 节将介绍如何在步骤中把 JMS 资源作为条目的目的地。

9.5.3 JmsItemWriter

Java 消息服务(Java Messaging Service,JMS)是一种用于在两个或多个端点之间进行通信的面向消息的方法。通过使用点对点通信(JMS 队列)或发布-订阅模型(JMS 主题),对于其他任何可能与消息实现进行交互的技术,Java 应用都可以与之通信。本小节将介绍如何使用 Spring Batch 的 JmsItemWriter 将消息添加到 JMS 队列中。

Spring 在简化一些常见的 Java 概念方面已经取得了巨大进步。JDBC 能够与不同的 ORM 框架进行集成就是这样的例子。但是,在简化 JMS 资源的接口方面,Spring 同样令人印象深刻。为了与 JMS 一起工作,你需要使用 JMS 代理(JMS Broker),这里使用了 Apache 的 ActiveMQ。

在使用 ActiveMQ 之前,需要把 ActiveMQ 和 Spring 的 JMS 依赖项添加到 POM 中。代码清单 9-5 列出了需要添加到 POM 中的依赖项。

代码清单 9-59　ActiveMQ 和 Spring JMS 所需的依赖项

```
...
<dependency>
    <groupId>org.springframework.boot</groupId>
    <artifactId>spring-boot-starter-activemq</artifactId>
</dependency>
<dependency>
    <groupId>org.apache.activemq</groupId>
    <artifactId>activemq-broker</artifactId>
</dependency>
...
```

现在就可以使用 ActiveMQ 了。不过,在编写代码之前,我们先看看作业的处理,因为这里与前面的示例略有不同。

在本章前面的示例中,有一个单独的步骤用于读取 customer.csv 文件并使用合适的 ItemWriter 写入输出。但是,对于现在这个示例来说,这还不够。如果读取条目,然后写入 JMS 队列,那么没有什么办法能够确认它们已正确进入队列,因为无法查看队列的内容。如图 9.6 所示,替代方式是在这个作业中使用两个步骤:第一个步骤读取 customer.csv 文件并写入 ActiveMQ 队列,第二个步骤读取队列并将记录写到 XML 文件中。

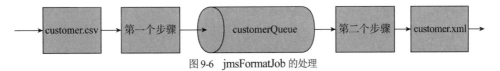

图 9-6　jmsFormatJob 的处理

特别需要注意的是：在真正的生产环境中不要这么做，因为在生产环境中不会在把消息全部放入队列后，才从队列中取出。这可能耗尽队列中的空间，具体取决于配置方式和可用的资源。不过，对于这个示例而言，由于要处理的客户数量很少，因此使用这种方式仅仅是为了演示要点。

- MessageConverter：负责把消息转换成 JSON 格式，用于在网络上传输。
- JmsTemplate：Spring Boot 尽管提供了自动配置的 JmsTemplate，但并没有提供使用 CachingConnectionFactory 的 JmsTemplate，后者才是使用 JmsTemplate 时的推荐方式。因此，我们需要配置自己的 JmsTemplate。

下面首先查看 MessageConverter。默认情况下，Spring Integration 会通过 Java 序列化自行处理消息(Message)对象的序列化。然而，这并不十分有用。作为替代，我们将配置 MessageConverter，并把传递的消息转换为 JSON，以便在 ActiveMQ 上进行传输。

JmsTemplate 是需要配置的下一个 Bean。虽然提供了 JmsTemplate，但是 Spring Boot 提供的 CachingConnectionFactory 并不能与 JmsTemplate 很好地协同工作。因此，我们需要配置自己的 JmsTemplate，从而利用 CachingConnectionFactory。

代码清单 9-60 展示了 JMS 资源的配置，它们位于 JmsJob.java 文件中。

代码清单 9-60　JMS 资源的配置

```
...
@Bean // Serialize message content to json using TextMessage
public MessageConverter jacksonJmsMessageConverter() {
    MappingJackson2MessageConverter converter = new MappingJackson2MessageConverter();
    converter.setTargetType(MessageType.TEXT);
    converter.setTypeIdPropertyName("_type");
    return converter;
}

@Bean
public JmsTemplate jmsTemplate(ConnectionFactory connectionFactory) {
    CachingConnectionFactory cachingConnectionFactory = new CachingConnectionFactory
        (connectionFactory);
    cachingConnectionFactory.afterPropertiesSet();

    JmsTemplate jmsTemplate = new JmsTemplate(cachingConnectionFactory);
    jmsTemplate.setDefaultDestinationName("customers");
    jmsTemplate.setReceiveTimeout(5000L);

    return jmsTemplate;
}
...
```

现在可以配置作业了。把到目前为止使用的那个读取器作为第一个步骤，而把本章前面的 XML 示例中使用的写入器作为第二个步骤。它们的配置如代码清单 9-61 所示。

代码清单 9-61　jmsFormatJob 作业的输入输出

```
...
@Bean
@StepScope
public FlatFileItemReader<Customer> customerFileReader(
```

第 9 章 ItemWriter

```
            @Value("#{jobParameters['customerFile']}")Resource inputFile) {

    return new FlatFileItemReaderBuilder<Customer>()
            .name("customerFileReader")
            .resource(inputFile)
            .delimited()
            .names(new String[] {"firstName",
                                "middleInitial",
                                "lastName",
                                "address",
                                "city",
                                "state",
                                "zip"})
            .targetType(Customer.class)
            .build();
}

@Bean
@StepScope
public StaxEventItemWriter<Customer> xmlOutputWriter(
        @Value("#{jobParameters['outputFile']}") Resource outputFile) {

    Map<String, Class> aliases = new HashMap<>();
    aliases.put("customer", Customer.class);

    XStreamMarshaller marshaller = new XStreamMarshaller();
    marshaller.setAliases(aliases);

    return new StaxEventItemWriterBuilder<Customer>()
            .name("xmlOutputWriter")
            .resource(outputFile)
            .marshaller(marshaller)
            .rootTagName("customers")
            .build();
}
...
```

JmsReader 与 JmsWriter 的配置相同。二者都是基本的 Spring Bean，并且引用了代码清单 9-60 中的 JmsTemplate。你在代码清单 9-62 中可以看到 JmsItemReader、JmsItemWriter 以及用来让所有的读取器和写入器一起工作的作业配置。

代码清单 9-62　JmsItemReader、JmsItemWriter 以及作业配置

```
...
@Bean
public JmsItemReader<Customer> jmsItemReader(JmsTemplate jmsTemplate) {

    return new JmsItemReaderBuilder<Customer>()
                    .jmsTemplate(jmsTemplate)
                    .itemType(Customer.class)
                    .build();
}

@Bean
public JmsItemWriter<Customer> jmsItemWriter(JmsTemplate jmsTemplate) {
```

```java
            return new JmsItemWriterBuilder<Customer>()
                    .jmsTemplate(jmsTemplate)
                    .build();
}

@Bean
public Step formatInputStep() throws Exception {

    return this.stepBuilderFactory.get("formatInputStep")
                .<Customer, Customer>chunk(10)
                .reader(customerFileReader(null))
                .writer(jmsItemWriter(null))
                .build();
}

@Bean
public Step formatOutputStep() throws Exception {

    return this.stepBuilderFactory.get("formatOutputStep")
                .<Customer, Customer>chunk(10)
                .reader(jmsItemReader(null))
                .writer(xmlOutputWriter(null))
                .build();
}

@Bean
public Job jmsFormatJob() throws Exception {

    return this.jobBuilderFactory.get("jmsFormatJob")
                .start(formatInputStep())
                .next(formatOutputStep())
                .build();
}
...
```

这就是所有的改动！在配置了这些资源之后，构建和运行这个作业的方式与前面执行过的其他作业相比并无二致。不过，在运行这个作业时需要注意，除了在第二个步骤执行之前查看作业存储库或浏览队列之外，并不会看到第一个步骤到底输出了什么。当查看第二个步骤生成的 XML 时，你可以看到消息已经按照预期成功传给了队列。代码清单 9-63 展示了这个作业生成的 XML 示例。

代码清单 9-63　JMS 版本的格式化作业生成的 XML 示例

```xml
<?xml version="1.0" encoding="UTF-8"?>
<customers>
    <customer>
        <id>0</id>
        <firstName>Ri1chard</firstName>
        <middleInitial>N</middleInitial>
        <lastName>Darrow</lastName>
        <address>5570 Isabella Ave</address>
        <city>St. Louis</city>
        <state>IL</state>
        <zip>58540</zip>
    </customer>
```

```
    <customer>
        <id>0</id>
        <firstName>Warren</firstName>
        <middleInitial>L</middleInitial>
        <lastName>Darrow</lastName>
        <address>4686 Mt. Lee Drive</address>
        <city>St. Louis</city>
        <state>NY</state>
        <zip>94935</zip>
    </customer>
    ...
</customers>
```

通过使用 Spring 的 JmsTemplate，Spring Batch 以最小的代价把 Spring 的 JMS 处理功能完全地暴露给了批处理过程。9.5.4 节将介绍一个你可能没有想到的写入器，它用于在批处理过程中发送 e-mail。

9.5.4　SimpleMailMessageItemWriter

邮件发送功能听起来很有用。当作业完成时，如果能收到一封电子邮件，表明事情进展顺利，可能会更方便。不过，这并不是 ItemWriter 的用法。你需要在步骤中为处理过的每个条目调用一次 ItemWriter。如果想执行自己的邮件群发操作，这么 ItemWriter 正好适用！本小节将介绍如何使用 Spring Batch 的 SimpleMailMessageItemWriter(简单邮件消息条目写入器)从作业中发送 e-mail。

尽管你可能不会使用这个条目写入器编写群发邮件的处理程序，但是你可以用它来做一些其他的事情。假设到现在为止我们处理过的文件都是真实的客户导入文件，你想在导入所有新的客户之后，给每一位新客户发送一封欢迎邮件。那么，使用 org.springframework.batch.item.mail. SimpleMailMessageItemWriter 将是最完美的方式。

这个示例与使用 JMS 的那个示例相同，也将使用一个包含两个步骤的作业。第一个步骤导入 customer.csv 文件到 CUSTOMER 表中。第二个步骤读取导入的所有客户，然后发送欢迎邮件。图 9-7 展示了这个作业的流程。

图 9-7　customerImport 作业的流程

在开始编码之前，我们先看看 SimpleMailMessageItemWriter。与其他条目写入器类似，这个条目写入器实现了 ItemWriter 接口，它的唯一方法 write(List<T> items)接收一个对象列表。然而，与你到目前为止见到的其他 ItemWriter 不同的是，SimpleMailMessageItemWriter 并不仅仅接收条目。除了邮件里的文本之外，发送 e-mail 时还需要更多的信息：主题(subject)、收件地址和发件地址。因此，SimpleMailMessageItemWriter 所需的对象列表中的对象必须扩展 Spring 的 SimpleMailMessage。这么做了以后，SimpleMailMessageItemWriter 就有了构建 e-mail 消息所需的所有信息。

但是，这是否意味着读入的任何条目都必须扩展 SimpleMailMessage 呢？这似乎是一个十分糟糕的耦合了 e-mail 功能和业务逻辑的作业——这也是你为什么不必这么做的原因。你是否还记得，第 8 章讨论了 ItemProcessor 返回的对象类型并不需要与接收的对象类型相同。例如，虽然接收到的是 Car

类型的对象，但可以返回 House 类型的对象。在这里，你可以创建一个 ItemProcessor 以接收 Customer 对象，然后返回所需的 SimpleMailMessage 对象。

为此，复用前面的客户输入文件，并在每一行添加一个新的字段：客户的电子邮件地址。代码清单 9-64 展示了将要处理的客户输入文件。

代码清单 9-64　customerWithEmail.csv

```
Ann,A,Smith,2501 Mt. Lee Drive,Miami,NE,62935,ASmith@yahoo.com
Laura,B,Jobs,9542 Isabella Ave,Aurora,FL,62344,LJobs@yahoo.com
Harry,J,Williams,1909 4th Street,Seatle,TX,48548,HWilliams@hotmail.com
Larry,Y,Minella,7839 S. Greenwood Ave,Miami,IL,65371,LMinella@hotmail.com
Richard,Q,Jobs,9732 4th Street,Chicago,NV,31320,RJobs@gmail.com
Ann,P,Darrow,4195 Jeopardy Lane,Aurora,CA,24482,ADarrow@hotmail.com
Larry,V,Williams,3075 Wall Street,St. Louis,NY,34205,LWilliams@hotmail.com
Michael,H,Gates,3219 S. Greenwood Ave,Boston,FL,24692,MGates@gmail.com
Harry,H,Johnson,7520 Infinite Loop Drive,Hollywood,MA,83793,HJohnson@hotmail.com
Harry,N,Ellison,6959 4th Street,Hollywood,MO,70398,HEllison@gmail.com
```

为了处理每个客户的电子邮件地址，你还需要为 Customer 对象添加 e-mail 字段。代码清单 9-65 展示了更新后的 Customer 类。

代码清单 9-65　添加了 e-mail 字段的 Customer.java

```java
package com.apress.springbatch.chapter9;

import java.io.Serializable;

import javax.persistence.Entity;
import javax.persistence.GeneratedValue;
import javax.persistence.GenerationType;
import javax.persistence.Id;
import javax.persistence.Table;

@Entity
@Table(name="CUSTOMER")
public class Customer implements Serializable {
    private static final long serialVersionUID = 1L;

    @Id
    @GeneratedValue(strategy = GenerationType.IDENTITY)
    private long id;
    private String firstName;
    private String middleInitial;
    private String lastName;
    private String address;
    private String city;
    private String state;
    private String zip;
    private String email;

    // Accessors go here
    ...
}
```

第 9 章 ItemWriter

由于作业把客户信息保存在了数据库中，因此我们快速地查看一下这个交互是如何进行的。首先，图 9-8 展示了本例使用的 CUSTOMER 表的数据模型。

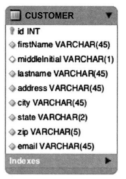

图 9-8　customer 表

在介绍任何与 email 相关的代码之前，我们需要先在 POM 中添加正确的依赖项。Spring Boot 使得给应用添加 e-mail 变得很简单，只需要添加一个依赖项，也就是代码清单 9-66 所示的邮件启动包。

代码清单 9-66　Java Mail 依赖项

```
...
<dependency>
        <groupId>org.springframework.boot</groupId>
        <artifactId>spring-boot-starter-mail</artifactId>
</dependency>
...
```

在添加了邮件启动包之后，还需要做一些配置，以便发送电子邮件。对于本例而言，可使用个人的 Gmail 账户，通过 Google 的 SMTP 服务器发送邮件[1]。代码清单 9-67 展示了需要在 application.properties 文件中配置的属性。

代码清单 9-67　Java Mail 配置

```
spring.mail.host=smtp.gmail.com
spring.mail.port=587
spring.mail.username=<SOME_USERNAME>
spring.mail.password=<SOME_PASSWORD>
spring.mail.properties.mail.smtp.auth=true
spring.mail.properties.mail.smtp.starttls.enable=true
```

在完成电子邮件的配置之后，就可以配置作业里的组件。首先是你应该很熟悉的三个 Bean，其中两个是 ItemReader，另一个是 ItemWriter，我们已经使用过它们。customerEmailFileReader 与我们使用过的其他 FlatFileItemReader 几乎相同，只是在每条记录中配置了一个额外的字段(email)。接下来是 customerBatchWriter，这个 Bean 是 JdbcBatchItemWriter，可配置为使用 JDBC 来写数据库。代码清

---

1 为了使用 Google 的 SMTP 服务器，可能需要开启"安全性较低的应用的访问权限"，更多信息详见 https://support.google.com/accounts/answer/6010255?hl=en(中文网址为 https://support.google.com/accounts/answer/6010255?hl=zh-Hans)。另外，虽然这里提到的是 Gmail，但只要支持 SMTP 服务器，任何邮件账户都可以使用。与 Gmail 类似，对于一些邮件服务商，可能需要开启一些安全性配置才能使用。

单 9-68 展示了为作业的第一个步骤所做的准备。

代码清单 9-68　第一个步骤的 ItemReader 和 ItemWriter

```java
...
@Bean
@StepScope
public FlatFileItemReader<Customer> customerEmailFileReader(
            @Value("#{jobParameters['customerFile']}")Resource inputFile) {

    return new FlatFileItemReaderBuilder<Customer>()
                .name("customerFileReader")
                .resource(inputFile)
                .delimited()
                .names(new String[] {"firstName",
                                     "middleInitial",
                                     "lastName",
                                     "address",
                                     "city",
                                     "state",
                                     "zip",
                                     "email"})
                .targetType(Customer.class)
                .build();
}

@Bean
public JdbcBatchItemWriter<Customer> customerBatchWriter(DataSource dataSource) {

    return new JdbcBatchItemWriterBuilder<Customer>()
                .namedParametersJdbcTemplate(new NamedParameterJdbcTemplate
                  (dataSource))
                .sql("INSERT INTO CUSTOMER (first_name, middle_initial, last_name, " +
                     "address, city, state, zip, email) " +
                     "VALUES(:firstName, :middleInitial, :lastName, " +
                     ":address, :city, :state, :zip, :email)")
                .beanMapped()
                .build();
}
...
```

在配置了第一个步骤的组件之后，接下来配置第二个步骤的条目读取器 customerCursorItemReader。customerCursorItemReader 是 JdbcCursorItemReader，用于返回 CUSTOMER 表中的所有数据。你将注意到，我们并没有开发自定义的 RowMapper，作为替代方式，我们使用 BeanPropertyRowMapper 把列名映射到 Bean 的 setter 方法，以映射数据库中的数据。

到目前为止，所有的配置对你来说都不陌生，因为你以前使用过它们。在配置第二个步骤的 ItemWriter 时，你将看到新的部分。在第二个步骤中，可使用 SimpleMailMessageItemWriter 作为 ItemWriter。代码清单 9-69 展示了第二个步骤所需的 Bean 配置以及作业配置。

代码清单 9-69　第二个步骤所需的 Bean 配置以及作业配置

```java
...
@Bean
public JdbcCursorItemReader<Customer> customerCursorItemReader(DataSource dataSource) {
```

第 9 章 ItemWriter

```java
        return new JdbcCursorItemReaderBuilder<Customer>()
                .name("customerItemReader")
                .dataSource(dataSource)
                .sql("select * from customer")
                .rowMapper(new BeanPropertyRowMapper<>(Customer.class))
                .build();
    }

    @Bean
    public SimpleMailMessageItemWriter emailItemWriter(MailSender mailSender) {
        return new SimpleMailMessageItemWriterBuilder()
                .mailSender(mailSender)
                .build();
    }

    @Bean
    public Step importStep() throws Exception {
        return this.stepBuilderFactory.get("importStep")
                .<Customer, Customer>chunk(10)
                .reader(customerEmailFileReader(null))
                .writer(customerBatchWriter(null))
                .build();
    }
    @Bean
    public Step emailStep() throws Exception {
        return this.stepBuilderFactory.get("emailStep")
                .<Customer, SimpleMailMessage>chunk(10)
                .reader(customerCursorItemReader(null))
                .processor((ItemProcessor<Customer, SimpleMailMessage>) customer -> {
                    SimpleMailMessage mail = new SimpleMailMessage();
                    mail.setFrom("prospringbatch@gmail.com");
                    mail.setTo(customer.getEmail());
                    mail.setSubject("Welcome!");
                    mail.setText(String.format("Welcome %s %s,\nYou were
                        imported into the system using Spring Batch!",
                            customer.getFirstName(), customer.
                                getLastName()));
                    return mail;
                })
                .writer(emailItemWriter(null))
                .build();
    }

    @Bean
    public Job emailJob() throws Exception {
        return this.jobBuilderFactory.get("emailJob")
                .start(importStep())
                .next(emailStep())
                .build();
    }
    ...
```

你将注意到，SimpleMailMessageItemWriter 只需要一个依赖项 MailSender，可由 Spring Boot 提供。在配置好 ItemWriter 之后，就可以配置步骤了。importStep 步骤与我们看到过的其他步骤类似，也提供了 ItemReader 和 ItemWriter。第二个步骤 emailStep 同样如此，不过，我们还使用了 lambda 表达式来配置 ItemProcessor，从而将 Customer 条目转换成 SimpleMailMessage(这是 ItemWriter 的输入类型)。这就是你要做的一切！在命令行中使用./mvnw clean install 命令构建作业，然后执行代码清单 9-70 中的命令，就可以处理输入文件并发出 e-mail。

代码清单 9-70　执行 e-mail 作业

```
java -jar itemWriters-0.0.1-SNAPSHOT.jar customerFile=/input/customerWithEmail.csv
```

在作业完成后，检查 e-mail 收件箱，如图 9-9 所示，就可以看到客户已经成功收到电子邮件。

图 9-9　e-mail 作业的结果

Spring Batch 提供了完整的 ItemWriter 集合，用于完成你所需要的绝大多数输出处理。9.6 节将介绍如何综合使用 ItemWriter 的各个特性，以处理复杂的输出场景，例如基于不同场景向多个位置写入数据。

9.6　复合的 ItemWriter

你的新系统的需求之一是抽取客户数据并生成两种不同的格式：销售部门的客户关系系统(Customer Relationship Management，CRM)应用需要 XML 文件，开票部门的数据库导入系统需要 CSV 文件。这里的问题是：需要抽取一百万条客户数据。

如果使用前面的工具，将不得不在这一百万个条目中循环两次(一次用于输出 XML 文件的步骤，另一次用于输出 CSV 文件的步骤)。也可创建自定义的 ItemWriter 实现，在处理了每个条目之后分别写入每个文件。前者耗时太长，会占用大量资源；后者需要编写和测试 Spring Batch 框架已经提供的某些功能。本节将介绍如何组合 Spring Batch 中的多种 ItemWriter，以处理更复杂的输出场景。

9.6.1　MultiResourceItemWriter

第 7 章介绍了 Spring Batch 支持在一个步骤中读取多个格式相同的文件。Spring Batch 在条目写入端也提供了类似的特性。本小节将介绍根据写入文件的条目生成多个资源的方式。

Spring Batch 为你提供了在处理一定数量的记录之后创建新资源的能力。例如，抽取所有的客户记录，并将它们写入 XML 文件，每个 XML 文件只能包含 10 个客户。为此，可使用 MultiResourceItemWriter(多资源的条目写入器)。

第 9 章　ItemWriter

MultiResourceItemWriter 能够基于已经处理的记录数量动态地创建输出资源,它把自己处理的每个条目传给了一个委托的写入器,真正的写入操作就由这个委托的写入器完成。MultiResourceItemWriter 只负责维护当前的计数,并在处理条目的过程中创建新的资源。图 9-10 展示了一个使用 org.springframework.batch.item.file.MultiResourceItemWriter 的步骤。

图 9-10　使用 MultiResourceItemWriter 进行处理

MultiResourceItemWriter 的 write(List<T>)方法用来验证当前的资源已经创建并且是打开的(否则就会创建资源并打开一个新的文件),然后把条目传给委托的写入器。一旦写入条目,就检查写入文件的条目数量是否已经达到已配置的要创建文件的阈值。如果达到了,那么关闭当前的文件。

注意,当 MultiResourceItemWriter 正在处理数据时,我们不会在创建处理块的中间创建资源,而是等待块的结束,然后创建新的资源。例如,如果把写入器配置为在处理 15 个条目后滚动创建文件,但是每个块的大小被配置为 20,那么 MultiResourceItemWriter 会在写入块中的 20 个条目之后,才会创建新的资源。

MultiResourceItemWriter 有 5 个依赖项可以进行配置,表 9-4 展示了它们的用法。

表 9-4　MultiResourceItemWriter 的配置选项

选项	类型	默认值	描述
delegate	ResourceAwareItemWriterItemStream	null(必选)	MultiResourceItemWriter 用于写入每个条目委托的写入器
itemCountLimitPerResource	int	Integer.MAX_VALUE	写入每个资源的最大条目数量
resource	Resource	null(必选)	MultiResourceItemWriter 创建的资源的原型
resourceSuffixCreator	ResourceSuffixCreator	null	MultiResourceItemWriter 可以使用这个类为创建的文件名附加后缀
saveState	boolean	true	如果为 false,那就不在作业存储库中维护 ItemWriter 的状态

下面从数据库中抽取客户信息并创建一些 XML 文件,每个 XML 文件包含 10 个客户。我们不需要开发任何新的代码(XML 已经创建好了),只需要把所有东西连接起来即可。让我们先看看作业的配置。

代码清单 9-71 展示了本例中条目读取器的配置。在本例中，条目读取器是配置为选取所有客户的简单 JdbcCursorItemReader。在此基础上，把来自数据库的客户传给接下来将要配置的条目写入器。

代码清单 9-71　用于多资源的格式化作业的条目读取器

```
...
@Bean
public JdbcCursorItemReader<Customer> customerJdbcCursorItemReader(DataSource dataSource) {

    return new JdbcCursorItemReaderBuilder<Customer>()
            .name("customerItemReader")
            .dataSource(dataSource)
            .sql("select * from customer")
            .rowMapper(new BeanPropertyRowMapper<>(Customer.class))
            .build();
}
...
```

条目写入器的配置是分层的。首先配置用于生成 XML 的 StaxEventItemWriter。然后把 MultiResourceItemWriter 放在顶层，用于生成 StaxEventItemWriter 写入的多个资源。代码清单 9-72 展示了条目写入器以及步骤和作业的配置。

代码清单 9-72　条目写入器以及步骤和作业的配置

```
...
@Bean
@StepScope
public StaxEventItemWriter<Customer> delegateItemWriter() throws Exception {

    Map<String, Class> aliases = new HashMap<>();
    aliases.put("customer", Customer.class);

    XStreamMarshaller marshaller = new XStreamMarshaller();

    marshaller.setAliases(aliases);

    marshaller.afterPropertiesSet();

    return new StaxEventItemWriterBuilder<Customer>()
            .name("customerItemWriter")
            .marshaller(marshaller)
            .rootTagName("customers")
            .build();
}

@Bean
public MultiResourceItemWriter<Customer> multiCustomerFileWriter() throws Exception {

    return new MultiResourceItemWriterBuilder<Customer>()
                .name("multiCustomerFileWriter")
                .delegate(delegateItemWriter())
                .itemCountLimitPerResource(25)
                .resource(new FileSystemResource("Chapter9/target/customer"))
                .build();
}
```

第 9 章 ItemWriter

```
@Bean
public Step multiXmlGeneratorStep() throws Exception {

    return this.stepBuilderFactory.get("multiXmlGeneratorStep")
                .<Customer, Customer>chunk(10)
                .reader(customerJdbcCursorItemReader(null))
                .writer(multiCustomerFileWriter())
                .build();
}
@Bean
public Job xmlGeneratorJob() throws Exception {

    return this.jobBuilderFactory.get("xmlGeneratorJob")
                .start(multiXmlGeneratorStep())
                .build();
}
...
```

我们首先配置的 delegateItemWriter 用于生成所需的 XML。尽管与其他示例中的 StaxEventItemWriter 类似,但是 delegateItemWriter 并没有直接引用输出文件。相反,multiResourceItemWriter 在后边需要时提供了对应的引用。

在这个示例中,multiCustomerFileWriter 使用了三个依赖项：作为文件写入位置(目录和文件名)模板的资源(Resource),在创建的文件中完成实际写入工作的 delegateItemWriter,以及这个条目写入器在每个文件中写入的客户数量(itemCountLimitPerResource)——在本例中是 25。要做的最后一部分工作是进行步骤和作业的配置,以便综合使用它们。作业本身的配置很直观,如代码清单 9-72 所示。为了使用这个作业,可以执行代码清单 9-73 中的命令。

代码清单 9-73 用于执行 multiResource 作业的命令

```
java -jar itemWriters-0.0.1-SNAPSHOT.jar
```

当查看这个作业的输出时,如果查看/output 目录,就会发现每个文件包含 10 个加载自数据库的客户。不过,Spring Batch 确实做了一些有意思的事情。首先要注意,我们并没有给作业传递输出文件的扩展名。这是有原因的。如果查看输出文件所在的目录(如代码清单 9-74 所示),就会看到 MultiResourceItemWriter 给每个文件名添加了 .X 后缀,X 就是所创建文件的序号。

代码清单 9-74 由这个作业创建的文件

```
michael-minellas-macbook-pro:temp mminella$ ls Chapter9/target/customer
customer.1  customer.2  customer.3  customer.4
```

尽管区分每个文件是有意义的,但是对于如何命名文件,这可能是一种可行的解决方法,也可能不是(很多编辑器在默认情况下并不能识别这些文件)。因此,Spring Batch 决定由你来配置创建的每个文件的后缀。通过实现 org.springframework.batch.item.file.ResourceSuffixCreator 接口,并将它作为 multiResourceItemWriterBean 的一个依赖项,就可以修改文件名的后缀。当使用 MultiResourceItemWriter 创建新的文件时,可使用 ResourceSuffixCreator 来生成后缀,后缀会附加到新文件的名称的末尾。代码清单 9-75 展示了用于本例的后缀创建器。

代码清单 9-75　CustomerOutputFileSuffixCreator

```
...
@Component
public class CustomerOutputFileSuffixCreator implements ResourceSuffixCreator {
    @Override
    public String getSuffix(int arg0) {
        return arg0 + ".xml";
    }
}
```

代码清单 9-75 实现了 ResourceSuffixCreator 的唯一方法 getSuffix,并返回了由传入的数字和.xml 扩展组成的后缀。传入的数字是正在创建的文件的序号。如果想返回默认的扩展名,那么可以返回一个英文句点加上传入的数字。

为了使用 CustomerOutputFileSuffixCreator,可将其配置为一个 Bean,并且通过设置属性 resourceSuffixCreator,把它添加为 multiResourceItemWriter 的依赖项,如代码清单 9-76 所示。

代码清单 9-76　配置 CustomerOutputFileSuffixCreator

```
...
@Bean
public MultiResourceItemWriter<Customer> multiCustomerFileWriter
(CustomerOutputFileSuffixCreator suffixCreator) throws Exception {

    return new MultiResourceItemWriterBuilder<Customer>()
                .name("multiCustomerFileWriter")
                .delegate(delegateItemWriter())
                .itemCountLimitPerResource(25)
                .resource(new FileSystemResource("Chapter9/target/customer"))
                .resourceSuffixCreator(suffixCreator)
                .build();
}
...
```

使用代码清单 9-76 所示的配置再次执行这个作业,就会得到略微不同的结果,如代码清单 9-77 所示。

代码清单 9-77　使用 ResourceSuffixCreator 执行这个作业的结果

```
michael-minellas-macbook-pro:output mminella$ ls Chapter9/target/customer
customer1.xml       customer2.xml       customer3.xml       customer4.xml
```

XML 文件中的头片段和尾片段

在创建文件时,无论是用于步骤或作业的单个文件,还是你在上一个示例中看到的多个文件,在文件中生成头片段(Header)和尾片段(Footer)是很常见的需求。可以使用头片段来定义平面文件的格式(文件中的字段及其顺序),或者在 XML 中包含单独的、与条目无关的片段。尾片段可能包含已处理过的文件中的记录数,或者包含在处理完文件后用于完整性检查的记录总数。下面将介绍如何使用 Spring Batch 的回调机制生成头片段和尾片段。

在打开或关闭文件时,Spring Batch 为你提供了向文件添加头片段或尾片段(无论使用哪种方式)的能力。根据文件是平面文件还是 XML 文件,为文件添加头片段或尾片段意味着做不同的事情。对于平面文件而言，添加头片段意味着在文件的顶部或底部添加一条或多条记录。由于为平面文件生成纯文本与为 XML 文件生成 XML 片段是不同的，Spring Batch 提供了两个不同的接口用于帮助达成这个目的。我们首先看看用于 XML 文件的回调接口 org.springframework.batch.item.xml.StaxWriterCallback。

StaxWriterCallback 接口包含唯一的 write(XMLEventWriter writer)方法，以便给当前的 XML 文档添加 XML 片段。基于具体的配置，Spring Batch 会在头片段或尾片段执行已经配置过的回调。在下面的示例中，我们将编写一个 StaxWriterCallback 实现，用它添加一个 XML 片段，其中包含编写这个作业的人的名字，如代码清单 9-78 所示。

代码清单 9-78　CustomerXmlHeaderCallback

```
...
@Component
public class CustomerXmlHeaderCallback implements StaxWriterCallback {

    @Override
    public void write(XMLEventWriter writer) throws IOException {
        XMLEventFactory factory = XMLEventFactory.newInstance();

        try {
            writer.add(factory.createStartElement("", "", "identification"));
            writer.add(factory.createStartElement("", "", "author"));
            writer.add(factory.createAttribute("name", "Michael Minella"));
            writer.add(factory.createEndElement("", "", "author"));
            writer.add(factory.createEndElement("", "", "identification"));
        } catch (XMLStreamException xmlse) {
            System.err.println("An error occured: " + xmlse.getMessage());
            xmlse.printStackTrace(System.err);
        }
    }
}
```

代码清单 9-78 展示了 CustomerXmlHeaderCallback，这个回调向 XML 添加了两个标签：标识(identification)和作者(author)。作者标签包含唯一的属性 name，属性值为 Michael Minella。为了创建作者标签，我们使用了 javax.xml.stream.XMLEventFactory 的 createStartElement 和 createEndElement 方法，其中的每个方法都接收三个参数：前缀、命名空间和标签名。由于我们并不使用前缀和命名空间，因此可为它们传入空的字符串。为了使用这个实现，你需要把这个回调配置为 StaxEventItemWriter 的 headerCallback，如代码清单 9-79 所示。

代码清单 9-79　CustomerXmlHeaderCallback 的 XML 配置

```
...
@Bean
@StepScope
public StaxEventItemWriter<Customer> delegateItemWriter(CustomerXmlHeaderCallback
headerCallback) throws Exception {

    Map<String, Class> aliases = new HashMap<>();
    aliases.put("customer", Customer.class);

    XStreamMarshaller marshaller = new XStreamMarshaller();
```

```
    marshaller.setAliases(aliases);

    marshaller.afterPropertiesSet();

    return new StaxEventItemWriterBuilder<Customer>()
            .name("customerItemWriter")
            .marshaller(marshaller)
            .rootTagName("customers")
            .headerCallback(headerCallback)
                .build();
}
```

当把代码清单 9-79 中的头片段配置添加到前面示例中的 MultiResourceJob 之后，如果执行这个作业，就会看到输出文件以代码清单 9-80 所示的 XML 片段开始。

代码清单 9-80　XML 头片段

```
<?xml version="1.0" encoding="UTF-8"?>
<customers>
<identification>
<author name="Michael Minella"/>
</identification>
<customer>
...
```

如你所见，在 XML 文件中添加头片段或尾片段非常简单：实现 StaxWriterCallback 接口，然后在 ItemWriter 中将其配置为头片段或尾片段即可。

平面文件中的头记录和尾记录

下面看看如何在平面文件中添加头记录和尾记录。与使用同一个接口生成 XML 头片段和尾片段不同，在平面文件中写入头记录和尾记录时使用的是不同的接口。对于头记录，需要实现 org.springframework.batch.item.file.FlatFileHeaderCallback 接口；对于尾记录，需要实现 org.springframework.batch.item.file.FlatFileFooterCallback 接口。这两个接口分别只包含一个方法，它们是 writeHeader(Writer writer)和writeFooter(Writer writer)。下面介绍如何编写尾记录，尾记录中的内容是你在当前文件中处理的记录总数。

这个示例将使用 MultiResourceItemWriter 在每个文件中写入 30 条格式化的记录和一条尾记录，这条尾记录声明了你在每个文件中写入的记录总数。为了保存写入文件的记录总数，可以使用几个切面(Aspect)。第一个切入点(Pointcut)位于对 FlatFileItemWriter.open(ExecutionContext ec)方法的任何调用之前。我们将使用这个切入点在打开每个文件时重置计数器。第二个切入点位于对 FlatFileItemWriter.write(List<T> items)方法的任何调用之前，在这里我们会增加计数器的值。

你现在可能想知道，为什么不使用 ItemWriteListener.beforeWrite(List<T> items)调用，而是突然使用切面呢？原因在于调用的顺序。对 beforeWrite(List<T> items)的调用发生在对 write(List<T> items)的调用之前。但是，对 open(ExecutionContext ec)的调用发生在方法 write(List<T> items)中。由于我们需要在调用 write(List<T> items)之前重置计数器，因此使用了切面。

接下来要做的是创建一个组件，它既是切面，也是 FlatFileFooterCallback 的实现。切面负责管理回调的状态(在当前文件中写入了多少条记录)，而 FlatFileFooterCallback.writeFooter(Writer writer)方法负责写出结果,如代码清单 9-81。

代码清单 9-81　CustomerRecordCountFooterCallback

```
...
@Component
@Aspect
public class CustomerRecordCountFooterCallback implements FlatFileFooterCallback {

    private int itemsWrittenInCurrentFile = 0;

    @Override
    public void writeFooter(Writer writer) throws IOException {
        writer.write("This file contains " +
                        itemsWrittenInCurrentFile + " items");
    }

    @Before("execution(* org.springframework.batch.item.file.FlatFileItemWriter.open(..))")
    public void resetCounter() {
        this.itemsWrittenInCurrentFile = 0;
    }

    @Before("execution(* org.springframework.batch.item.file.FlatFileItemWriter.write(..))")
    public void beforeWrite(JoinPoint joinPoint) {
        List<Customer> items = (List<Customer>) joinPoint.getArgs()[0];

        this.itemsWrittenInCurrentFile += items.size();
    }
}
```

如代码清单 9-81 所示,CustomerRecordCountFooterCallback 使用了 @Component 和 @Aspect 注解:前者用于将这个回调装配为 Spring 的 Bean,后者用于指明它是一个切面。这个回调实现了 FlatFileFooterCallback 及其 writeFooter(Writer writer)方法,这个方法用于写入真正的尾记录。接下来的两个方法是切面方法。其中:resetCounter 方法会在调用 FlatFileItemWriter 的 open(ExecutionContext ec)方法之前调用,用于把当前的计数重置为 0,并且对于每个文件会被调用一次;beforeWrite(List<T> items)方法则使用传给 FlatFileItemWriter.write(List<T> items)方法的记录数量来增加计数。

为了使用这个回调,我们需要更新 MultiResourceJob 的两个地方。首先,需要把 FlatFileItemWriter 作为委托,以替代前面使用的基于 XML 的 ItemWriter;然后,需要配置 ItemWriter 以使用这个回调,如代码清单 9-82 所示。

代码清单 9-82　delegateCustomerItemWriter

```
@Bean
@StepScope
public FlatFileItemWriter<Customer> delegateCustomerItemWriter
(CustomerRecordCountFooterCallback footerCallback) throws Exception {
    BeanWrapperFieldExtractor<Customer> fieldExtractor = new
        BeanWrapperFieldExtractor<>();
    fieldExtractor.setNames(new String[] {"firstName", "lastName", "address", "city",
        "state", "zip"});
```

```
fieldExtractor.afterPropertiesSet();

FormatterLineAggregator<Customer> lineAggregator = new FormatterLineAggregator<>();

lineAggregator.setFormat("%s %s lives at %s %s in %s, %s.");
lineAggregator.setFieldExtractor(fieldExtractor);

FlatFileItemWriter<Customer> itemWriter = new FlatFileItemWriter<>();

itemWriter.setName("delegateCustomerItemWriter");
itemWriter.setLineAggregator(lineAggregator);
itemWriter.setAppendAllowed(true);
itemWriter.setFooterCallback(footerCallback);

return itemWriter;
}
```

MultiResourceItemWriter 使得基于每个文件中的记录数量把条目写入多个文件变得十分容易。Spring 为你提供了添加头记录或尾记录的能力,也可以通过使用合适的接口和配置,以简单和实用的方式对它们进行管理。9.6.2 节将介绍如何把相同的条目写入多个写入器,并且不用添加任何代码。

9.6.2 CompositeItemWriter

虽然看起来可能不是这样,但是到目前为止,我们介绍的示例都很简单:它们把输出写入同一个位置。这个位置可能是数据库、文件、e-mail 等。然而,实际情况并不总是这么简单。企业可能需要写入 Web 应用使用的数据库以及数据仓库。在处理条目时,可能需要记录各种各样的商业指标。Spring Batch 允许在处理步骤中的每个条目时,将其写到多个地方。本小节将介绍 CompositeItemWriter(组合条目写入器)如何让步骤把条目写入多个 ItemWriter。

与 Spring Batch 中的大部分事情类似,为处理的每个条目调用多个 ItemWriter 非常容易。不过,在编写代码之前,我们先看看把相同的条目写入多个 ItemWriter 的流程,如图 9-11 所示。

读取操作都是一个一个地进行,处理也是这样。此外,图 9-11 展示了写入操作是一个块一个块地进行的(跟预期的一样)。在写入时,使用当前块中的条目调用每个 ItemWriter,并且按照它们的配置顺序进行调用。

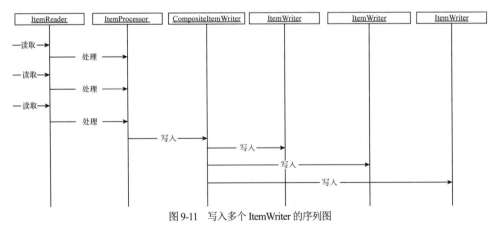

图 9-11 写入多个 ItemWriter 的序列图

第 9 章 ItemWriter

下面创建一个作业，读取本章前面使用的 customerWithEmail.csv 文件。代码清单 9-83 展示了如何读取 customerWithEmail.csv 文件。

代码 9-83　读取 customerWithEmail.csv 文件

```
...
@Bean
@StepScope
public FlatFileItemReader<Customer> compositewriterItemReader(
            @Value("#{jobParameters['customerFile']}") Resource inputFile) {

    return new FlatFileItemReaderBuilder<Customer>()
                .name("compositewriterItemReader")
                .resource(inputFile)
                .delimited()
                .names(new String[] {"firstName",
                                "middleInitial",
                                "lastName",
                                "address",
                                "city",
                                "state",
                                "zip",
                                "email"})
                .targetType(Customer.class)
                .build();
}
...
```

你对代码清单 9-83 中的代码应该不会感到陌生。本章前面的示例已经使用过相同的输入文件。代码清单 9-83 中的配置还包括了 FlatFileItemReader 的配置，以使用 DelimitedLineTokenizer(通过调用.delimited()方法)和 BeanWrapperFieldSetMapper(通过调用.targetType(Customer.class)方法)读取文件。

在输出端，需要创建三个 ItemWriter：XML 写入器及其依赖项、JDBC 写入器及其依赖项，以及封装了前面两者的 CompositeItemWriter 写入器。代码清单 9-84 展示了输出以及步骤和作业的配置。

代码清单 9-84　输出以及步骤和作业的配置

```
...
@Bean
@StepScope
public StaxEventItemWriter<Customer> xmlDelegateItemWriter(
            @Value("#{jobParameters['outputFile']}") Resource outputFile) throws
            Exception {
    Map<String, Class> aliases = new HashMap<>();
    aliases.put("customer", Customer.class);

    XStreamMarshaller marshaller = new XStreamMarshaller();

    marshaller.setAliases(aliases);

    marshaller.afterPropertiesSet();

    return new StaxEventItemWriterBuilder<Customer>()
                .name("customerItemWriter")
                .resource(outputFile)
```

289

```
                    .marshaller(marshaller)
                    .rootTagName("customers")
                    .build();
}

@Bean
public JdbcBatchItemWriter<Customer> jdbcDelgateItemWriter(DataSource dataSource) {
    return new JdbcBatchItemWriterBuilder<Customer>()
            .namedParametersJdbcTemplate(new NamedParameterJdbcTemplate
            (dataSource))
            .sql("INSERT INTO CUSTOMER (first_name, " +
                    "middle_initial, " +
                    "last_name, " +
                    "address, " +
                    "city, " +
                    "state, " +
                    "zip, " +
                    "email) " +
                    "VALUES(:firstName, " +
                    ":middleInitial, " +
                    ":lastName, " +
                    ":address, " +
                    ":city, " +
                    ":state, " +
                    ":zip, " +
                    ":email)")
            .beanMapped()
            .build();
}

@Bean
public CompositeItemWriter<Customer> compositeItemWriter() throws Exception {
    return new CompositeItemWriterBuilder<Customer>()
            .delegates(Arrays.asList(xmlDelegateItemWriter(null),
                    jdbcDelgateItemWriter(null)))
            .build();
}

@Bean
public Step compositeWriterStep() throws Exception {
    return this.stepBuilderFactory.get("compositeWriterStep")
            .<Customer, Customer>chunk(10)
            .reader(compositewriterItemReader(null))
            .writer(compositeItemWriter())
            .build();
}

@Bean
public Job compositeWriterJob() throws Exception {
    return this.jobBuilderFactory.get("compositeWriterJob")
            .start(compositeWriterStep())
            .build();
}
...
```

这些 ItemWriter 的配置应该已经符合你的预期。首先是 XML 写入器(xmlDelegateItemWriter)的配置，这里与本章前面示例中的配置相同。接下来是 JDBC 写入器，这里使用命名参数配置了 PreparedStatement，并注入了 NamedParameterTemplate 以处理命名参数。在调用.beanMapped()方法后，Spring Batch 就会把条目中的字段名称映射为 SQL 语句中的字段名称。最后是 CompositeItemWriter 的定义(compositeItemWriter)。对于 compositeItemWriter，这里配置了一个 ItemWriter 列表，以便由封装器调用。需要注意的是，ItemWriter 是按照它们的配置顺序进行调用的，并在调用时使用块中的所有条目。所以，如果块中有 10 个条目，那么将使用所有的 10 个条目调用第一个 ItemWriter，然后调用第二个 ItemWriter，以此类推。你还需要注意，尽管写操作是串行执行的(每次只有一个写入器)，但是所有 ItemWriter 的所有写操作都发生在相同的事务中。因此，如果块中的任何条目在任何时间写入失败，那么整个块都会回滚。

当使用命令 java -jar itemWriters-0.0.1-SNAPSHOT.jar customerFile=/input/customerWithEmail.csv outputFile=/output/xmlCustomer.xml 执行配置后的作业时，就可以在输出中看到，所有记录同时被写入数据库和 XML 文件。你应该会想到，如果写入 100 个客户，Spring Batch 会认为写入了 200 个条目。但是，如果在作业存储库中查看 Spring Batch 的处理记录，就会看到写入了 100 个条目。

原因是 Spring Batch 计算的是写入的条目，而不关心把条目写到了什么地方。如果作业失败，重启地点取决于读入和处理了多少条目，而不是在每个位置写入了多少条目(因为它们最后都会回滚)。

CompositeItemWriter 简化了将所有条目写入多个位置的实现。但是，有时候你可能想把一些东西写到一个地方，而把另一些东西写到另一个地方。你在本章中将要了解的最后一个条目处理器应对的正是这种情况，它就是 ClassifierCompositeItemWriter。

9.6.3　ClassifierCompositeItemWriter

在第 7 章，你看到过在一个文件中包含多种记录类型的场景。将不同类型的行映射到不同的转换器和映射器，以便得到正确的对象，这并不是一项简单的任务。但是，在写入端，Spring Batch 让事情变得容易了。本小节将介绍如何基于预定的条件，使用 ClassifierCompositeItemWriter 选择写入条目的位置。

org.springframework.batch.item.support.ClassifierCompositeItemWriter(分类复合 ItemWriter)可用于查看不同类型的条目，决定写入哪个 ItemWriter，然后进行相应的转发。以上功能主要基于 ClassifierCompositeItemWriter 和 org.springframework.batch.classify.Classifier 接口的实现。下面首先看看 Classifier 接口。

Classifier 接口只有一个方法 classify(C item)，如代码清单 9-85 所示。当 ClassifierCompositeItemWriter 使用Classifier接口的实现时，classify(C item)方法接收一个条目作为输入，并返回用于写入这个条目的 ItemWriter。在本质上，Classifier 接口的实现充当了上下文，这些 ItemWriter 则充当了策略的实现。

代码清单 9-85　Classifier 接口

```
package org.springframework.batch.classify;
public interface Classifier<C, T> {

    T classify(C classifiable);
}
```

ClassifierCompositeItemWriter 只接收一个依赖项——Classifier 接口的一个实现,在处理过程中,将从这个实现获取每个条目所需的 ItemWriter。

与常见的把所有条目写入所有 ItemWriter 的 CompositeItemWriter 不同,ClassifierCompositeItemWriter 最后会在每个 ItemWriter 中写入不同数量的条目。举个例子,根据居住地所在州的缩写,以字母 A~M 的顺序把客户写入平面文件,以字母 N~Z 的顺序把客户写入数据库。

你可能已经猜到,Classifier 接口的实现是使 CompositeItemWriter 工作的关键,所以我们先从 Classifier 接口开始。为了实现代码清单 9-86 所示的 CustomerClassifier,我们需要给 classify(C item)方法传入一个客户对象作为唯一的参数。然后,我们可以使用正则表达式来决定应该把条目写入平面文件还是数据库,之后返回所需的 ItemWriter。

代码清单 9-86　CustomerClassifier

```
...
public class CustomerClassifier implements
        Classifier<Customer, ItemWriter<? super Customer>> {

    private ItemWriter<Customer> fileItemWriter;
    private ItemWriter<Customer> jdbcItemWriter;

    public CustomerClassifier(StaxEventItemWriter<Customer> fileItemWriter,
    JdbcBatchItemWriter<Customer> jdbcItemWriter) {
        this.fileItemWriter = fileItemWriter;
        this.jdbcItemWriter = jdbcItemWriter;
    }

    @Override
    public ItemWriter<Customer> classify(Customer customer) {

        if(customer.getState().matches("^[A-M].*")) {
            return fileItemWriter;
        } else {
            return jdbcItemWriter;
        }
    }
}
```

在编写了 CustomerClassifier 之后,就可以配置作业和 ItemWriter 了。复用之前的 CompositeItemWriter 示例中使用的输入和每个独立的 ItemWriter,然后只需要配置 ClassifierCompositeItemWriter 即可。ClassifierCompositeItemWriter 和 CustomerClassifier 的配置如代码清单 9-87 所示。

代码清单 9-87　ClassifierCompositeItemWriter 和 CustomerClassifier 的配置

```
...
@Bean
public ClassifierCompositeItemWriter<Customer> classifierCompositeItemWriter()
throws Exception {
        Classifier<Customer, ItemWriter<? super Customer>> classifier = new CustomerClassifier
            (xmlDelegate(null), jdbcDelgate(null));
        return new ClassifierCompositeItemWriterBuilder<Customer>()
                .classifier(classifier)
                .build();
```

```
}

@Bean
public Step classifierCompositeWriterStep() throws Exception {
    return this.stepBuilderFactory.get("classifierCompositeWriterStep")
            .<Customer, Customer>chunk(10)
            .reader(classifierCompositeWriterItemReader(null))
            .writer(classifierCompositeItemWriter())
            .build();
}

@Bean
public Job classifierCompositeWriterJob() throws Exception {
    return this.jobBuilderFactory.get("classifierCompositeWriterJob")
            .start(classifierCompositeWriterStep())
            .build();
}
...
```

构建 classifierFormatJob 作业，并使用命令 java -jar itemWriters-0.0.1-SNAPSHOT.jar jobs/formatJob.xml formatJob customerFile=/input/customerWithEmail.csv outputFile=/output/xmlCustomer.xml 执行作业，你会遇到一些意外。Spring 并没有输出作业已完成的信息，而是抛出了异常，如代码清单 9-88 所示。

代码清单 9-88　classifierFormatJob 作业的输出

```
2018-05-10 22:51:23.691 INFO 11102 --- [main] o.s.b.c.l.support.SimpleJobLauncher:
Job: [SimpleJob: [name=classifierCompositeWriterJob]] launched with the following
parameters: [{customerFile=/data/customerWithEmail.csv, outputFile=file:/Users/
mminella/Documents/IntelliJWorkspace/def-guide-spring-batch/Chapter9/target/
formattedCustomers.xml}]
2018-05-10 22:51:23.701 INFO 11102 --- [main] o.s.batch.core.job.SimpleStepHandler:
Executing step: [classifierCompositeWriterStep]
2018-05-10 22:51:23.900 ERROR 11102 --- [main] o.s.batch.core.step.AbstractStep:
Encountered an error executing step classifierCompositeWriterStep in job
classifierCompositeWriterJob

org.springframework.batch.item.WriterNotOpenException: Writer must be open before it can be
written to
  at org.springframework.batch.item.xml.StaxEventItemWriter.write(StaxEventItemWriter.
  java:761) ~[spring-batch-infrastructure-4.0.1.RELEASE.jar:4.0.1.RELEASE]
  at org.springframework.batch.item.xml.
  StaxEventItemWriter$$FastClassBySpringCGLIB$$d105dd1.invoke(<generated>)~[spring-
  batch-infrastructure-4.0.1.RELEASE.jar:4.0.1.RELEASE]
  at org.springframework.cglib.proxy.MethodProxy.invoke(MethodProxy.java:204)~[springcore-
  5.0.5.RELEASE.jar:5.0.5.RELEASE]
  at org.springframework.aop.framework.CglibAopProxy$CglibMethodInvocation.
  invokeJoinpoint(CglibAopProxy.java:747)~[spring-aop-5.0.5.RELEASE.jar:5.0.5.RELEASE]
  at org.springframework.aop.framework.ReflectiveMethodInvocation.proceed
  (ReflectiveMethodInvocation.java:163) [spring-aop-5.0.5.RELEASE.jar:5.0.5.RELEASE]
```

到底是哪里出错了呢？所做的改动仅仅是把之前使用的 CompositeItemWriter 换成新的 ClassifierCompositeItemWriter 而已。问题的关键在于 ItemStream 接口。

ItemStream 接口

ItemStream 接口充当用于定期存储和恢复状态的契约，其中包括三个方法：open(ExecutionContext ec)、update(ExecutionContext ec)和 close。Spring Batch 中任何有状态的组件都实现了 ItemStream 接口。在这种情况下，输入输出都用到了文件，open(ExecutionContext ec)方法打开所需的文件，close 方法关闭所需的文件，update(ExecutionContext ec)方法则记录每个块在完成处理时的当前状态(写入的记录数量等)。

CompositeItemWriter 和 ClassifierCompositeItemWriter 的区别在于 CompositeItemWriter 实现了 org.springframework.batch.item.ItemStream 接口。CompositeItemWriter 的 open(ExecutionContext ec)方法将循环遍历委托的 ItemWriter，并且根据需求调用它们各自的 open(ExecutionContext ec)方法；close 和 update(ExecutionContext ec)方法采用与之相同的工作方式。然而，ClassifierCompositeItemWriter 并没有实现 ItemStream 的方法，因此没有打开 XML 文件，也没有创建 XMLEventFactory(或底层的 XML 写操作)，于是抛出代码清单 9-88 中的异常。

如何修复这个问题呢？在 Spring Batch 中，可以手动在步骤中注册 ItemStream。如果有 ItemReader 或 ItemWriter 实现了 ItemStream，就可以使用这种方法；如果没有实现(就像 ClassifierCompositeItemWriter 那样)，那么当 ItemReader 或 ItemWriter 维护状态时，就必须把它们注册为流(Stream)，以便能够工作。代码清单 9-89 展示了更新后的作业配置，其中把 xmlOutputWriter 注册成了 ItemStream[1]。

代码清单 9-89　更新后的作业配置：注册合适的 ItemStream

```
...
@Bean
public Step classifierCompositeWriterStep() throws Exception {
    return this.stepBuilderFactory.get("classifierCompositeWriterStep")
            .<Customer, Customer>chunk(10)
            .reader(classifierCompositeWriterItemReader(null))
            .writer(classifierCompositeItemWriter())
            .stream(xmlDelegate(null))
            .build();
}

@Bean
public Job classifierCompositeWriterJob() throws Exception {
    return this.jobBuilderFactory.get("classifierCompositeWriterJob")
            .start(classifierCompositeWriterStep())
            .build();
}
...
```

如果使用更新后的作业配置重新构建和执行作业，就会看到所有的记录都已按预期得到处理。

9.7　本章小结

Spring Batch 的众多 ItemWriter 实现为你提供了广泛的输出选择，包括从写入简单的平面文件到

---

1 你只需要将 xmlDelegate 注册为流。JdbcBatchItemWriter 没有实现 ItemStream 是因为它不需要保存任何状态。

在运行时选择将哪些条目写入哪些 ItemWriter。Spring Batch 提供了一些开箱即用的组件,几乎涵盖了所有的场景。

本章介绍了 Spring Batch 中大部分可用的 ItemWriter。你已经了解了如何使用 Spring Batch 框架提供的不同 ItemWriter 来完成示例应用。在第 10 章,你将了解如何使用 Spring Batch 框架的可伸缩特性,以便让作业按需伸缩和执行。

第 10 章
示例应用

互联网上的技术教程都很巧妙。对于任何新的概念，大部分教程的复杂程度很少会超过"Hello, World!"示例。虽然这有助于对一种技术的理解，但是你也知道，实际情况从来都不像教程中描述的那样简单。因此，本章将介绍一个更加真实的 Spring Batch 作业。

本章将要讨论的话题如下。
- 回顾银行对账单作业：在开发新功能之前，回顾这个作业的开发目标，如第 3 章所述。
- 配置新项目：从 Spring Initializr 创建一个崭新的 Spring Batch 项目。
- 开发作业：逐步完成银行对账单作业的整个开发过程。

我们首先从回顾银行对账单作业的需求开始。

10.1 回顾银行对账单作业

本章要开发的作业用于一家虚拟的银行，名为 Apress Banking。Apress Banking 有大量的客户，并且每位客户有多个交易账户。在每个月末，客户会收到一份组合的对账单，上面会列出客户所有的账户、过去一个月的所有交易、贷方总额、借方总额以及账户的当前余额。

为了完成这些需求，我们将使用一个包含 4 个步骤的作业，如图 10-1 所示。

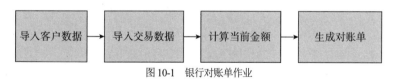

图 10-1 银行对账单作业

这个作业首先在第一个步骤中导入客户数据。我们导入的是一个 CSV 文件，其中包含很多不同的记录格式，每一条记录都是对客户的不同类型的更新。我们把这些更新应用到数据库中的客户记录，然后在第二个步骤中导入交易数据。

已经发生过的交易由一个 XML 文档提供，我们将这些交易作为新记录读入数据库。一旦成功地把所有交易记录导入数据库，我们就将它们应用到账户的当前余额：计算贷方总额，然后从中减去借方总额，这些发生在作业的第三个步骤中。

银行对账单作业的最后一个步骤是生成对账单。为每个客户创建一个文件，这个文件的开头包含了客户地址。对于客户的每个账户(可能有不止一个账户)，打印户头信息：一个包括所有交易的列表、贷方总额、借方总额以及当前余额。在本章，我们将实现这些步骤，并且详细说明这样设计作业的原因。

下面我们将通过从 Spring Initializr 创建一个新的空项目来开始编写这个作业。

10.2 配置新项目

在开始任何新的基于 Spring Boot 的项目时，最好的起点是 https://start.spring.io，也可以选择使用任何 IDE(例如 Spring Tool Suite 或 IntelliJ IDEA)，从而直接访问项目。下面我们使用 IntelliJ IDEA 创建新项目。

首先选择 File | New Project。在打开的窗口中，从左侧选择 Spring Initializr。然后，选择 Project SDK 和 Spring Initializr 的 Service URL(服务地址)。对于这个项目，我们使用 Java 8，因为 Spring Boot 2 默认使用的就是 Java 8，我们还使用了默认的服务地址[1]，如图 10-2 所示。

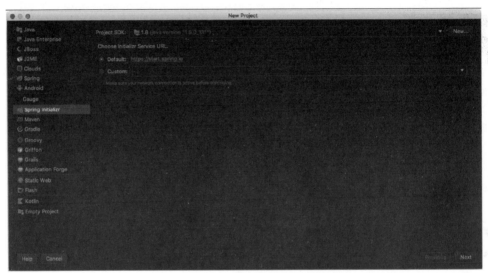

图 10-2　在 IntelliJ IDEA 中使用 Spring Initializr

单击 Next 按钮，输入项目的元信息。本书一直在使用 Maven，所以这里保持不变。我们可以填入 Group ID(组 Id)、Artifact ID(工件 ID)，然后选择 Maven Project (Maven 项目)，这样就有了 Maven 项目的目录结构，这里选择 Java 作为编程语言。我们选择使用 JAR 进行打包，这样生成的 POM 文件就会配置用于生成超级 Jar(über jar)的 Spring Boot 插件。我们选择的 Java 版本为 8。为什么要选择 Java 版本两次？第一个是 IDE 构建和运行项目时使用的版本。第二个是配置 Maven POM 时使用的编译版本。最后，配置 version(版本)、name(项目名)、description(描述)和默认的包。在我们的示例中，它们的值如下。

---

[1] 一些企业选择运行它们自己的 Spring Initializr 实例，从而能够进行定制或者避免开发人员的机器访问互联网。对于这种情况，可以在 IDE 中输入自定义的 URL。

- Group ID：com.apress.batch。
- Artifact ID：chapter10。
- Version：0.0.1-SNAPSHOT。
- Name：Statement Batch Job。
- Description：Apress Banking statement generation batch job。
- Default Package：com.apress.batch.chapter10。

在单击 Next 按钮后，就会进入选择依赖项的界面。这些依赖项都是将要在项目中引入的 Spring Boot 启动包。对于这个项目，需要选择 Batch、JDBC 和 HSQLDB，其中每一个都可以通过在窗口顶部的搜索框中输入名称并按回车键来添加。在加载项目时，我们还需要添加一些其他的依赖项，但它们并没有 Spring Boot 启动包。图 10-3 展示了选择结果。

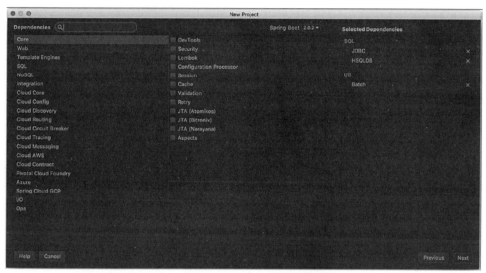

图 10-3　在 IntelliJ IDEA 中选择项目的依赖项

单击 Next 按钮，最后这个界面用于设置项目的名称和下载目录。输入最适合环境的值，单击 Finish 按钮。一个新的窗口将会打开，用于配置项目。

这个项目与其他 Maven 项目一样。Spring Initializr 在默认包的根路径下提供了一个基本的类，其中包含了 Spring Boot 所需的 main 方法。Spring Initializr 还提供了一个测试，该测试仅仅启动了 ApplicationContext，其他什么也不做。由于 Spring Boot 默认会启动所有能够找到的 Spring Batch 作业，因此这个测试并没有多大用处。

准备工作已经完成。在项目的根目录下，在命令行中使用./mvnw clean install 命令进行 Maven 构建。构建应该会成功。回忆一下如何构建批处理作业。我们需要为主类添加 @EnableBatchProcessing 注解。代码清单 10-1 展示了更新后的主类。

代码清单 10-1　Chapter10Application

```
@EnableBatchProcessing
@SpringBootApplication
public class Chapter10Application {
```

```
public static void main(String[] args) {
    SpringApplication.run(Chapter10Application.class, realArgs);
}
}
```

在配置了主类之后,就可以开始编写作业和第一个步骤了。参考前面的图 10-1,第一个步骤是导入客户数据。

10.3　导入客户数据

正如前面提到的,银行对账单作业中的第一个步骤是导入客户数据。我们将接收一个 CSV 文件,其中包含三种不同的记录格式。本节将介绍如何转换这些记录,并把相关的最新更新应用到数据库中。

不过,在深入代码之前,我们先回顾一下银行对账单作业的数据模型。与大多数的企业数据模型相比,这个作业的数据模型非常简单。这个数据模型从 CUSTOMER 表开始。这个表中包含了姓名、地址以及其他联系人信息(包括 e-mail 地址和不同类型的电话号码)。CUSTOMER 表中还有一个字段,用于标识用户首选的通信类型。下一个表是 ACCOUNT 表。CUSOMTER 表与 ACCOUNT 表之间是一对多的关系(一个客户可以有多个账户)。ACCOUNT 表非常简单,其中只包含 id、当前余额以及上次签发对账单时的日期。最后,TRANSACTION 表包含了业务数据,也就是账户中每笔交易的详细信息。如你所料,ACCOUNT 表与 TRANSACTION 表之间是一对多的关系。这个数据模型中的第四个,也是最后一个表是 CUSTOMER_ACCOUNT 表,它用于充当 CUSTOMER 表和 ACCOUNT 表之间的关联表。图 10-4 展示了这个数据模型。

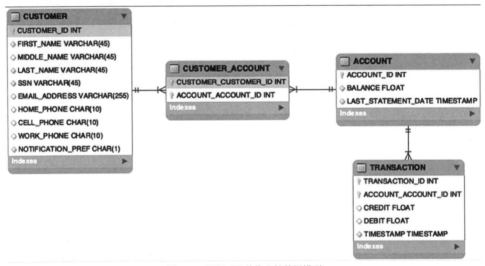

图 10-4　银行对账单作业的数据模型

为了从客户文件中导入数据,你需要理解文件的内容格式。客户文件包含三种记录格式:一种用于客户姓名字段的更新,另一种用于客户地址相关字段的更新,第三种用于客户通信方式的更新。代码清单 10-2 中的示例展示了这三种记录格式。

代码清单 10-2　customer_update.csv 文件中的一些记录

```
2,2,,,Montgomery,Alabama,36134
2,2,,,Montgomery,Alabama,36134
3,441,,,,316-510-9138,2
3,174,trothchild3o@pinterest.com,,785-790-7373,467-631-6632,5
2,287,,,Rochester,New York,14646
2,287,,,Rochester,New York,14646
1,168,Rozelle,Heda,Farnill
2,204,2 Warner Junction,,Akron,Ohio,44305
```

记录中的第一个字段标识了记录的类型。在转换文件时，我们将用到这个字段。现在，我们看看每种记录类型。类型 1(以数字 1 开始)用于更新客户姓名字段。这种记录类型有 5 个字段。

- Record type (记录类型)：对于记录类型 1 来说，总是 1。
- Customer ID (客户 ID)：用于更新的客户记录的 ID。
- First Name (名字)：用于更新的客户记录中的名字。如果为空，那么不更新名字。
- Middle Name (中间名)：用于更新的客户记录中的中间名。如果为空，那么不更新中间名。
- Last Name (姓氏)：用于更新的客户记录中的姓氏。如果为空，那么不更新姓氏。

需要注意的是，在更新之前，客户记录应该已经位于数据库中。我们将在导入阶段进行校验。

下一种记录格式是记录类型 2，这种记录类型有 7 个字段。

- Record type (记录类型)：对于记录类型 2，总是 2。
- Customer ID (客户 ID)：用于更新的客户记录的 ID。
- Address 1 (地址的第一行)：用于更新地址的第一行。如果为空，那么不应该更新。
- Address 2 (地址的第二行)：用于更新地址的第二行(可选)。如果为空，那么不应该更新。
- City (城市)：客户所在的城市。如果为空，那么不应该更新。
- State (州)：客户所在的州。如果为空，那么不应该更新。
- Postal Code (邮政编码)：客户的邮政编码。如果为空，那么不应该更新。

最后一种记录格式是记录类型 3。这种记录类型用于更新客户的联系信息，也有 7 个字段。

- Record type (记录类型)：对于记录类型 3，总是 3。
- Customer ID (客户 ID)：用于更新的客户记录的 ID。
- e-mail address (e-mail 地址)：用于更新的客户记录的电子邮件地址。如果为空，那么不应该更新。
- Home phone (家庭电话)：用于更新的客户记录的家庭电话号码。如果为空，那么不应该更新。
- Cell phone (手机号码)：用于更新的客户记录的手机号码。如果为空，那么不应该更新。
- Work phone (工作电话)：用于更新的客户记录的工作电话号码。如果为空，那么不应该更新。
- Notification preference (通知的首选项)：用于标识使用什么机制来联系客户。如果为空，那么不应该更新。

为了处理客户文件，下面定义作业和第一个步骤。作业的配置位于 com.apress.batch.chapter10.configuration.ImportJobConfiguration 类中。代码清单 10-3 展示了作业和第一个步骤的配置。

代码清单 10-3　导入作业的定义

```
...
@Configuration
public class ImportJobConfiguration {
```

```
    @Autowired
    private JobBuilderFactory jobBuilderFactory;

    @Autowired
    private StepBuilderFactory stepBuilderFactory;

    @Bean
    public Job job() throws Exception {
        return this.jobBuilderFactory.get("importJob")
                    .start(importCustomerUpdates())
                    .build();
    }

    @Bean
    public Step importCustomerUpdates() throws Exception {
        return this.stepBuilderFactory.get("importCustomerUpdates")
                    .<CustomerUpdate, CustomerUpdate>chunk(100)
                    .reader(customerUpdateItemReader(null))
                    .processor(customerValidatingItemProcessor(null))
                    .writer(customerUpdateItemWriter())
                    .build();
}
...
```

代码清单 10-3 首先定义了一个 Spring 配置类(使用了@Configuration 注解)，然后使用@Autowired 自动装配了 JobBuilderFactory 和 StepBuilderFactory，它们由主类的@EnableBatchProcessing 提供。有了这些构建器，就可以定义作业和步骤。作业的定义使用了 jobBuilderFactory，并且从步骤 importCustomerUpdates 开始，然后调用 build 方法以构造作业。

在步骤 importCustomerUpdates 中，可使用 StepBuilderFactor 得到一个 StepBuilder，然后将其配置为基于块进行处理。每个块包含 100 个数据项。这个步骤将使用一个名为 customerUpdateItemReader 的 ItemReader、一个名为 customerValidatingItemProcessor 的 ItemProcessor，以及一个名为 customerUpdateItemWriter 的 ItemWriter。所有这些都会读取、处理和写入 CustomerUpdate 类型的对象。这听起来不错，但是那些 ItemReader、ItemProcessor 和 ItemWriter 会是什么样呢？下面就从 ItemReader 开始。

这个步骤中的 ItemReader 是一个 FlatFileItemReader。我们可以使用 FlatFileItemReaderBuilder 配置名称(以支持重启)、Resource(将要读入的文件)、用于转换文件中记录的 LineTokenizer(分词器)，以及用于把转换结果映射到领域对象 FieldSetMapper(字段集合映射器)。代码清单 10-4 展示了这个 FlatFileItemReader 的代码。

代码清单 10-4　读入客户文件

```
...
@Bean
@StepScope
public FlatFileItemReader<CustomerUpdate> customerUpdateItemReader(
@Value("#{jobParameters['customerUpdateFile']}") Resource inputFile) throws Exception {

    return new FlatFileItemReaderBuilder<CustomerUpdate>()
                .name("customerUpdateItemReader")
```

```
                .resource(inputFile)
                .lineTokenizer(customerUpdatesLineTokenizer())
                .fieldSetMapper(customerUpdateFieldSetMapper())
                .build();
    }
...
```

你将注意到,customerUpdateItemReader 是步骤作用域内的一个对象,我们可以使用作业参数来指定读取文件的什么位置。

尽管我们定义了 ItemReader,但是还需要定义 LineTokenizer 和 FieldSetMapper。我们已经讨论过,客户文件有三种记录格式,那么如何在 LineTokenizer 中进行分词呢?很简单,我们不会这么做。我们可以创建一个组合分词器,并根据每个文件的模式将其委托给正确的 LineTokenizer。Spring Batch 提供的 PatternMatchingCompositeLineTokenizer 正好用于这种情况,但是需要一个 Map<String, LineTokenizer> 类型的对象。这个映射中的字符串键定义了记录必须匹配的模式,以便使用对应的 LineTokenzier。所以,我们真正需要做的事情是定义 LineTokenizer 的三个实现,并定义一种模式来识别使用哪个 LineTokenizer,如代码清单 10-5 所示。

代码清单 10-5　用于客户文件的 LineTokenizer 配置

```
...
@Bean
public LineTokenizer customerUpdatesLineTokenizer() throws Exception {
    DelimitedLineTokenizer recordType1 = new DelimitedLineTokenizer();

    recordType1.setNames("recordId", "customerId", "firstName",
            "middleName", "lastName");

    recordType1.afterPropertiesSet();

    DelimitedLineTokenizer recordType2 = new DelimitedLineTokenizer();

    recordType2.setNames("recordId", "customerId", "address1",
            "address2", "city", "state", "postalCode");

    recordType2.afterPropertiesSet();

    DelimitedLineTokenizer recordType3 = new DelimitedLineTokenizer();

    recordType3.setNames("recordId", "customerId", "emailAddress",
            "homePhone", "cellPhone", "workPhone", "notificationPreference");

    recordType3.afterPropertiesSet();

    Map<String, LineTokenizer> tokenizers = new HashMap<>(3);
    tokenizers.put("1*", recordType1);
    tokenizers.put("2*", recordType2);
    tokenizers.put("3*", recordType3);

    PatternMatchingCompositeLineTokenizer lineTokenizer =
            new PatternMatchingCompositeLineTokenizer();

    lineTokenizer.setTokenizers(tokenizers);
```

```
        return lineTokenizer;
    }
...
```

如代码清单 10-5 所示，我们配置了 3 个 DelimitedLineTokenizer 实例，其中的每一个实例都定义了每种记录类型的字段。然后我们将每个实例映射到标识了每种记录类型的前缀模式。PatternMatchingCompositeLineTokenizer 将负责基于这些模式将它们委托给正确的 LineTokenizer。

读取客户文件的最后一部分操作是将记录映射到领域对象。现在，我们可以使用一个包含三种记录类型中所有字段的超级模型对象(über domain object)以及一个简单的用于填充对象的 FieldSetMapper。不过，这是一种相当混乱的数据处理方式。我们可以为这三种记录类型分别创建领域对象。记录类型 1 使用 CustomerNameUpdate 对象，记录类型 2 使用 CustomerAddressUpdate 对象，记录类型 3 使用 CustomerContactUpdate 对象。它们都扩展了共同的父类 CustomerUpdate。CustomerUpdate 类包含了 customerId 字段，这个父类服务于两个目的。第一是存放共有的 customerId 字段。不过，它更重要的目的是允许我们在用于步骤配置的泛型中使用 CustomerUpdate(可以在代码清单 10-3 中查看实际用法)。代码清单 10-6 展示了这个步骤用到的领域对象。

代码清单 10-6　用于客户更新步骤的领域对象

```
...
public class CustomerUpdate {
    protected final long customerId;

    public CustomerUpdate(long customerId) {
        this.customerId = customerId;
    }
// accessors removed
}

...
public class CustomerNameUpdate extends CustomerUpdate {

    private final String firstName;

    private final String middleName;

    private final String lastName;
    public CustomerNameUpdate(long customerId, String firstName,
            String middleName, String lastName) {

        super(customerId);
        this.firstName = StringUtils.hasText(firstName) ? firstName : null;
        this.middleName = StringUtils.hasText(middleName) ? middleName : null;
        this.lastName = StringUtils.hasText(lastName) ? lastName : null;
    }
// accessors removed
}

...
public class CustomerAddressUpdate extends CustomerUpdate {
```

第 10 章 示例应用

```java
        private final String address1;

        private final String address2;

        private final String city;

        private final String state;

        private final String postalCode;

        public CustomerAddressUpdate(long customerId, String address1,
                String address2, String city, String state, String postalCode) {

            super(customerId);
            this.address1 = StringUtils.hasText(address1) ? address1 : null;
            this.address2 = StringUtils.hasText(address2) ? address2 : null;
            this.city = StringUtils.hasText(city) ? city : null;
            this.state = StringUtils.hasText(state) ? state : null;
            this.postalCode = StringUtils.hasText(postalCode) ? postalCode : null;
        }
// accessors removed
}

...
public class CustomerContactUpdate extends CustomerUpdate {

        private final String emailAddress;

        private final String homePhone;

        private final String cellPhone;

        private final String workPhone;

        private final Integer notificationPreferences;

        public CustomerContactUpdate(long customerId, String emailAddress, String homePhone,
                String cellPhone, String workPhone, Integer notificationPreferences) {
            super(customerId);
            this.emailAddress = StringUtils.hasText(emailAddress) ? emailAddress : null;
            this.homePhone = StringUtils.hasText(homePhone) ? homePhone : null;
            this.cellPhone = StringUtils.hasText(cellPhone) ? cellPhone : null;
            this.workPhone = StringUtils.hasText(workPhone) ? workPhone : null;
            this.notificationPreferences = notificationPreferences;
        }
// accessors removed
}
```

为了决定返回哪个对象，需要使用 FieldSetMapper，从而基于记录类型创建和返回正确的领域对象。代码清单 10-7 展示了如何保持简单并使用 lambda 表达式来创建用于处理这种映射的 FieldSetMapper。

代码清单 10-7　用于客户文件的 FieldSetMapper 的配置

```java
...
@Bean
public FieldSetMapper<CustomerUpdate> customerUpdateFieldSetMapper() {
```

```
            return fieldSet -> {
                switch (fieldSet.readInt("recordId")) {
                    case 1: return new CustomerNameUpdate(
                                fieldSet.readLong("customerId"),
                                fieldSet.readString("firstName"),
                                fieldSet.readString("middleName"),
                                fieldSet.readString("lastName"));
                    case 2: return new CustomerAddressUpdate(
                                fieldSet.readLong("customerId"),
                                fieldSet.readString("address1"),
                                fieldSet.readString("address2"),
                                fieldSet.readString("city"),
                                fieldSet.readString("state"),
                                fieldSet.readString("postalCode"));
                    case 3:
                        String rawPreference =
                                fieldSet.readString("notificationPreference");

                        Integer notificationPreference = null;

                        if(StringUtils.hasText(rawPreference)) {
                            notificationPreference = Integer.
                                            parseInt(rawPreference);
                        }

                        return new CustomerContactUpdate(fieldSet.
                        readLong("customerId"),
                                fieldSet.readString("emailAddress"),
                                fieldSet.readString("homePhone"),
                                fieldSet.readString("cellPhone"),
                                fieldSet.readString("workPhone"),
                                notificationPreference);
                    default: throw new IllegalArgumentException(
                            "Invalid record type was found:" +
                                    fieldSet.readInt("recordId"));
                }
            };
    }
    ...
```

代码清单 10-7 中的 lambda 表达式用于查找每条记录中的记录类型字段，然后为每条记录创建合适的领域对象。如果找不到记录的类型，就会抛出异常以表明记录类型是无效的。

读取只是应用客户更新的第一部分。这个步骤的目标是把数据保存到数据库中。为此，需要首先验证每条记录是否有合法的客户 ID。

10.3.1 验证客户 ID

在步骤的定义中，我们定义了一个 ItemProcessor，名为 customerValidatingItemProcessor。这个组件用于在接收的 CustomerUpdate 对象中查找客户 ID。如果位于数据库中，就跳过这条记录继续处理；如果不在数据库中，就过滤掉这些记录。在真实的场景中，我们可能会把这些数据项写入新的文件，以便将来进行调试。但是对于这个作业，仅仅过滤掉它们就可以了。

为此，我们可以使用 Spring Batch 的 ValidatingItemProcessor。ValidatingItemProcessor 将接收

org.springframework.batch.item.validator.Validator 接口(它与 Spring 框架中的 Validator 接口不同)的一个实现。在这个示例中，我们将创建自定义的 Validator 实现，用于在数据库中查找客户 ID，并且在没有找到时抛出 ValidationException 异常。代码清单 10-8 展示了 CustomerItemValidator 的代码。

代码清单 10-8　CustomerItemValidator

```
...
@Component
public class CustomerItemValidator implements Validator<CustomerUpdate> {

    private final NamedParameterJdbcTemplate jdbcTemplate;

    private static final String FIND_CUSTOMER =
        "SELECT COUNT(*) FROM CUSTOMER WHERE customer_id = :id";

    public CustomerItemValidator(DataSource dataSource) {
        this.jdbcTemplate = new NamedParameterJdbcTemplate(dataSource);
    }

    @Override
    public void validate(CustomerUpdate customer) throws ValidationException {
        Map<String, Long> parameterMap =
            Collections.singletonMap("id", customer.getCustomerId());

        Long count = jdbcTemplate.queryForObject(FIND_CUSTOMER, parameterMap,
                                        Long.class);
        if(count == 0) {
            throw new ValidationException(
                String.format("Customer id %s was not able to be found",
                        customer.getCustomerId() ));
        }
    }
}
```

在定义了验证器之后，就可以配置 ItemProcessor 了。代码清单 10-9 提供了 ValidatingItemProcessor 的配置。

代码清单 10-9　customerValidatingItemProcessor

```
...
@Bean
public ValidatingItemProcessor<CustomerUpdate> customerValidatingItemProcessor
(CustomerItemValidator validator) {

    ValidatingItemProcessor<CustomerUpdate> customerValidatingItemProcessor =
                new ValidatingItemProcessor<>(validator);
    customerValidatingItemProcessor.setFilter(true);

    return customerValidatingItemProcessor;
}
...
```

在配置了 ItemProcessor 之后，剩下要做的事情就是配置步骤的 ItemWriter。不过，如果我们有三种类型的数据需要写入，那么如何在步骤的写入端进行管理呢？详见 10.3.2 节。

10.3.2 写入客户更新

第一个步骤的最后一部分是将更新应用到 CUSTOMER 表。由于我们有三种记录类型，也就是说有三种不同类型的更新，因此我们需要能够委托给三个不同的 ItemWriter。我们可以使用 Spring Batch 的 ClassifierCompositeItemWriter，基于 Classifier 的实现，委托 ItemWriter 的实现。

我们首先看看需要配置的三个 ItemWriter 的实现。除了用于更新数据的 SQL 之外，它们都是相同的。使用 JdbcBatchItemWriterBuilder，只需要几行代码就可以配置 SQL 和数据源，并且让 Spring 使用 Bean 的名称映射语句参数。代码清单 10-10 展示了这三个 JdbcBatchItemWriter 的配置。

代码清单 10-10　customerValidatingItemProcessor

```
...
@Bean
public JdbcBatchItemWriter<CustomerUpdate> customerNameUpdateItemWriter(DataSource
dataSource) {
    return new JdbcBatchItemWriterBuilder<CustomerUpdate>()
            .beanMapped()
            .sql("UPDATE CUSTOMER " +
                "SET FIRST_NAME = COALESCE(:firstName, FIRST_NAME), " +
                "MIDDLE_NAME = COALESCE(:middleName, MIDDLE_NAME), " +
                "LAST_NAME = COALESCE(:lastName, LAST_NAME) " +
                "WHERE CUSTOMER_ID = :customerId")
            .dataSource(dataSource)
            .build();
}

@Bean
public JdbcBatchItemWriter<CustomerUpdate> customerAddressUpdateItemWriter(DataSource
dataSource) {
    return new JdbcBatchItemWriterBuilder<CustomerUpdate>()
            .beanMapped()
            .sql("UPDATE CUSTOMER SET " +
                "ADDRESS1 = COALESCE(:address1, ADDRESS1), " +
                "ADDRESS2 = COALESCE(:address2, ADDRESS2), " +
                "CITY = COALESCE(:city, CITY), " +
                "STATE = COALESCE(:state, STATE), " +
                "POSTAL_CODE = COALESCE(:postalCode, POSTAL_CODE) " +
                "WHERE CUSTOMER_ID = :customerId")
            .dataSource(dataSource)
            .build();
}

@Bean
public JdbcBatchItemWriter<CustomerUpdate> customerContactUpdateItemWriter(DataSource
dataSource) {
    return new JdbcBatchItemWriterBuilder<CustomerUpdate>()
            .beanMapped()
            .sql("UPDATE CUSTOMER SET " +
                "EMAIL_ADDRESS = COALESCE(:emailAddress, EMAIL_ADDRESS), " +
                "HOME_PHONE = COALESCE(:homePhone, HOME_PHONE), " +
                "CELL_PHONE = COALESCE(:cellPhone, CELL_PHONE), " +
                "WORK_PHONE = COALESCE(:workPhone, WORK_PHONE), " +
                "NOTIFICATION_PREF = COALESCE(:notificationPreferences,
```

```
                    NOTIFICATION_PREF) " +
                    "WHERE CUSTOMER_ID = :customerId")
            .dataSource(dataSource)
            .build();
}
```
...

代码清单 10-10 中的每个 ItemWriter 配置都将完成同样的事情：使用合适的值设置不同的列。对 SQL 语句中的每个值使用 COALESCE 的原因是，我们只想更新输入文件提供的值。如果输入文件提供的值是 null，那么不应该更新。

在配置了这 3 个 ItemWriter 之后，我们需要能够基于接收到的数据类型选择正确的 ItemWriter(因为数据类型基于输出文件中的记录类型)。为了进行选择，我们实现了一个 org.springframework.classify.Classifier 类型的分类器，用于评估给定的数据，并返回合适的 ItemWriter，如代码清单 10-11 所示。

代码清单 10-11　CustomerUpdateClassifier

```
...
public class CustomerUpdateClassifier implements
Classifier<CustomerUpdate, ItemWriter<? super CustomerUpdate>> {

    private final JdbcBatchItemWriter<CustomerUpdate> recordType1ItemWriter;
    private final JdbcBatchItemWriter<CustomerUpdate> recordType2ItemWriter;
    private final JdbcBatchItemWriter<CustomerUpdate> recordType3ItemWriter;

    public CustomerUpdateClassifier(
        JdbcBatchItemWriter<CustomerUpdate> recordType1ItemWriter,
        JdbcBatchItemWriter<CustomerUpdate> recordType2ItemWriter,
        JdbcBatchItemWriter<CustomerUpdate> recordType3ItemWriter) {

        this.recordType1ItemWriter = recordType1ItemWriter;
        this.recordType2ItemWriter = recordType2ItemWriter;
        this.recordType3ItemWriter = recordType3ItemWriter;
    }

    @Override
    public ItemWriter<? super CustomerUpdate> classify(CustomerUpdate classifiable) {

        if(classifiable instanceof CustomerNameUpdate) {
            return recordType1ItemWriter;
        }
        else if(classifiable instanceof CustomerAddressUpdate) {
            return recordType2ItemWriter;
        }
        else if(classifiable instanceof CustomerContactUpdate) {
            return recordType3ItemWriter;
        }
        else {
            throw new IllegalArgumentException("Invalid type: " +
                classifiable.getClass().getCanonicalName());
        }
    }
}
```

如你所见，这个分类器会将每一个 ItemWriter 实例作为构造方法的参数，然后基于传入的数据返回正确的 ItemWriter。银行对账单作业的第一个步骤的最后一部分是配置 ClassifierCompositeItemWriter。这个 ItemWriter 的配置非常简单，因为所有的工作都在分类器以及委托的 ItemWriter 实例中完成了。代码清单 10-12 展示了 customerUpdateItemWriter 的配置。

代码清单 10-12　customerUpdateItemWriter

```
...
@Bean
public ClassifierCompositeItemWriter<CustomerUpdate> customerUpdateItemWriter() {

    CustomerUpdateClassifier classifier =
            new CustomerUpdateClassifier(customerNameUpdateItemWriter(null),
                    customerAddressUpdateItemWriter(null),
                    customerContactUpdateItemWriter(null));

    ClassifierCompositeItemWriter<CustomerUpdate> compositeItemWriter =
            new ClassifierCompositeItemWriter<>();

    compositeItemWriter.setClassifier(classifier);

    return compositeItemWriter;
}
...
```

在编写和配置了用于第一个步骤的所有组件之后，就可以运行这个批处理作业，查看第一个步骤的运行情况。我们可以使用代码清单 10-13 所示的 Spring Boot 属性来配置一个"真实的数据库"，以便查看结果。

代码清单 10-13　application.properties

```
spring.datasource.driverClassName=com.mysql.jdbc.Driver
spring.datasource.url=jdbc:mysql://localhost:3306/statement
spring.datasource.username=<USERNAME>
spring.datasource.password=<PASSWORD>
spring.datasource.schema=schema-mysql.sql
spring.datasource.initialization-mode=always
spring.batch.initialize-schema=always
```

运行作业和测试第一个步骤所需的最后一部分是添加 MySQL 驱动(这个示例使用的是 MySQL。如果想要使用其他类型的数据库，那么需要替换相应的配置值和驱动)。代码清单 10-14 展示了 MySQL 所需的 Maven 依赖项(版本由 Spring Boot 提供)。

代码清单 10-14　MySQL 依赖项

```xml
<dependency>
    <groupId>mysql</groupId>
    <artifactId>mysql-connector-java</artifactId>
</dependency>
```

有了这些配置，就可以在命令行中使用 ./mvnw clean install 命令构建项目。构建完成后，就可以在项目的 target 目录下执行命令 java -jar chapter10-0.0.1-SNAPSHOT.jar customerUpdateFile=<PATH_TO_

CUSTOMER_FILE>。之后就可以验证数据是否已正确应用到 CUSTOMER 表。

接下来，我们将完成银行对账单作业的第二个步骤：导入交易数据。

10.4 导入交易数据

在更新了客户文件之后，就可以从交易文件中读取交易数据。尽管客户文件由于处理了多种记录类型而显得复杂，但是交易文件实际上非常简单。交易文件是简单的 XML 文件，可直接导入数据库的 TRANSACTION 表中。代码清单 10-15 展示了将要导入的交易文件。

代码清单 10-15　交易文件

```xml
<?xml version='1.0' encoding='UTF-8'?>
<transactions>
    <transaction>
        <transactionId>2462744</transactionId>
        <accountId>405</accountId>
        <description>Skinix</description>
        <credit/>
        <debit>-438</debit>
        <timestamp>2018-06-01 19:39:53</timestamp>
    </transaction>
    <transaction>
        <transactionId>4243424</transactionId>
        <accountId>584</accountId>
        <description>Yakidoo</description>
        <credit>8681.98</credit>
        <debit/>
        <timestamp>2018-06-12 18:39:09</timestamp>
    </transaction>
...
</transactions>
```

交易文件的开头是 transations 元素，其中包含所有独立的 transaction 元素。每个 transaction 元素代表一条银行交易记录，从而在作业中生成一条数据条目。这些块已被映射到代码清单 10-16 中列出的 Transaction 领域对象。

代码清单 10-16　Transaction 领域对象

```java
...
@XmlRootElement(name = "transaction")
public class Transaction {

    private long transactionId;

    private long accountId;

    private String description;

    private BigDecimal credit;

    private BigDecimal debit;
```

```
    private Date timestamp;

    public Transaction() {
    }

    public Transaction(long transactionId,
                       long accountId,
                       String description,
                       BigDecimal credit,
                       BigDecimal debit,
                       Date timestamp) {

        this.transactionId = transactionId;
        this.accountId = accountId;
        this.description = description;
        this.credit = credit;
        this.debit = debit;
        this.timestamp = timestamp;
    }

    // accessors removed for brevity

    @XmlJavaTypeAdapter(JaxbDateSerializer.class)
    public void setTimestamp(Date timestamp) {

        this.timestamp = timestamp;
    }

    public BigDecimal getTransactionAmount() {
        if(credit != null) {
            if(debit != null) {
                return credit.add(debit);
            }
            else {
                return credit;
            }
        }
        else if(debit != null) {
            return debit;
        }
        else {
            return new BigDecimal(0);
        }
    }
}
```

如你所见，在 Transaction 领域对象中，有些字段已经直接映射到了输入文件的 XML 块。Transaction 类有三个值得注意的地方：类级别的 @XmlRootElement、用于 timestamp 字段的设置方法的 @XmlJavaTypeAdapter，以及额外的方法 getTransactionAmount()。@XmlRootElement 作为 JAX-B 注解定义了领域对象的根标签。在本例中，这个根标签是 transaction。我们为 timestamp 字段的 setter 方法使用了 @XmlJavaTypeAdapter，这是因为 JAX-B 并没有提供用于处理字符串类型与日期类型之间转换的简单方法。因此，我们需要提供少许代码，供 JAX-B 用来完成转换(也就是使用 JaxbDateSerializer)。代码清单 10-17 展示了 JaxbDateSerializer。

代码清单 10-17　JaxbDateSerializer

```java
...
public class JaxbDateSerializer extends XmlAdapter<String, Date> {

    private SimpleDateFormat dateFormat = new SimpleDateFormat("yyyy-MM-dd hh:mm:ss");

    @Override
    public String marshal(Date date) throws Exception {
        return dateFormat.format(date);
    }

    @Override
    public Date unmarshal(String date) throws Exception {
        return dateFormat.parse(date);
    }
}
```

JaxbDateSerializer 扩展了 XmlAdapter，在本例中，JAX-B 使用前者来完成 String 类型到 java.util.Date 类型的转换。Transaction 类的最后一部分是 getTransactionAmount() 方法。交易要么包含贷方，要么包含借方。不过，当我们进行数学运算时，我们并不关心数值是贷方还是借方，我们仅仅关心数值是多少。所以，这个方法将返回交易的实际数额。

在定义了领域对象之后，就可以配置第二个步骤 importTransactions 了。

10.4.1　读取交易

我们首先配置这个步骤并将其添加到作业中，以导入交易数据。代码清单 10-18 展示了第二个步骤 importTransactions 的配置。

代码清单 10-18　importTransactions

```java
...
@Bean
public Job job() throws Exception {
    return this.jobBuilderFactory.get("importJob")
            .start(importCustomerUpdates())
            .next(importTransactions())
            .build();
}

@Bean
public Step importTransactions() {
    return this.stepBuilderFactory.get("importTransactions")
            .<Transaction, Transaction>chunk(100)
            .reader(transactionItemReader(null))
            .writer(transactionItemWriter(null))
            .build();
}
...
```

importTransactions 很简单。在这里，我们使用了名为 transactionItemReader 的条目读取器和名为 transactionItemWriter 的条目写入器。由于读取的是 XML 文件，因此使用了 StaxEventItemReader。对

于如何解析 XML，领域对象的 JAX-B 注解应该已经提示了一些信息，但是我们将使用 JAX-B 来解析 XML。在这种方式下，条目读取器的配置将变得非常简单，如代码清单 10-19 所示。

代码清单 10-19　transactionItemReader

```
@Bean
@StepScope
public StaxEventItemReader<Transaction> transactionItemReader(
    @Value("#{jobParameters['transactionFile']}") Resource transactionFile) {

    Jaxb2Marshaller unmarshaller = new Jaxb2Marshaller();
    unmarshaller.setClassesToBeBound(Transaction.class);

    return new StaxEventItemReaderBuilder<Transaction>()
            .name("fooReader")
            .resource(transactionFile)
            .addFragmentRootElements("transaction")
            .unmarshaller(unmarshaller)
            .build();
}
...
```

步骤作用域内的 transactionItemReader 方法将接收输入文件的位置作为作业参数，参数的名称为 transactionFile。在这个方法中，我们创建了 Jaxb2Marshaller，并将其绑定到了代码清单 10-16 定义的 Transaction 领域对象。最后，我们使用 StaxEventItemReaderBuilder 配置了 transactionItemWriter。我们传入了名称(用于重启)、通过作业参数注入的资源，还定义了每个 XML 片段的根元素(transaction 元素)以进行解析，并且把 Jaxb2Marshaller 传给了 StaxEventItemReaderBuilder 以在解析 XML 时使用。最后，我们调用了 build 方法以返回所需的 StaxEventItemReader。

有了 ItemReader 之后，就需要 ItemWriter。对此你应该很熟悉，因为使用的配置与前一个步骤中的相同(仅仅进行了简化)。10.4.2 节将介绍 transactionItemWriter 的配置。

10.4.2　写入交易

transactionItemWriter 负责把交易数据写入数据库的 TRANSACTION 表中。为此，我们将再次使用 JdbcBatchItemWriter，如代码清单 10-20 所示。

代码清单 10-20　transactionItemWriter

```
...
@Bean
public JdbcBatchItemWriter<Transaction> transactionItemWriter(DataSource dataSource) {
    return new JdbcBatchItemWriterBuilder<Transaction>()
            .dataSource(dataSource)
            .sql("INSERT INTO TRANSACTION (TRANSACTION_ID, " +
                    "ACCOUNT_ACCOUNT_ID, " +
                    "DESCRIPTION, " +
                    "CREDIT, " +
                    "DEBIT, " +
                    "TIMESTAMP) VALUES (:transactionId, " +
                    ":accountId, " +
                    ":description, " +
```

```
                            ":credit, " +
                            ":debit, " +
                            ":timestamp)")
                .beanMapped()
                .build();
    }
...
```

JdbcBatchItemWriterBuilder 被用来配置 JdbcBatchItemWriter，方法是接收数据源和 SQL 语句，并且让 ItemWriter 使用数据对象的属性名作为键来设置 SQL 语句的值。

以上就是用于第二个步骤的所有配置。在配置了 ItemReader 和 ItemWriter 之后，就可以使用./mvnw clean install 命令构建作业了，然后在如下命令中添加新的文件参数来运行作业：java -jar chapter10-0.0.1-SNAPSHOT.jar customerUpdateFile=<PATH_TO_CUSTOMER_FILE>transactionFile=<PATH_TO_TRANSACTION_FILE>。在作业成功运行之后，就可以在数据库中验证 XML 交易文件的内容已写入 TRANSACTION 表。

在导入交易数据之后，现在需要将它们应用于 ACCOUNT 表中的余额。

10.5　计算当前余额

银行对账单作业中的下一个步骤是把刚刚导入的交易数据应用到账户余额。这实际上是最容易配置的一个步骤。与导入交易数据类似，这个步骤也有一个简单的 ItemReader 和一个简单的 ItemWriter。我们将以与前面两个步骤相同的方式使用 JdbcBatchItemWriter。但是，我们的输入也是来自数据库，并且与前面导入交易数据的形式相同。代码清单 10-21 展示了 applyTransactions 步骤的配置。

代码清单 10-21　applyTransactions 步骤的配置

```
...
@Bean
public Job job() throws Exception {
    return this.jobBuilderFactory.get("importJob")
                .start(importCustomerUpdates())
                .next(importTransactions())
                .next(applyTransactions())
                .build();
}

...
@Bean
public Step applyTransactions() {
    return this.stepBuilderFactory.get("applyTransactions")
                .<Transaction, Transaction>chunk(100)
                .reader(applyTransactionReader(null))
                .writer(applyTransactionWriter(null))
                .build();
}
...
```

代码清单 10-21 把这个新的步骤添加到了作业的配置中，然后定义了步骤的 Bean。这个步骤是使用构造函数创建的，并且配置了读取和写入 Transaction 领域对象时的块大小为 100。读取器是

applyTransactionReader，用于接收数据源；写入器是 applyTransactionWriter，也用于接收数据源。10.5.1 节将介绍如何在作业中定义 applyTransactionReader。

10.5.1 读取交易

多亏有了 JdbcCursorItemReader 的帮助，从数据表中读取前面导入的交易数据变得很简单。为了使用这个条目读取器，我们为它配置了名称(用于重启)、数据源、一条 SQL 语句以及本例中 RowMapper 的实现。我们将使用 lambda 表达式来完成这些配置。代码清单 10-22 展示了配置这个条目读取器所需的代码。

代码清单 10-22　applyTransactionsReader

```
...
@Bean
public JdbcCursorItemReader<Transaction> applyTransactionReader(DataSource dataSource) {
    return new JdbcCursorItemReaderBuilder<Transaction>()
            .name("applyTransactionReader")
            .dataSource(dataSource)
            .sql("select transaction_id, " +
                    "account_account_id, " +
                    "description, " +
                    "credit, " +
                    "debit, " +
                    "timestamp " +
                    "from TRANSACTION " +
                    "order by timestamp")
            .rowMapper((resultSet, i) ->
                    new Transaction(
                            resultSet.getLong("transaction_id"),
                            resultSet.getLong("account_account_id"),
                            resultSet.getString("description"),
                            resultSet.getBigDecimal("credit"),
                            resultSet.getBigDecimal("debit"),
                            resultSet.getTimestamp("timestamp")))
            .build();
}
...
```

以上真的就是读取交易数据所需的一切。10.5.2 节将介绍如何通过 JdbcBatchItemWriter 把交易应用到账户余额。

10.5.2　更新账户余额

在读取了每个数据条目之后，就可以把交易结果应用到对应的账户。好消息是：我们只需要一条 SQL 语句即可完成。当然，还有许多更有效的方法可以做到这一点。代码清单 10-23 展示了将每笔交易应用到账户余额的实现代码。

代码清单 10-23　applyTransactionsWriter

```
...
@Bean
public JdbcBatchItemWriter<Transaction> applyTransactionWriter(DataSource dataSource) {
```

```
        return new JdbcBatchItemWriterBuilder<Transaction>()
                .dataSource(dataSource)
                .sql("UPDATE ACCOUNT SET " +
                        "BALANCE = BALANCE + :transactionAmount " +
                        "WHERE ACCOUNT_ID = :accountId")
                .beanMapped()
                .assertUpdates(false)
                .build();
}
...
```

代码清单 10-23 在配置 JdbcBatchItemWriter 时,使用了 DataSource(数据源)和一条 SQL 语句。后者用于把交易数据应用到账户的当前余额,SQL 语句中的参数可通过调用 Bean 的属性来填充。

在构建了读取器和写入器之后,就可以构建和运行作业了。使用与构建和运行 importTransactions 步骤时相同的命令(./mvnw clean install 用于构建项目,java -jar chapter10-0.0.1-SNAPSHOT.jar customerUpdateFile=<PATH_TO_CUSTOMER_FILE>transactionFile=<PATH_TO_TRANSACTION_FILE>用于执行作业,就可以验证步骤是否已经生效,以及是否把交易数据正确应用到了账户的当前余额。

接下来我们将进入银行对账单作业的最后一个步骤:生成对账单。尽管这个步骤表面上看很简单,但涉及的代码更多,让我们拭目以待吧!

10.6 生成对账单

这个批处理作业的终极目标是为每个客户生成一份包含账户汇总信息的银行对账单。到目前为止,所有处理都是为了给写入对账单做准备。本节介绍与写入对账单相关的处理。

10.6.1 读取对账单数据

当查看最后一个步骤期望的输出时,你会很快意识到,为了生成对账单,你需要拉取大量的数据。在拉取这些数据之前,我们先看看用于呈现数据的领域对象 Statement,如代码清单 10-24 所示。

代码清单 10-24　Statement.java

```
...
public class Statement {

    private final Customer customer;
    private List<Account> accounts = new ArrayList<>();

    public Statement(Customer customer, List<Account> accounts) {
        this.customer = customer;
        this.accounts.addAll(accounts);
    }

    // accessors removed for brevity
...
}
```

Statement 对象包含一个 Customer 实例(对账单为谁生成)和一个 Account 对象列表,后者代表客户拥有的所有账户。每个 Customer 对象包含了数据库中 CUSTOMER 表的所有数据。如你所料，Account 对象被直接映射到了数据库中的 ACCOUNT 表。代码清单 10-25 展示了这些领域对象的代码。

代码清单 10-25　Customer.java 和 Account.java

```
...
public class Customer {

    private final long id;
    private final String firstName;
    private final String middleName;
    private final String lastName;
    private final String address1;
    private final String address2;
    private final String city;
    private final String state;
    private final String postalCode;
    private final String ssn;
    private final String emailAddress;
    private final String homePhone;
    private final String cellPhone;
    private final String workPhone;
    private final int notificationPreferences;

    public Customer(long id, String firstName, String middleName, String lastName,
    String address1, String address2, String city, String state, String postalCode,
    String ssn, String emailAddress, String homePhone, String cellPhone, String
    workPhone, int notificationPreferences) {
        this.id = id;
        this.firstName = firstName;
        this.middleName = middleName;
        this.lastName = lastName;
        this.address1 = address1;
        this.address2 = address2;
        this.city = city;
        this.state = state;
        this.postalCode = postalCode;
        this.ssn = ssn;
        this.emailAddress = emailAddress;
        this.homePhone = homePhone;
        this.cellPhone = cellPhone;
        this.workPhone = workPhone;
        this.notificationPreferences = notificationPreferences;
    }

    // accessors removed

    ...
}

...
public class Account {
    private final long id;
    private final BigDecimal balance;
    private final Date lastStatementDate;
```

```
    private final List<Transaction> transactions = new ArrayList<>();

    public Account(long id, BigDecimal balance, Date lastStatementDate) {
        this.id = id;
        this.balance = balance;
        this.lastStatementDate = lastStatementDate;
    }

    // accessors removed
    ...
}
```

尽管领域对象包含了银行对账单作业的所有组件,但是读取器并不会填充所有这些组件。对于这个步骤,我们将使用所谓的驱动查询模式(Driving Query Pattern)。这意味着 ItemReader 只读取基本的数据(在本例中是 Customer);而 ItemProcessor 将使用账户信息充实 Statement 对象,然后最终在 ItemWriter 中生成对账单。代码清单 10-26 展示了最后这个步骤的配置,然后将其添加到了作业中。

代码清单 10-26　generateStatements 步骤

```
...
@Bean
public Job job() throws Exception {
    return this.jobBuilderFactory.get("importJob")
                .start(importCustomerUpdates())
                .next(importTransactions())
                .next(applyTransactions())
                .next(generateStatements(null))
                .build();
}
...
@Bean
public Step generateStatements(AccountItemProcessor itemProcessor) {
    return this.stepBuilderFactory.get("generateStatements")
                .<Statement, Statement>chunk(1)
                .reader(statementItemReader(null))
                .processor(itemProcessor)
                .writer(statementItemWriter(null))
                .build();
}
...
```

作为银行对账单作业中的最后一个步骤,generateStatements 包括一个简单的 ItemReader、一个 ItemProcessor 和一个 ItemWriter。你将注意到最后这个步骤的块大小是 1。因为对于每个对账单将生成一个文件,所以我们使用了 MultiResourceItemWriter。不过,这只是对每个块生成一个文件。如果期望对每个数据项生成一个文件,就需要把块大小设置为 1。

在配置了这个步骤之后,就可以配置条目读取器。用于 generateStatements 的条目读取器是一个简单的 JdbcCursorItemReader。我们为这个 JdbcCursorItemReader 配置了名称(用于重启)、数据源、一条 SQL 语句以及一个 RowMapper 实现(本例将使用一个 lambda 表达式)。代码清单 10-27 展示了这个条目读取器的配置。

代码清单 10-27　statementItemReader

```
...
@Bean
public JdbcCursorItemReader<Statement> statementItemReader(DataSource dataSource) {
    return new JdbcCursorItemReaderBuilder<Statement>()
                .name("statementItemReader")
                .dataSource(dataSource)
                .sql("SELECT * FROM CUSTOMER")
                .rowMapper((resultSet, i) -> {
                    Customer customer =
                        new Customer(resultSet.getLong("customer_id"),
                            resultSet.getString("first_name"),
                            resultSet.getString("middle_name"),
                            resultSet.getString("last_name"),
                            resultSet.getString("address1"),
                            resultSet.getString("address2"),
                            resultSet.getString("city"),
                            resultSet.getString("state"),
                            resultSet.getString("postal_code"),
                            resultSet.getString("ssn"),
                            resultSet.getString("email_address"),
                            resultSet.getString("home_phone"),
                            resultSet.getString("cell_phone"),
                            resultSet.getString("work_phone"),
                            resultSet.getInt("notification_pref"));

                    return new Statement(customer);
                }).build();
}
...
```

在配置了这个条目读取器之后，你将需要编写 ItemProcessor，以通过添加客户的账户列表来充实 ItemReader 返回的 Statement 对象。

10.6.2　为对账单添加账户信息

在读取了客户数据之后，就可以读取账户和交易数据来生成对账单。这可以通过 Spring 的 JdbcTemplate 来实现。不过，由于查询会导致父子关系(一个账户对应多笔交易)，因此我们不使用 RowMapper，而是使用 ResultSetExtractor。RowMapper 接口用于把单行记录映射到对象，ResultSetExtractor 则把 ResultSet 作为整体(当使用 RowMapper 时，如果自己提前执行 ResultSet，那么会抛出异常)。我们需要从 ResultSet 中读取多行以创建账户。下面首先深入研究一下代码清单 10-28 中的代码，它们用于执行查询并充实对账单对象。

代码清单 10-28　AccountItemProcessor

```
...
@Component
public class AccountItemProcessor implements ItemProcessor<Statement, Statement> {

    @Autowired
    private final JdbcTemplate jdbcTemplate;
```

```java
public AccountItemProcessor(JdbcTemplate jdbcTemplate) {
    this.jdbcTemplate = jdbcTemplate;
}

@Override
public Statement process(Statement item) throws Exception {

    item.setAccounts(this.jdbcTemplate.query("select a.account_id," +
                    "       a.balance," +
                    "       a.last_statement_date," +
                    "       t.transaction_id," +
                    "       t.description," +
                    "       t.credit," +
                    "       t.debit," +
                    "       t.timestamp " +
                    "from account a left join " + //HSQLDB
                    "   transaction t on a.account_id = t.account_account_id "+
                    "where a.account_id in " +
                    "   (select account_account_id " +
                    "   from customer_account " +
                    "   where customer_customer_id = ?) " +
                    "order by t.timestamp",
            new Object[] {item.getCustomer().getId()},
            new AccountResultSetExtractor()));
    return item;
}
}
```

AccountItemProcessor 运行了一个查询,以查找特定客户的所有账户和交易。除 SQL 查询本身外,其他的代码非常简单。真正要做的"工作"是在代码清单 10-29 所示的 AccountResultSetExtractor 中完成的。

代码清单 10-29　AccountResultSetExtractor

```java
...
public class AccountResultSetExtractor implements ResultSetExtractor<List<Account>> {

    private List<Account> accounts = new ArrayList<>();
    private Account curAccount;

    @Nullable
    @Override
    public List<Account> extractData(ResultSet rs) throws SQLException, DataAccessException {

        while (rs.next()) {

            if(curAccount == null) {
                curAccount = new Account(
                        rs.getLong("account_id"),
                        rs.getBigDecimal("balance"),
                        rs.getDate("last_statement_date"));
            }
            else if (rs.getLong("account_id") != curAccount.getId()) {
                accounts.add(curAccount);
```

```java
            curAccount = new Account(rs.getLong("account_id"),
                    rs.getBigDecimal("balance"),
                    rs.getDate("last_statement_date"));
        }

        if(StringUtils.hasText(rs.getString("description"))) {
            curAccount.addTransaction(
                new Transaction(rs.getLong("transaction_id"),
                        rs.getLong("account_id"),
                        rs.getString("description"),
                        rs.getBigDecimal("credit"),
                        rs.getBigDecimal("debit"),
                        new Date(rs.getTimestamp("timestamp").getTime())));
        }
    }

    if(curAccount != null) {
        accounts.add(curAccount);
    }

    return accounts;
}
```

如代码清单 10-29 所示，可通过迭代 ResultSet 来构建 Account 对象。如果当前的账户为 null，或者记录的账户 ID(account_id)与当前的账户 ID 不同，就创建一个新的 Account 对象。有了 Account 对象后，对于每一条交易记录，都添加一个 Transaction 对象。这将允许我们构建一个返回给 AccountItemProcessor 的账户对象列表，可以把这个列表添加到对账单中以进行写入。最后一部分是配置 ItemWriter。10.6.3 节将深入研究用于写对账单的 ItemWriter。

10.6.3 写对账单

在真正生成最终的银行对账单之前，进行很多预处理是很常见的。无论如何，就在此时，我们将为每个文件写一份对账单。对于这个条目写入器，我们需要使用 MultiResourceItemWriter，并把写操作委托给 FlatFileItemWriter。对于每个文件，我们需要生成客户的信息作为头记录，还需要生成每个账户的信息。下面我们从各个细节部分开始，然后将它们组合起来。首先需要创建自定义的 LineAggregator，用于输出每个对账单中的账户。代码清单 10-30 展示了 StatementLineAggregator 的代码。

代码清单 10-30　StatementLineAggregator

```java
public class StatementLineAggregator implements LineAggregator<Statement> {

    private static final String ADDRESS_LINE_ONE =
                String.format("%121s\n", "Apress Banking");
    private static final String ADDRESS_LINE_TWO =
                String.format("%120s\n", "1060 West Addison St.");
    private static final String ADDRESS_LINE_THREE =
                String.format("%120s\n\n", "Chicago, IL 60613");
    private static final String STATEMENT_DATE_LINE =
                String.format("Your Account Summary %78s ", "Statement Period") +
```

第 10 章 示例应用

```java
                    "%tD to %tD\n\n";

    public String aggregate(Statement statement) {
        StringBuilder output = new StringBuilder();
        formatHeader(statement, output);
        formatAccount(statement, output);

        return output.toString();
    }

    private void formatAccount(Statement statement, StringBuilder output) {
        if(!CollectionUtils.isEmpty(statement.getAccounts())) {

            for (Account account : statement.getAccounts()) {

                output.append(
                    String.format(STATEMENT_DATE_LINE,
                        account.getLastStatementDate(),
                        new Date()));

                BigDecimal creditAmount = new BigDecimal(0);
                BigDecimal debitAmount = new BigDecimal(0);
                for (Transaction transaction : account.getTransactions()) {
                    if(transaction.getCredit() != null) {
                        creditAmount =
                        creditAmount.add(transaction.getCredit());
                    }

                    if(transaction.getDebit() != null) {
                        debitAmount =
                            debitAmount.add(transaction.getDebit());
                    }

                    output.append(
                        String.format("        %tD         %-50s       %8.2f\n",
                            transaction.getTimestamp(),
                            transaction.getDescription(),
                            transaction.getTransactionAmount()));
                }

                output.append(
        String.format("%80s %14.2f\n", "Total Debit:" , debitAmount));
                output.append(
        String.format("%81s %13.2f\n", "Total Credit:", creditAmount));
                output.append(
        String.format("%76s %18.2f\n\n", "Balance:", account.getBalance()));
            }
        }
    }

    private void formatHeader(Statement statement, StringBuilder output) {
        Customer customer = statement.getCustomer();
        String customerName =
            String.format("\n%s %s",
                    customer.getFirstName(),
                    customer.getLastName());
```

```
            output.append(customerName +
                ADDRESS_LINE_ONE.substring(customerName.length()));

            output.append(customer.getAddress1() +
                ADDRESS_LINE_TWO.substring(customer.getAddress1().length()));

            String addressString =
                String.format("%s, %s %s",
                            customer.getCity(),
                            customer.getState(),
                            customer.getPostalCode());
            output.append(addressString +
                ADDRESS_LINE_THREE.substring(addressString.length()));
    }
}
```

这里的代码有点多,不过其中大部分都是 String.format 调用和定义良好的表达式。formatHeader (Statement statement, StringBuilder output)方法负责格式化字符串并将其追加到输出中。formatAccount (Statement statement, StringBuilder output)方法完成同样的事情,只不过处理的是每个账户以及其中的交易。

这个条目写入器的下一个组件是 StatementHeaderCallback,用于提供每份对账单的通用元素。代码清单 10-31 展示了 StatementHeaderCallback 的代码。

代码清单 10-31　StatementHeaderCallback

```
...
public class StatementHeaderCallback implements FlatFileHeaderCallback {

    public void writeHeader(Writer writer) throws IOException {
        writer.write(String.format("%120s\n", "Customer Service Number"));
        writer.write(String.format("%120s\n", "(800) 867-5309"));
        writer.write(String.format("%120s\n", "Available 24/7"));
        writer.write("\n");
    }
}
```

StatementHeaderCallback 在本质上与 StatementLineAggregator 所做的事情相同:格式化输出并将其追加到当前的流。不过,这些数据是静态的,所以不会变化。

以上这些就是 FlatFileItemWriter 为生成对账单所需的组件。为了配置真正的 FlatFileItemWriter,只需要将前面两个实例传给构建器,如代码清单 10-32 所示。

代码清单 10-32　individualStatementItemWriter

```
...
@Bean
public FlatFileItemWriter<Statement> individualStatementItemWriter() {
    FlatFileItemWriter<Statement> itemWriter = new FlatFileItemWriter<>();

    itemWriter.setName("individualStatementItemWriter");
    itemWriter.setHeaderCallback(new StatementHeaderCallback());
    itemWriter.setLineAggregator(new StatementLineAggregator());

    return itemWriter;
```

}
...

此处配置了FlatFileItemWriter的名称、StatementHeaderCallback以及前面创建的StatementLineAggregator。完成这些配置后，最后一部分是MultiResourceItemWriter组件。这个组件用于为每个客户生成一个文件，如代码清单10-33所示。

代码清单10-33　statementItemWriter

```
...
@Bean
@StepScope
public MultiResourceItemWriter<Statement> statementItemWriter(@Value("#{jobParameters[
'outputDirectory']}") Resource outputDir) {
    return new MultiResourceItemWriterBuilder<Statement>()
                .name("statementItemWriter")
                .resource(outputDir)
                .itemCountLimitPerResource(1)
                .delegate(individualStatementItemWriter())
                .build();
}
```

此处配置了MultiResoureItemWriter的名称、写入目录的资源、每个资源中数据条目的数量(设置为1)以及委托。

在配置了作业的这些组件之后，可通过./mvnw clean install命令构建作业，通过 java -jar chapter10-0.0.1-SNAPSHOT.jar customerUpdateFile=<PATH_TO_CUSTOMER_FILE>transactionFile=<PATH_TO_TRANSACTION_FILE> outputDirectory=L<PATH_TO_OUTPUT_DIR>运行作业。输出是完整的对账单，如代码清单10-34所示。

代码清单10-34　对账单示例

```
                                              Customer Service Number
                                                       (800) 867-5309
                                                       Available 24/7
Elliot Winslade                                       Apress Banking
3 Clyde Gallagher Parkway                        1060 West Addison St.
San Antonio, Texas 78250                             Chicago, IL 60613

Your Account           Summary Statement Period 05/08/18 to 06/20/18

                                       Total Debit:             0.00
                                       Total Credit:            0.00
                                       Balance:             24082.61

Your Account           Summary Statement Period 05/06/18 to 06/20/18

            06/05/18   Quinu                              10733.88
            06/15/18   Jabbercube                         -1061.00
                                       Total Debit:        -1061.00
                                       Total Credit:      10733.88
                                       Balance:           11413.68
```

10.7 本章小结

如果没有合适的环境，就很难将学到的知识应用到真实世界中。本章介绍了如何把 Spring Batch 中的常见元素综合应用到真实的批处理作业中。

在介绍了基本知识之后，接下来的几章将深入 Spring Batch 中的一些高级话题。第 11 章将介绍如何在单线程执行方式(我们到目前为止使用的方式)之外，对作业的规模进行伸缩。

第 11 章
伸缩和调优

在 2018 年，IRS(美国国家税务局)处理了超过 1.15 亿张的独立纳税申报表。亚特兰大的 Hartsfield-Jackson 机场在 2017 年接待了接近 1.04 亿人次的旅客。每天会有超过 3 亿张照片上传到 Facebook。苹果公司在 2017 年卖出了超过 2 亿台的 iPhone 手机。世界上每天产生的数据量是惊人的。在过去，随着数据量的增长，处理器的处理能力也在提升。如果应用不够快，那么可以在一年之后换台新的服务器，这一切都很好。

但是现在情况不一样了。CPU 不再像以前那样变得更快。但是，计算的总体成本在下降。不是通过获得更快的处理速度，而是通过在单颗芯片上集成更多的核心，或者通过分布式系统获得更多的芯片，从而获得更强的计算能力。Spring Batch 的开发人员深刻理解了这一点，并且将并行处理作为 Spring Batch 框架的主要关注点之一。本章主要介绍如下内容。

- 分析批处理作业的性能：通过学习如何分析批处理作业的性能，使性能优化决策产生积极的影响。
- 评估 Spring Batch 中的各个伸缩选项：Spring Batch 提供了很多不同的伸缩选项，其中的每个都值得详细查看。

11.1 分析批处理作业的性能

Michael A. Jackson 在 1975 年出版的 *Principals fo Program Design* 一书中提出了两条最佳的优化规则。
- 规则 1：不要优化。
- 规则 2：现在优化还不是时候(仅供专家使用)。

背后的想法很简单。软件在开发过程中会发生变化。因此，在系统开发出来之前，几乎不可能对系统的设计做出准确的决定。在系统开发完之后，可以测试系统的性能瓶颈，并根据需求解决性能问题。

为了分析 Java 应用的性能，可以使用很多工具，有免费的，也有收费的。其中，最佳的免费工具是 VisualVM。这个工具可以用来分析批处理作业。在开始分析批处理作业的性能之前，我们先快速地了解一下 VisualVM。

11.1.1 VisualVM 之旅

Oracle 提供的 VisualVM 可以让你了解 JVM 中正在发生的事情。作为 JConsole 的 "大哥"，VisualVM 提供的不仅有像 JConsole 一样的 JMX 管理，还包括 CPU 和内存使用、方法执行次数，以及线程管理和垃圾回收方面的信息。本小节将介绍 VisualVM 工具的这些功能。

在尝试之前，首先必须安装 VisualVM。在 Java 9 之前，JVM 提供了 VisualVM。不过，Java 9 之后情况发生了变化，现在只能从 GitHub 获取。访问 https://visualvm.github.io/index.html，就可以获取最新版本的 VisualVM 和安装说明。

安装好之后，就可以启动 VisualVM 了。在 VisualVM 的欢迎界面中，左边是菜单，右边是开始页面(Start Page)，如图 11-1 所示。

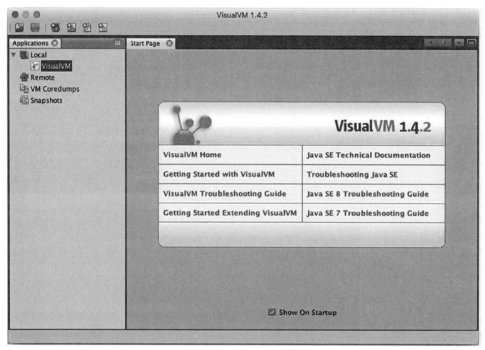

图 11-1　VisualVM 的欢迎界面

左边的菜单有四部分：Local(本地)和 Remote(远程)是用来找到可以连接并进行性能分析的应用的地方。在启动 VisualVM 时，由于 VisualVM 本身就是 Java 应用，因此它也出现在了 Local 部分。在 Local 和 Remote 部分的下面，可以在 VM Coredumps 中加载前面已经收集的 Java VM Coredumps 并进行分析，也可以在 Snapshots 中加载快照，快照是使用 VirtualVM 在某个时间点捕获的 VM 的状态。为了让你了解 VirtualVM 的一些功能，我们用它连接一个 Eclipse 实例。

当首次连接到运行的 JVM 时，VisualVM 会展示如图 11-2 所示的界面。

第 11 章 伸缩和调优

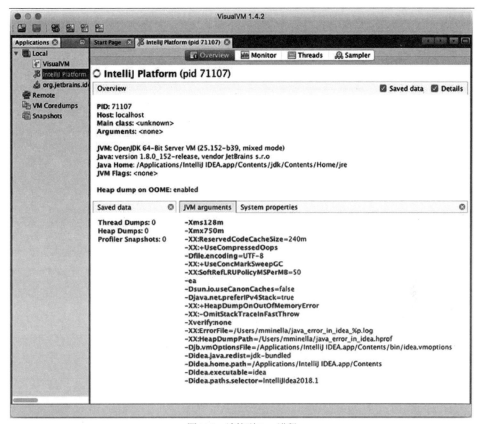

图 11-2　连接到 Java 进程

图 11-2 所示界面的顶部有四个选项卡。

- Overview(概览)：提供 Java 应用的运行概览，包括主类、应用的名称、进程 ID 以及在启动时传入 JVM 的参数。
- Monitor(监视器)：通过图表展示 CPU 占用率、内存占用率(包括堆和 PermGen)、已加载的类的数量以及活动线程和守护线程的数量。在 Monitor 选项卡中，你还可以执行垃圾回收，以及生成堆转储(heap dump)以进行后续分析。
- Threads(线程)：显示应用启动的所有线程的信息以及线程当前的状态(运行、休眠、等待还是监视)。这些数据将以时间线、表格或详细表单的形式展示。
- Sampler(采样器)：允许对应用的 CPU 占用率和内存分配进行采样和创建快照。CPU 展示了正在执行哪个方法以及它们运行了多长时间。内存占用率展示了类占用的内存。

在 Monitor 选项卡中，可以把内存和 CPU 作为整体来查看 JVM 的状态。在从 Monitor 选项卡中识别出问题(例如持续的内存不足或者 CPU 由于某种原因而被挂起)后，其他选项卡在确定问题的原因时将更有用。Monitor 选项卡中的所有图表都是可调整大小的，也可以根据需要隐藏它们，如图 11-3 所示。

329

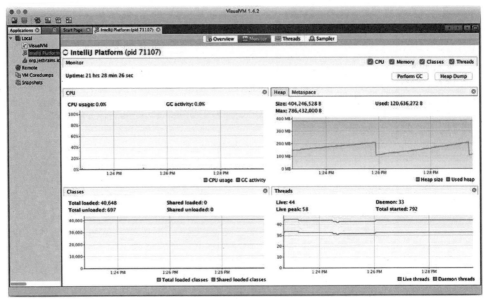

图 11-3　分析 Eclipse 实例时的 Monitor 选项卡

Threads 选项卡如图 11-4 所示。

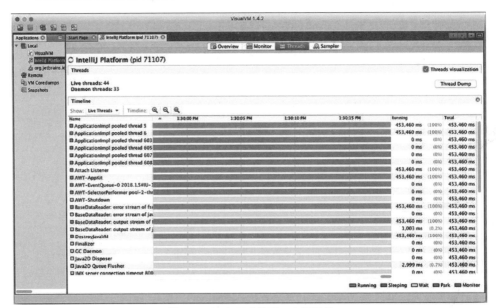

图 11-4　VitualVM 中的 Threads 选项卡

所有的 Java 应用都是多线程的。Java 应用至少有一个主执行线程和一个用于垃圾回收的普通线程。然而，大部分 Java 应用出于不同的原因会衍生出更多的线程。在 Threads 选项卡中，你可以查看 Java 应用衍生出的不同线程的信息，以及它们正在做什么。图 11-4 以时间线展示了这些数据，你也可以用表格展示数据。每个线程的数据也可以展示为详情图。

Sampler 选项卡如图 11-5 所示。

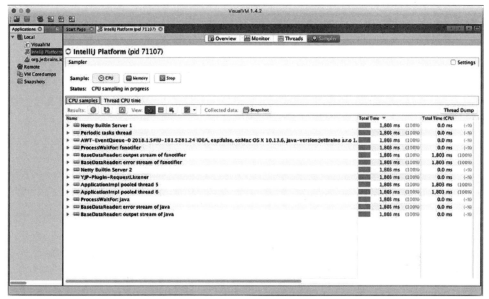

图 11-5　VisualVM 中的 Sampler 选项卡

Sampler 选项卡提供了 CPU、Memory(内存)和 Stop(停止)按钮。无论是根据方法对 CPU 进行采样，还是根据类对内存的足迹进行采样，都可以单击合适的按钮，下方的列表会根据 VisualVM 正在研究的 VM 的当前状态定期更新。

VisualVM 不仅功能强大，而且可扩展。市面上有许多可用的插件提供了开箱即用的特性。可添加插件来扩展 VirtualVM，例如，Thread Inspector (线程检测器)可用于查看当前执行的线程的栈跟踪信息、Visual GC 可用于可视化垃圾回收、可通过 MBean 浏览器访问 MBeans，等等。

现在，你已经了解了 VisualVM 能做什么，下面看看如何用它分析 Spring Batch 应用的性能。

11.1.2　分析 Spring Batch 应用的性能

在分析应用时，通常需要考虑以下两个方面：CPU 的工作强度以及工作在哪里；使用了多少内存以及用到了什么地方。第一个方面与 CPU 正在处理的内容有关。作业在计算方面很难吗？CPU 是否在业务逻辑之外的地方还有大量的处理事项——例如，CPU 是否在解析文件方面花费了比进行实际计算更多的时间？第二个方面是关于内存的。应用是否使用了大部分(如果不是全部的话)可用内存？如果是，那么是什么占用了这些内存？是否有 Hibernate 对象没有延迟加载集合，从而导致内存压力？本小节将介绍如何查看 Spring 批处理应用中的资源都用在了哪里。

1. CPU 分析

在分析应用的性能时，如果有一份直观的检查清单，那就再好不过了。有时候，分析应用的性能更像是艺术活动而不是科学活动。下面将介绍如何获取与应用性能有关的数据以及它们的 CPU 占用率。

当查看 CPU 在应用中的性能表现时，你通常会通过时间度量来确定热点(Hot Pot，不符合预期性能表现的区域)。CPU 工作最多的区域是哪些？例如，如果代码中有一个无限循环，那么在触发这个无限循环之后，CPU 将在它上面花费大量的时间。然而，如果一切正常，就可以查看是否有处理瓶颈(输入输出操作通常是大部分现代系统的瓶颈所在)。

下面以第 10 章使用的银行对账单作业为例。这个作业包括四个步骤，并且与文件和数据库进行了交互，如图 11-6 所示。

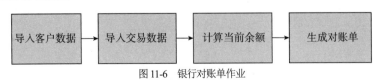

图 11-6　银行对账单作业

为了执行作业，可使用命令 java -jar chapter10-0.0.1-SNAPSHOT.jar customerUpdateFile=<PATH_TO_CUSTOMER_FILE> transactionFile=<PATH_TO_TRANSACTION_FILE> outputDirectory=<PATH_TO_OUTPUT_DIR>。在启动了作业之后，作业就会出现在 VirtualVM 左边菜单的 Local 部分，只需要双击即可连接。

只要连接到正在运行的银行对账单作业，就可以开始查看里边的东西是如何运作的。首先在 Monitor 选项卡中查看 CPU 有多忙。在使用包含 100 个客户和 20 000 笔以上交易的客户交易文件运行银行对账单作业时，就会发现作业的 CPU 占用率非常低。图 11-7 展示了在运行作业之后 Monitor 选项卡中的各种图表。

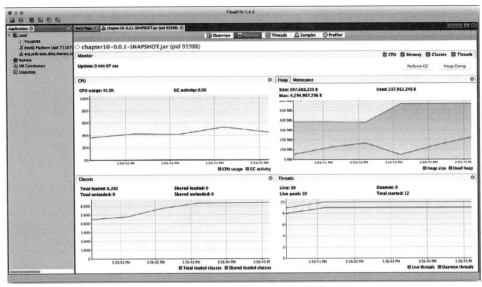

图 11-7　银行对账单作业的资源占用率

如图 11-7 所示，银行对账单作业并不是 CPU 密集型进程。事实上，如果查看内存分析，就会发现这个作业也不是内存密集型进程。当然，只要在第四个步骤的 ItemProcessor 中添加一个小小的循环，很快就能让 CPU 忙碌起来。代码清单 11-1 展示了添加的这个循环。

第 11 章　伸缩和调优

代码清单 11-1　使用 PricingTiersItemProcessor 计算质数

```java
...
@Component
public class AccountItemProcessor implements ItemProcessor<Statement, Statement> {

    @Autowired
    private final JdbcTemplate jdbcTemplate;

    public AccountItemProcessor(JdbcTemplate jdbcTemplate) {
        this.jdbcTemplate = jdbcTemplate;
    }

    @Override
    public Statement process(Statement item) throws Exception {

        int threadCount = 10;
        CountDownLatch doneSignal = new CountDownLatch(threadCount);
        for(int i = 0; i < threadCount; i++) {
            Thread thread = new Thread(() -> {
                    for (int j = 0; j < 1000000; j++) {
                        new BigInteger(String.valueOf(j)).isProbablePrime(0);
                    }
                    doneSignal.countDown();
            });
            thread.start();
        }

        doneSignal.await();

        item.setAccounts(this.jdbcTemplate.query("select a.account_id," +
                        "     a.balance," +
                        "     a.last_statement_date," +
                        "     t.transaction_id," +
                        "     t.description," +
                        "     t.credit," +
                        "     t.debit," +
                        " t.timestamp " +
                        " from account a left join " + //HSQLDB
                        "   transaction t on a.account_id = t.account_account_id " +
//                      " from account a left join " + //MYSQL
//                      "   transaction t on a.account_id = t.account_account_id " +
                        " where a.account_id in " +
                        "    (select account_account_id " +
                        "     from customer_account " +
                        "     where customer_customer_id = ?) " +
                        " order by t.timestamp",
                new Object[] {item.getCustomer().getId()},
                new AccountResultSetExtractor()));
        return item;
    }
}
```

在代码清单 11-1 中，这个循环启动了大量的线程，并在每个线程中计算 0 和 1 000 000 之间的质数。这种意外的循环可能会在处理数百万个事务的过程中对批处理作业的性能产生灾难性的影响。图 11-8 展示了这个小小的循环对 CPU 利用率的影响。

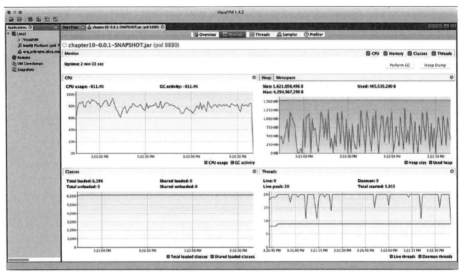

图 11-8　更新后的银行对账单作业的资源占用率

如你所料，银行对账单作业的前三个步骤执行得很快，它们甚至都没有出现在图 11-8 中，但是当作业执行到第四个步骤时，将消耗更多的线程、内存和 CPU。你并不知道是什么导致这样的峰值出现。

在识别出这样的峰值之后，下一个要查看的地方是 Sampler 选项卡。通过在相同的条件下重新运行作业，就可以看到哪些个别方法成了热点。在本例中，用于执行计算的 lambda 表达式在以 CPU 时间排序的列表中位列第三。在作业结束时，这占用执行作业所需的所有 CPU 时间的 24.2%，如图 11-9 所示。

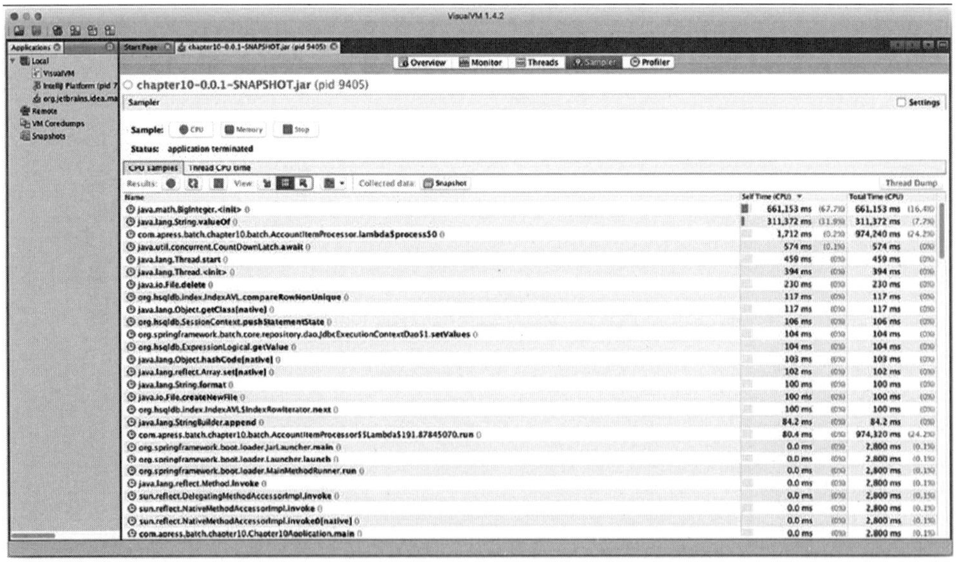

图 11-9　AccountItemProcessor 占用了大量的 CPU 时间

第 11 章 伸缩和调优

当遇到类似的场景时，为了查看是什么在消耗 CPU 执行时间，更好的方法是根据代码使用的包名来筛选列表。在本例中，可以通过 com.apress.batch.chapter10.*过滤列表，以查看每个类占用了总 CPU 利用率的多少。在这个示例中，罪魁祸首变得非常清楚：AccountItemProcessor. lambda$process$0，它占用 24.2%的 CPU 时间。你现在已经得到了从这个工具中所能获取的所有信息，是时候深入研究代码，以确定 AccountItemProcessor 中的哪个部分占用了这么多 CPU 时间。

这很简单，不是吗？不一定。尽管这里使用的流程可以在任何系统中用来缩小问题的范围，但是问题并不总是这么容易追踪。不过，可以使用 VirtualVM 逐步缩小问题在作业中的范围。CPU 利用率并不是性能的唯一部分。接下来介绍如何使用 VirtualVM 分析内存。

2. 内存分析

尽管 CPU 利用率可能是最容易遇到问题的地方，但根据笔者的经验，内存问题更可能出现在软件中。原因就在于我们使用的框架在幕后做了很多事情。当不正确地使用这些框架时，就可能会创建大量对象，并且没有任何迹象表明这种状况已经发生，直到完全耗尽内存为止。下面介绍如何使用 VisualVM 分析内存使用情况。

为了了解如何分析内存，我们稍微调整一下前面所做的修改。不过，这一次并不是占用处理时间，而是模拟创建字符串，并且是不加控制地创建。尽管这里的代码示例可能不会出现在真实的系统中，但是过度的字符串操作是导致内存问题的常见原因。代码清单 11-2 展示了更新后的 AccountItemProcessor。

代码清单 11-2　会发生内存泄漏的 PricingTierItemProcessor

```java
@Component
public class AccountItemProcessor implements ItemProcessor<Statement, Statement> {

    @Autowired
    private final JdbcTemplate jdbcTemplate;

    public AccountItemProcessor(JdbcTemplate jdbcTemplate) {
        this.jdbcTemplate = jdbcTemplate;
    }

    @Override
    public Statement process(Statement item) throws Exception {

        String memoryBuster = "memoryBuster";

        for (int i = 0; i < 200; i++) {
            memoryBuster += memoryBuster;
        }

        item.setAccounts(this.jdbcTemplate.query("select a.account_id," +
                "       a.balance," +
                "       a.last_statement_date," +
                "       t.transaction_id," +
                "       t.description," +
                "       t.credit," +
                "       t.debit," +
                "       t.timestamp " +
                " from account a left join " + //HSQLDB
```

```
//                          "      transaction t on a.account_id = t.account_account_id " +
//                          "  from account a left join " + //MYSQL
                            "      transaction t on a.account_id = t.account_account_id " +
                            " where a.account_id in " +
                            "    (select account_account_id " +
                            "     from customer_account " +
                            "     where customer_customer_id = ?) " +
                            " order by t.timestamp",
                            new Object[] {item.getCustomer().getId()},
                            new AccountResultSetExtractor());
        return item;
    }
}
```

代码清单 11-2 不加控制地创建了字符串。通过采用类似的做法，你会看到内存的增加轨迹也失控了。

当运行银行对账单作业并使用 VisualVM 进行分析时，你将从内存角度看到，事情很快变得失控；在步骤执行到一半时抛出了 OutOfMemoryException 异常。图 11-10 显示了在运行带有内存泄漏错误的银行对账单作业时的 VisualVM Inspector 选项卡。

图 11-10　带有内存泄漏错误的银行对账单作业的监视结果

注意在图 11-10 中，内存图表的右上角出现了内存峰值，进而导致 OutOfMemoryException 异常。但是，如何才能知道导致内存峰值的原因呢？如果不知道，那么 Sampler 选项卡可能会为你提供一些信息。

你在前面已经看到过，Sampler 选项卡可以展示导致 CPU 利用率上升的方法调用，并且能告诉你是哪些对象在占用珍贵的内存。为此，你需要像前面一样执行作业。在执行时，使用 VirtualVM 连接到作业进程，然后打开 Sampler 选项卡。为了确定产生内存泄漏的原因，需要确定发生的内存使用变化。如图 11-11 所示，每一个块代表一个类的实例。每一列中堆积的块越多，意味着内存中的实例越

第 11 章　伸缩和调优

多。每一列代表 JVM 在某一时刻的快照。当程序开始时，已创建实例的数量很小(在本例中是 1)；这个数字将缓慢增加，偶尔在发生垃圾回收时降低。最后，末尾出现了峰值——9 个实例。

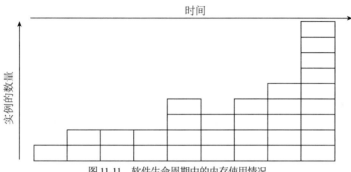

图 11-11　软件生命周期中的内存使用情况

为了在批处理作业中查看这种变化类型，可以使用 VisualVM 的快照特性。在作业运行时，单击屏幕中间的 Snapshot(快照)按钮。VirtualVM 会记录创建快照时的 JVM 确切状态。可以通过对比不同的快照来确定变化。通常，这种变化表明了产生问题的原因所在。即使这不是确凿的证据，也绝对应该是你开始寻找证据的地方。

本章在前面讨论过，在解决性能漏洞时，并不要求能够对批处理作业进行伸缩。相反，对于前面讨论过的带有漏洞的作业，无论做什么都是不能伸缩的。你需要在使用 Spring Batch 或其他框架提供的可伸缩特性之前，在应用中解决这种问题。当你的系统没有这些问题时，Spring Batch 提供的可伸缩特性超越了单线程、单 JVM 方式，在所有框架中最为强大。本章的剩余部分将介绍如何使用 Spring Batch 的可伸缩特性。

11.2　伸缩作业

随着业务的顺利开展，数据也变多了，企业有了更多的客户、更多的交易以及更大的站点访问量。批处理作业也需要与时俱进才行。Spring Batch 在最初设计时就具有很强的可伸缩性，以适应小的批处理作业和巨大的企业级规模的批处理基础设施。本节将介绍 Spring Batch 提供的默认流程之外的伸缩批处理作业的四种方法：多线程步骤、并行步骤、分区以及远程分块。

11.2.1　多线程步骤

默认情况下，步骤是在单个线程中处理的。尽管多线程步骤是并行化作业执行的最简单方式，但是对于所有的多线程环境而言，在使用的过程中仍需要慎重考虑一些方面。下面将介绍 Spring Batch 的多线程步骤，以及如何在批处理作业中安全地使用多线程步骤。

Spring Batch 的多线程步骤的理念是允许批处理作业使用 Spring 的 org.springframework.core.task.TaskExecutor(任务执行器)抽象在自己的线程中执行每个块。图 11-12 展示了如何使用多线程步骤处理块。

如图 11-12 所示，作业中的任何步骤都可以配置为在线程池中执行，并独立地处理每一个块。在处理块时，Spring Batch 跟踪相应的操作。如果任何一个线程发生了错误，那么按照 Spring Batch 的

337

常规步骤，作业的处理就会回滚或终止。

图 11-12　使用多线程步骤处理块

为了让步骤以多线程方式执行，只需要为给定的步骤配置指向 TaskExecutor 的引用。如果以银行对账单作业为例，那么代码清单 11-3 展示了如何使用多线程步骤配置单步骤的作业。

代码清单 11-3　使用了多线程步骤的 MultithreadedJobApplication

```java
...
@EnableBatchProcessing
@SpringBootApplication
public class MultithreadedJobApplication {
    @Autowired
    private JobBuilderFactory jobBuilderFactory;

    @Autowired
    private StepBuilderFactory stepBuilderFactory;

    @Bean
    @StepScope
    public FlatFileItemReader<Transaction> fileTransactionReader(
            @Value("#{jobParameters['inputFlatFile']}") Resource resource) {

        return new FlatFileItemReaderBuilder<Transaction>()
                .name("transactionItemReader")
                .resource(resource)
                .saveState(false)
                .delimited()
                .names(new String[] {"account", "amount", "timestamp"})
                .fieldSetMapper(fieldSet -> {
                    Transaction transaction = new Transaction();
                    transaction.setAccount(fieldSet.readString("account"));
                    transaction.setAmount(fieldSet.readBigDecimal("amount"));
                    transaction.setTimestamp(fieldSet.
                        readDate("timestamp", "yyyy-mm-dd hh:mm:ss"));

                    return transaction;
                })
                .build();
    }

    @Bean
    @StepScope
    public JdbcBatchItemWriter<Transaction> writer(DataSource dataSource) {
        return new JdbcBatchItemWriterBuilder<Transaction>()
                .dataSource(dataSource)
                .sql("INSERT INTO TRANSACTION (ACCOUNT, AMOUNT, TIMESTAMP) " +
                    "VALUES (:account, :amount, :timestamp)")
```

```
                    .beanMapped()
                    .build();
}

@Bean
public Job multithreadedJob() {
    return this.jobBuilderFactory.get("multithreadedJob")
                    .start(step1())
                    .build();
}

@Bean
public Step step1() {
    return this.stepBuilderFactory.get("step1")
                    .<Transaction, Transaction>chunk(100)
                    .reader(fileTransactionReader(null))
                    .writer(writer(null))
                    .taskExecutor(new SimpleAsyncTaskExecutor())
                    .build();
}

public static void main(String[] args) {
    String [] newArgs = new String[] {"inputFlatFile=/data/csv/bigtransactions.csv"};
    SpringApplication.run(MultithreadedJobApplication.class, newArgs);
}
}
```

如代码清单 11-3 所示，为了让作业中的步骤拥有 Spring 的多线程处理能力，只需要定义 TaskExecutor 的实现(在本例中是 org.springframework.core.task.SimpleAsyncTaskExecutor)，并且在步骤中引用即可。当执行作业时，Spring 将为步骤中的每个块创建一个新的线程，并且并行地执行每个块。可以想象，这对大多数作业来说都是强大的补充。

但是，使用多线程步骤时存在一个问题。Spring Batch 提供的大部分 ItemReader 都是有状态的。Spring Batch 在重启作业时只有通过状态，才能知道作业停止的位置。然而，在多线程环境中，如果对象维护的状态可被多个线程访问(但不是同步的)，那么可能会出现线程之间互相覆盖状态的问题。因此，我们将关闭条目读取器的状态保存特性，这样作业就不能重启了。

添加任务执行器是提高性能的第一步。不过在很多情况下，这实际上并不影响性能(如果输入机制已经占用了网络、磁盘总线等资源的话)。下一种用于伸缩 Spring Batch 作业的机制是并行步骤。

11.2.2 并行步骤

多线程步骤为你提供了对同一步骤中的块进行并行处理的能力，但有时候并行执行整个步骤也很有帮助。例如，当导入互相之间没有关联的多个文件时，我们没有理由让一个文件的导入等待另一个文件导入完之后才开始。Spring Batch 为你提供了以并行的方式执行步骤甚至流(可复用的步骤组)的能力，从而改善作业的整体吞吐量。本小节将介绍如何使用 Spring Batch 的并行步骤和流来改善作业的整体性能。

下面考虑一个从多个源读取文件的示例，假如每个客户都提供了一个用于导入系统的文件。一些客户喜欢 CSV 文件，而另一些客户喜欢 XML 文件。数据是相同的，但格式不同。在这种情况下，我们可以用不同的方式来实现。然而，由于每个文件都是独立的，因此一种相对容易的方式是通过执行并行步骤来实现。图 11-13 展示了并行步骤的执行方式。

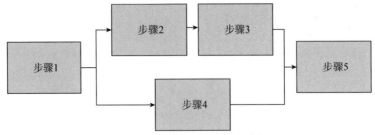

图 11-13　订单处理作业的执行流程

图 11-13 所示的订单处理作业从单个步骤开始，然后并行地分成两个流进行处理。上面的流执行步骤 2，完成后执行步骤 3。下面的流执行步骤 4。一旦两个流都完成，就执行步骤 5。

下面查看一个包含两个步骤的简单作业。每个步骤负责使用不同的输入格式导入数据。步骤 1 从 XML 文件读取数据。步骤 2 从平面文件读取数据。这两个步骤将并行地执行。每种格式的数据都是相同的。交易对象则包含了账户、时间戳和交易金额。注意，这里将使用一个简单的领域模型，这是为了专注于伸缩组件而不是领域对象本身。

1. 配置并行步骤

为了并行地执行步骤，Spring Batch 使用了 Spring 的 TaskExecutor。在这个示例中，每个流都在自己的线程中执行，这将允许并行地执行多个流。为了进行配置，可使用 FlowBuilder 的 split 方法，该方法接收一个 TaskExecutor 作为参数，返回一个 SplitBuilder，使用这个 SplitBuilder 可以添加任意多个流对象。这些流都会在它们自己的线程中执行(基于底层的 TaskExecutor 规则)，从而为你提供了并行执行步骤或流的能力。

注意，使用了 split 方法的作业的执行顺序与常规作业相同。在常规作业中，步骤在处理完自身的所有数据之后才算完成，而下一个步骤只有在前一个步骤完成后才开始。在使用了 split 方法之后，只有在配置的所有流都已完成时，才会执行之后的步骤。

■ **注意**　拆分之后的步骤并不会执行，除非拆分出来的所有流都已经完成。

对于每个输入源，我们都会配置步骤，以便并行执行。代码清单 11-4 展示了整个作业的配置。

代码清单 11-4　并行步骤的配置

```
@EnableBatchProcessing
@SpringBootApplication
public class ParallelStepsJobApplication {

    @Autowired
    private JobBuilderFactory jobBuilderFactory;

    @Autowired
    private StepBuilderFactory stepBuilderFactory;
```

第 11 章 伸缩和调优

```java
@Bean
public Job parallelStepsJob() {
    Flow secondFlow = new FlowBuilder<Flow>("secondFlow")
                    .start(step2())
                    .build();

    Flow parallelFlow = new FlowBuilder<Flow>("parallelFlow")
                    .start(step1())
                    .split(new SimpleAsyncTaskExecutor())
                    .add(secondFlow)
                    .build();

    return this.jobBuilderFactory.get("parallelStepsJob")
                    .start(parallelFlow)
                    .end()
                    .build();
}

@Bean
@StepScope
public FlatFileItemReader<Transaction> fileTransactionReader(
            @Value("#{jobParameters['inputFlatFile']}") Resource resource) {

return new FlatFileItemReaderBuilder<Transaction>()
                    .name("flatFileTransactionReader")
                    .resource(resource)
                    .delimited()
                    .names(new String[] {"account", "amount", "timestamp"})
                    .fieldSetMapper(fieldSet -> {
                            Transaction transaction = new Transaction();
                            transaction.setAccount(fieldSet.readString
                                ("account"));
                            transaction.setAmount(fieldSet.readBigDecimal
                                ("amount"));
                            transaction.setTimestamp(fieldSet.
                                readDate("timestamp", "yyyy-mm-dd hh:mm:ss"));
                            return transaction;
                    })
                    .build();
}

@Bean
@StepScope
public StaxEventItemReader<Transaction> xmlTransactionReader(
            @Value("#{jobParameters['inputXmlFile']}") Resource resource) {
    Jaxb2Marshaller unmarshaller = new Jaxb2Marshaller();
    unmarshaller.setClassesToBeBound(Transaction.class);

    return new StaxEventItemReaderBuilder<Transaction>()
                    .name("xmlFileTransactionReader")
                    .resource(resource)
                    .addFragmentRootElements("transaction")
                    .unmarshaller(unmarshaller)
                    .build();
}
```

```java
@Bean
@StepScope
public JdbcBatchItemWriter<Transaction> writer(DataSource dataSource) {
    return new JdbcBatchItemWriterBuilder<Transaction>()
                .dataSource(dataSource)
                .beanMapped()
                .sql("INSERT INTO TRANSACTION (ACCOUNT, AMOUNT, TIMESTAMP) " +
                        "VALUES (:account, :amount, :timestamp)")
                .build();
}

@Bean
public Step step1() {
    return this.stepBuilderFactory.get("step1")
                .<Transaction, Transaction>chunk(100)
                .reader(xmlTransactionReader(null))
                .writer(writer(null))
                .build();
}

@Bean
public Step step2() {
    return this.stepBuilderFactory.get("step2")
                .<Transaction, Transaction>chunk(100)
                .reader(fileTransactionReader(null))
                .writer(writer(null))
                .build();
}

public static void main(String[] args) {
    String[] newArgs = new String[] {
        "inputFlatFile=/data/csv/bigtransactions.csv",
        "inputXmlFile=/data/xml/bigtransactions.xml"};
    SpringApplication.run(ParallelStepsJobApplication.class, newArgs);
}
}
```

代码清单 11-4 展示了配置作业所需的一切。代码中的关键部分已加粗显示，其中使用 FlowBuilder 创建了两个流。其中一个流(secondFlow)用于提取 CSV 文件，其中包含了前面提到的三个字段(account、amount 和 timestamp)。另一个流(parallelFlow)用于实现真正的拆分，从而执行用于提取 XML 文件的步骤，同时并行地运行前面定义的流。

代码清单 11-4 的剩余部分定义了用于每个步骤的读取器、写入器以及步骤本身。代码清单 11-4 的最后是 Spring Boot 使用的 main 方法。有了输入才可以运行作业，代码清单 11-5 展示了一个 CSV 输入文件。

代码清单 11-5　bigtransactions.csv

```
5113971498870901,-546.68,2018-02-08 17:46:12
4041373995909987,-37.06,2018-02-02 21:10:33
3573694401052643,-784.93,2018-02-04 13:01:30
3543961469650122,925.44,2018-02-05 23:41:50
3536921428140325,507.57,2018-02-13 02:09:08
4905167183996244409,-575.81,2018-02-15 20:43:12
```

```
201904179222112,-964.21,2018-02-08 15:50:21
5602221470889083,23.71,2018-02-14 10:23:41
5038678280559913,979.94,2018-02-05 04:28:31
```

代码清单 11-6 展示了一个 XML 输入文件。

代码清单 11-6　bigtransactions.xml

```xml
<transactions>
        <transaction>
                <account>633110684460535475</account>
                <amount>961.93</amount>
                <timestamp>2018-02-03 18:30:51</timestamp>
        </transaction>
        <transaction>
                <account>3555221131716404</account>
                <amount>759.62</amount>
                <timestamp>2018-02-12 20:02:01</timestamp>
        </transaction>
        <transaction>
                <account>30315923571992</account>
                <amount>648.92</amount>
                <timestamp>2018-02-12 23:16:45</timestamp>
        </transaction>
        <transaction>
                <account>5574851814767258</account>
                <amount>-90.11</amount>
                <timestamp>2018-02-04 10:01:04</timestamp>
        </transaction>
</transactions>
```

当执行作业时，你可以在日志中看到两个步骤是同时开始执行的，并且作业是在两个步骤都完成后才完成的。代码清单 11-7 显示的日志片段说明了这一点。

代码清单 11-7　parallelStepsJob 作业的日志

```
2018-12-03 15:46:09.575 INFO 44705 --- [main] o.s.b.c.l.support.SimpleJobLauncher:
Job: [FlowJob: [name=parallelStepsJob]] launched with the following parameters:
[{inputXmlFile=/data/xml/bigtransactions.xml, inputFlatFile=/data/csv/bigtransactions.csv}]
2018-12-03 15:46:09.661 INFO 44705 --- [cTaskExecutor-2] o.s.batch.core.job.
SimpleStepHandler: Executing step: [step1]
2018-12-03 15:46:09.670 INFO 44705 --- [cTaskExecutor-1] o.s.batch.core.job.
SimpleStepHandler: Executing step: [step2]
2018-12-03 15:46:09.819 INFO 44705 --- [cTaskExecutor-2] o.s.oxm.jaxb.Jaxb2Marshaller:
Creating JAXBContext with classes to be bound [class io.spring.batch.scalingdemos.domain.
Transaction]
2018-12-03 15:46:29.960 INFO 44705 --- [main] o.s.b.c.l.support.SimpleJobLauncher:
Job: [FlowJob: [name=parallelStepsJob]] completed with the following parameters:
[{inputXmlFile=/data/xml/bigtransactions.xml, inputFlatFile=/data/csv/bigtransactions.
csv}] and the following status: [COMPLETED]
```

当执行独立的步骤并希望改善它们的性能时，并行地执行步骤可能非常有用。Spring Batch 提供的另一种机制仅依赖于单个 JVM 中的线程来进行伸缩，并且组合使用了 AsyncItemProcessor(异步条目处理器)和 AsyncItemWriter(异步条目写入器)。11.2.3 节将介绍如何改善 ItemProcessor 阶段的性能。

11.2.3 组合使用 AsyncItemProcessor 和 AsyncItemWriter

在某些处理中，ItemProcessor 会成为步骤的瓶颈。例如，ItemProcessor 因为涉及复杂的计算，所以减慢了步骤的整体执行速度。改善性能的一种方法是在新的线程中进入步骤的 ItemProcessor 阶段。AsyncItemProcessor 和 AsyncItemWriter 允许我们这么做。

AsyncItemProcessor 修饰器用于封装已有的 ItemProcessor 实现。当把数据传给这个修饰器时，对底层委托的调用将在新的线程中执行。返回的 Future 代表了 ItemProcessor 的执行结果，然后就被传给 AsyncItemWriter。就像 AsyncItemProcessor，AsyncItemWriter 是 ItemWriter 的封装器。AsyncItemWriter 最终打开了返回的 Future，并把结果传给委托的 ItemWriter。注意，应该成对儿使用 AsycnItemProcessor 和 AsyncItemWriter，否则你就只能自己打开 AsyncItemProcessor 返回的 Future 了。

在使用 AsyncItemProcessor 和 AsyncItemWriter 之前，需要为项目导入 spring-batch-integration 模块。代码清单 11-8 展示了用于将这个模块添加到 pom.xml 中的 Maven 配置。

代码清单 11-8　spring-batch-integration

```
...
<dependency>
        <groupId>org.springframework.batch</groupId>
        <artifactId>spring-batch-integration</artifactId>
</dependency>
...
```

对于这个示例，我们将使用与并行步骤相同的用例。不过，这次我们只导入 CSV 文件。但是，区别在于我们将添加 ItemProcessor，对于每个数据，都会执行 Thread.sleep(5)。尽管 5 毫秒似乎不长，但是如果按顺序处理一百万条记录，就将给作业增加一个多小时的处理时间。另外，如果这是并行化的，那么效果将立竿见影。

在定义好了 ItemProcessor 之后，我们将定义 AsyncItemProcessor 来对它进行修饰。我们将在另一个线程中使用 TaskExecutor 调用底层的 ItemProcessor#process。代码清单 11-9 展示了用于本例的 ItemProcessor 的配置。

代码清单 11-9　异步的条目处理器

```
...
  @Bean
  public AsyncItemProcessor<Transaction, Transaction> asyncItemProcessor() {
      AsyncItemProcessor<Transaction, Transaction> processor = new
      AsyncItemProcessor<>();

      processor.setDelegate(processor());
      processor.setTaskExecutor(new SimpleAsyncTaskExecutor());

      return processor;
  }

  @Bean
  public ItemProcessor<Transaction, Transaction> processor() {
      return (transaction) -> {
```

```
            Thread.sleep(5);
            return transaction;
        };
    }
...
```

有了这些 ItemProcessor，就可以添加 AsyncItemWriter 并配置步骤以使用它们。代码清单 11-10 的开头是执行真正任务的条目写入器，在这个示例中，具体配置与并行步骤中的条目写入器相同。代码清单 11-10 中的第二个 Bean 是 AsyncItemWriter，它用来修饰将要委托的 JdbcBatchItemWriter。然后，我们将步骤配置为使用 AsyncItemProcessor 和 AsyncItemWriter，而不是使用已经配置的委托。在步骤的配置中，需要注意的另一个细节是调用 chunk 方法时使用的类型。这里的范型并不是通常的 <Transaction, Transaction>，由于第二个范型用于指明 ItemWriter 的输入，因此需要将其更新为 <Transaction, Future<Transaction>>，这是因为 AsyncItemProcessor 实际返回的是 Future<Transaction>。代码清单 11-10 的最后一部分是作业的配置，这里使用了刚刚配置的步骤。

代码清单 11-10　异步的条目写入器

```
...
@Bean
public JdbcBatchItemWriter<Transaction> writer(DataSource dataSource) {
    return new JdbcBatchItemWriterBuilder<Transaction>()
                    .dataSource(dataSource)
                    .beanMapped()
                    .sql("INSERT INTO TRANSACTION (ACCOUNT, AMOUNT, TIMESTAMP) " +
                        "VALUES (:account, :amount, :timestamp)")
                    .build();
}

@Bean
public AsyncItemWriter<Transaction> asyncItemWriter() {
    AsyncItemWriter<Transaction> writer = new AsyncItemWriter<>();

    writer.setDelegate(writer(null));

    return writer;
}

@Bean
public Step step1async() {
    return this.stepBuilderFactory.get("step1async")
                    .<Transaction, Future<Transaction>>chunk(100)
                    .reader(fileTransactionReader(null))
                    .processor(asyncItemProcessor())
                    .writer(asyncItemWriter())
                    .build();
}

@Bean
public Job asyncJob() {
    return this.jobBuilderFactory.get("asyncJob")
                    .start(step1async())
                    .build();
}
```

......

如果现在运行使用了 AsyncItemProcessor 和 AsyncItemWriter 的作业，你就能看到，就算有一百万条记录，性能也会有显著的提升。你还要注意一点，这里的示例使用了 SimpleAsyncTaskExecutor，它对于每个请求都会使用一个新的线程。在生产环境中，应该使用一些更安全的方式，例如使用 ThreadPoolTaskExecutor(线程池任务执行器)。

到目前为止，所有的伸缩选项都在使用单个 JVM 中的线程。然而，并不是所有的工作负载都能放入单个 JVM 中。在 Spring Batch 工作负载的下一个伸缩选项中，可以选择是通过单个 JVM 中的线程来使用，还是通过远程的 JVM Worker 来使用。下面我们看看分区。

11.2.4 分区

大多数基于批处理的工作负载都是 I/O 密集型的。在与数据库交互或读取文件时，通常会涉及性能和可伸缩性问题。为了应对这种状况，Spring Batch 提供了让多个 Worker 执行完整步骤的能力。可以把所有的 ItemReader、ItemProcessor 和 ItemWriter 交互卸载给 Worker。本小节将介绍什么是分区，以及如何让作业利用分区这一强大特性。

分区是一种概念，其中的主步骤(Master Step)会将工作转交给任意数量的从步骤(Work Step)进行处理。在分区步骤中，大的数据集(例如包含一百万行记录的数据表)被划分成小的分区。每个分区都由 Worker 并行处理。每个 Worker 都是一个完整的 Spring Batch 步骤，负责各自的读取、处理、写入等。这种模型有很大的优势，例如天生具有重启特性。Worker 的实现很自然，因为 Worker 仅仅是另一个步骤。

在 Spring Batch 中使用分区步骤时，你需要理解两个重要的抽象。首先是 Partitioner(分区器)接口，这个接口负责理解进行分区的数据，以及如何将它们拆分到不同的分区。回到包含一百万条记录的那个数据表，Partitioner 接口的实现可能通过执行查询来确定每个分区中有哪些 ID。Spring Batch 提供了一个开箱即用的 Partitioner 实现——MultiResourcePartitioner，用于查看一个资源(Resource)数组，然后为其中的每个资源创建一个文件。

这里暂停一下，思考如下问题：在 Spring Batch 中，分区是什么？它们是如何表示的？其实很简单。分区由包含相关数据的 ExecutionContext 表示，这些数据用于标识分区的组成。当使用 MultiResourcePartitioner 时，Spring Batch 将在 ExecutionContext 中为每个分区设置资源的名称。这些信息存储在作业存储库中，以便将来被 Worker 引用。

Partitioner 接口只有一个方法 partition(int gridSize)，该方法将返回一个 Map<String, ExecutionContext>对象。参数 gridSize 仅仅表示以对整个集群有效的方式划分数据的 Worker 数量，也就是说，Spring Batch 并不会动态地设置这个参数，而是由你来计算或设置。在该方法返回的键值映射中，键是分区的名称，并且是唯一的。前面已经提到过的 ExecutionContext 用于表示分区的元数据以及标识要处理的内容。

Spring Batch 的分区步骤中的另一个重要抽象是 PartitionHandler(分区处理器)，这个接口能帮助你理解如何与 Worker 通信、如何告诉每个 Worker 该做什么，以及如何表明所有的工作已经完成。在使用 Spring Batch 时，尽管你可能要编写自己的 Partitioner 实现，但可能不会编写自己的 PartitionHandler。

在 Spring Portfolio 中，PartitionHandler 接口有三个实现；而在 Spring Batch 中有两个，分别是

TaskExecutorPartitionHandler 和 MessageChannelPartitionHandler。TaskExecutorPartitionHandler 在同一个 JVM 中以线程的方式启动 Worker，这样就可以在单个 JVM 中使用分区的概念。MessageChannelPartitionHandler 使用 Spring Integration 把元数据发送到远程的 JVM 以进行处理。Spring 中的最后一个 PartitionHandler 实现由 Spring Cloud Task 项目提供，这个项目提供了 DeployerPartitionHandler 实现，并委托 Spring Cloud Deployer 实现在支持的平台上按需启动 Worker。这些 Worker 在启动后，将执行它们的分区，然后在运行时关闭提供的动态扩展功能。在探索 Spring Batch 的分区时，我们需要了解这三个 Partitioner 实现。图 11-14 展示了分区步骤中各个组件之间的关系。

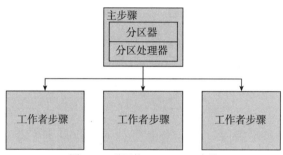

图 11-14　分区的 Spring Batch 步骤

在使用分区步骤时，需要考虑一些问题。例如，步骤的状态是在作业存储库中维护的，包括主步骤和所有从步骤的状态，所以必须配置集群的所有组件以访问同一个作业存储库的数据库实例。另外，如果想要使用 MessageChannelPartitionHandler，那么需要能够访问远程的 JVM。这意味着需要使用消息中间件或者 Spring Integration 支持的其他机制进行通信。下面先从使用单个 JVM 的代码角度深入研究一下分区。

1. TaskExecutorPartitionHandler

TaskExecutorPartitionHandler 允许分区步骤在单个 JVM 中使用线程执行多个 Worker。我们首先从分区开始，然后再考虑因编排远程 JVM 而增加的复杂性，这在很多情况下是一种不错的方式，但这种方式受限于单个 JVM 的限制。当以这种方式使用线程时，在单台机器上所能完成的工作(从磁盘获取数据的速度、一次能够拥有的网络连接的速度以及数量等)是有限制的。

我们将要查看的分区步骤的用例是提取多个文件到数据库中，其中的每个文件都可以独立地进行处理。这种用例风格(I/O密集型并且输入之间没有依赖)是使用分区的经典示例。下面将使用与拆分示例相同的 CSV 输入文件。领域对象也是相同的。事实上，这个作业的从步骤只是在代码清单 11-4 所示拆分作业的第二个步骤的基础上做了一些微小的修改。fileTransactionReader 的方法签名通过作业参数得到了相同的用于读取的文件。当使用分区时，可以从步骤执行上下文中获取有关分区的任何信息(例如将要处理的文件)。代码清单 11-11 展示了更新后的用来从步骤中读取文件的代码。

代码清单 11-11　fileTransactionReader

```
...
@Bean
@StepScope
public FlatFileItemReader<Transaction> fileTransactionReader(
        @Value("#{stepExecutionContext['file']}") Resource resource) {
```

```
        return new FlatFileItemReaderBuilder<Transaction>()
                .name("flatFileTransactionReader")
                .resource(resource)
                .delimited()
                .names(new String[] {"account", "amount", "timestamp"})
                .fieldSetMapper(fieldSet -> {
                    Transaction transaction = new Transaction();

                    transaction.setAccount(fieldSet.readString("account"));
                    transaction.setAmount(fieldSet.readBigDecimal("amount"));
                    transaction.setTimestamp(fieldSet.readDate("timestamp",
                                            "yyyy-mm-dd hh:mm:ss"));

                    return transaction;
                })
                .build();
    }
...
```

Spring Batch 的可伸缩特性的一大优点就是能够以迭代的方式添加伸缩性，而不需要每次都进行大量的重写。这个作业的单线程版本与分区版本之间的唯一区别在于添加了分区步骤的配置，并修改了读取器，以便从步骤执行上下文(stepExecutionContext)而不是作业参数中获取文件位置。

现在需要注意的是，前面定义的并且正在复用的步骤是现在的从步骤而不是作业中的主步骤。因此，作业不会直接引用这个步骤。相反，我们需要用两个组件来定义分区步骤：一个 Partitioner 实现和一个 PartitionHandler 实现。对于本例，二者都可以从 Spring Batch 中获得。我们需要做的就是配置它们。下面从 Partitioner 接口开始。

前面已经讨论过，Partitioner 接口负责理解数据集以及如何将数据划分成各个分区。Spring Batch 只提供了一个开箱即用的 Partitioner 实现——MultiResourcePartitioner。这个 Partitioner 实现接收一个资源数组，并为其中的每个资源创建一个新的分区。代码清单 11-12 展示了 MultiResourcePartitioner 的配置。

代码清单 11-12　MultiResourcePartitioner

```
...
@Bean
@StepScope
public MultiResourcePartitioner partitioner(
        @Value("#{jobParameters['inputFiles']}") Resource[] resources) {

    MultiResourcePartitioner partitioner = new MultiResourcePartitioner();

    partitioner.setKeyName("file");
    partitioner.setResources(resources);

    return partitioner;
}
```

这个 Bean 接收的唯一参数，就是当我们把路径传给 inputFiles 时由 Spring 提供的资源(Resource)对象数组。MultiResourcePartitioner 设置了两个值。第一个值是键的名称，由 Worker 用来在

ExecutionContext 中查找所要读取的资源的名称。在本例中，这个值 file 与代码清单 11-11 在定义 FlatFileItemReader 时使用的键相匹配。另一个需要设置的值是资源数组。完成上面的配置后，就可以返回这个实例。

在这个分区步骤中，我们将要使用的另一个组件是 PartitionHandler 实现。针对第一次查看分区的情况，我们将使用 TaskExecutorPartitionHandler。我们为其设置了两个值：第一个值是将要执行的步骤，在这里名为 step1；第二个值是一个 TaskExecutor。默认情况下，如果不给 PartitionStepBuilder 提供 PartitionHandler，或者不给 TaskExecutorPartitionHandler 提供 TaskExecutor，那么默认会使用 SyncTaskExecutor(同步任务执行器)。如果你正在寻找并行化方案，那么这可能看起来与你的目的不符。但实际上，这里并没有其他合理的方案。所以，你需要确保为 PartitionHandler 实现设置一个使用了多线程的 TaskExecutor。为了测试，可使用 SimpleAsyncTaskExecutor。代码清单 11-13 展示了用于作业的 TaskExecutorPartitionHandler 的配置[1]。

代码清单 11-13　TaskExecutorPartitionHandler

```
...
@Bean
public TaskExecutorPartitionHandler partitionHandler() {
    TaskExecutorPartitionHandler partitionHandler =
        new TaskExecutorPartitionHandler();

    partitionHandler.setStep(step1());
    partitionHandler.setTaskExecutor(new SimpleAsyncTaskExecutor());

    return partitionHandler;
}
...
```

在定义了这两个 Bean 之后，就以创建新的分区步骤，并更新作业以使用它。partitionedMaster 步骤使用通常的 StepBuilderFactory 来获取构建器，然后设置了分区器(partitioner)，并提供了将要执行的步骤名称和 Partitioner 实例。我们需要使用步骤名称来为每个分区创建步骤执行上下文。接下来设置的是分区处理器(partitionHandler)。代码清单 11-14 展示了 partitionedMaster 步骤和作业。

代码清单 11-14　partitionedMaster 步骤和作业

```
...
@Bean
public Step partitionedMaster() {
    return this.stepBuilderFactory.get("step1")
                .partitioner(step1().getName(), partitioner(null))
                .partitionHandler(partitionHandler())
                .build();
}

@Bean
public Job partitionedJob() {
    return this.jobBuilderFactory.get("partitionedJob")
                .start(partitionedMaster())
                .build();
```

---

1 不要在生成环境中使用 SimpleAsyncTaskExecutor，否则既不会回收线程，也不会限制创建的线程数量。

```
}
...
```

在定义了所有这些之后，就可以构建和执行作业。在构建了 Jar 之后，可通过命令 java -jar partition-demo-0.0.1-SNAPSHOT.jar inputFiles=/data/csv/transactions*.csv 来执行作业。当在日志中查看输出时，与前面相比并没有多少变化。然而，这里有两个关键不同。首先，位于/data/csv 目录且前缀为 transactions、扩展名为 .csv 的文件都被导入了 TRANSACTION 表。其次，如果查看 BATCH_STEP_EXECUTION 表，你会发现有一条关于分区步骤的记录，并且对于每个执行过的分区，也都有一条额外的记录。代码清单 11-15 展示了运行分区步骤后的 BATCH_STEP_EXECUTION 表。

代码清单 11-15　运行分区步骤后的 BATCH_STEP_EXECUTION 表

```
mysql> select step_name, status, commit_count, read_count, write_count from
SCALING.BATCH_STEP_EXECUTION;
+------------------+-----------+--------------+------------+-------------+
| step_name        | status    | commit_count | read_count | write_count |
+------------------+-----------+--------------+------------+-------------+
| step1            | COMPLETED | 303          | 30000      | 30000       |
| step1:partition1 | COMPLETED | 101          | 10000      | 10000       |
| step1:partition2 | COMPLETED | 101          | 10000      | 10000       |
| step1:partition0 | COMPLETED | 101          | 10000      | 10000       |
+------------------+-----------+--------------+------------+-------------+
4 rows in set (0.01 sec)
```

使用 TaskExecutorPartitionHandler 是为 Spring Batch 步骤添加分区的最简单方式。但是，考虑到被绑定到了单个 JVM，这种方式也是最受限制的。稍后将介绍分布式的批处理过程：通过 MessageChannelPartitionHandler，将分区负载分布到多个 JVM 上。

2．MessageChannelPartitionHandler

Spring Integration 作为 Spring Portfolio 中的项目之一，实现了 Gregor Hohpe 和 Bobby Woolf 的 *Enterprise Integration Patterns* 中所讲的企业集成模式。当在 Spring 项目中寻求 JVM 之间的通信时，通常就会想到 MessageChannelPartitionHandler。

MessageChannelPartitionHandler 是一个 PartitionHandler 实现，它使用 Spring Integration 的 MessageChannel 抽象，以某种方式与外部的 JVM 进行通信。我们的示例将使用 RabbitMQ。RabbitMQ 是开源的消息代理，在生产环境和本地环境中十分易于使用。

当以分布式形式查看分区步骤时，拓扑图显然会发生一些变化。此时，不是通过线程在相同的 JVM 中运行从步骤，而是在每个 Worker JVM 中都提供一个监听器，以监听从步骤的执行请求。图 11-15 展示了远程分区的 Spring Batch 步骤。

如你所见，主步骤通过消息与从步骤通信，以执行从步骤。每个 Worker JVM 都有一个监听器，用于监听队列上的消息。当请求经过网络到达从步骤时，监听器执行从步骤并返回结果。注意，这种架构的一些细节是很重要的。

首先，所有 JVM 都必须配置为使用同样的作业存储库。由于这是维护状态的地方，因此每个步骤维护处理的结果。如果没有这个共享的状态，就不可能重启分区作业。其次，每个从步骤都在作业的上下文之外执行。所以，在从步骤中，无法从 JobExecution 或作业的 ExecutionContext 中获取任何东西。

第 11 章 伸缩和调优

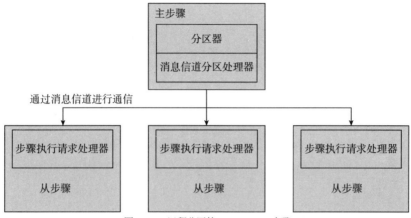

图 11-15 远程分区的 Spring Batch 步骤

现在，除了三个地方之外，这种分区步骤的配置与主步骤的配置是相同的。首先是需要配置通信机制，以便让应用与 RabbitMQ 交互。多亏有了 Spring Boot 和 Spring Integration，这很容易实现。其次是配置新的 PartitionHandler，也就是 MessageChannelPartitionHandler。幸运的是，Spring Batch 提供了构建器以帮助你完成配置。最后，为了简化部署，我们使用 Spring Profile 对从步骤和主步骤的组件进行分组，从而允许为二者使用单独的 JAR，并且通过一个简单的参数来指定正在执行哪一个步骤。

下面从主步骤的配置开始。我们将创建一个配置类，名为 MasterConfiguration；并添加新的注解 @EnableBatchIntegration，这个注解以简单的方式提供了用于构建远程分区步骤所需的构建器。在主步骤的配置中，实际上需要配置两个流：一个出站流(Outbound Flow，用于发送消息给从步骤)和一个入站流(Inbound Flow，用于接收来自从步骤的消息)。代码清单 11-16 展示了 MasterConfiguration 和出站流的配置。

代码清单 11-16　MasterConfiguration 和出站流的配置

```
...
@Configuration
@Profile("master")
@EnableBatchIntegration
public class MasterConfiguration {

    private final JobBuilderFactory jobBuilderFactory;

    private final RemotePartitioningMasterStepBuilderFactory
        masterStepBuilderFactory;

    public MasterConfiguration(JobBuilderFactory jobBuilderFactory,
            RemotePartitioningMasterStepBuilderFactory masterStepBuilderFactory) {

        this.jobBuilderFactory = jobBuilderFactory;
        this.masterStepBuilderFactory = masterStepBuilderFactory;
    }

    /*
```

```
         * Configure outbound flow (requests going to workers)
         */
        @Bean
        public DirectChannel requests() {
           return new DirectChannel();
        }

        @Bean
        public IntegrationFlow outboundFlow(AmqpTemplate amqpTemplate) {
           return IntegrationFlows.from(requests())
                                  .handle(Amqp.outboundAdapter(amqpTemplate)
                                  .routingKey("requests"))
                                  .get();
        }
    ...
```

代码清单 11-16 的开头是多个注解，它们分别用于将 MasterConfiguration 标记为配置类(@Configuration)并且在主配置中使用(@Profile("master"))以及启用 Spring Batch Integration (@EnableBatchIntegration)。这个配置类中的第一个构建器工厂与定义其他作业的构建器工厂相同。但是，第二个构建器工厂(RemotePartitioningMasterStepBuilderFactory)是新的，作用是获取用于构建主远程分区步骤的步骤构建器。

接下来是一个构造函数，用于注入这两个构建器工厂。再接下来是出站流的配置，这里把一条直接信道的一端连接到了 AMQP 模板(由 Spring Boot 的自动配置提供)。IntegrationFlow 使用 Spring Integration 的 Java DSL(Domain Specific Language，领域特定语言)来配置流。换种通俗的说法，就是"当消息进入请求信道时，将它们传递给配置过的一个处理器。这个处理器被配置为 AMQP 的出站适配器，并把消息发送给名为 requests 的 RabbitMQ 队列"。

在配置了出站管道之后，就应该配置入站管道。现在你应该注意到，有两种方式可用于从具有远程分区的从步骤那里获取结果。第一种方式是在接收到每个从步骤发回给主步骤的消息后，进行聚合并评估结果，以确定步骤是否成功(这是我们将要使用的方式)。另一种方式是轮询作业存储库，检查发出去的每个 StepExecution 的状态。一旦它们的状态在数据库中是"完成"，就可以在此基础之上评估状态。如前所述，本例使用的机制选择接收来自每个从步骤的结果。代码清单 11-17 展示了入站主流的配置。

代码清单 11-17 入站流的配置

```
    ...
        /*
         * Configure inbound flow (replies coming from workers)
         */
        @Bean
        public DirectChannel replies() {
            return new DirectChannel();
        }

        @Bean
        public IntegrationFlow inboundFlow(ConnectionFactory connectionFactory) {
            return IntegrationFlows
                        .from(Amqp.inboundAdapter(connectionFactory,"replies"))
                        .channel(replies())
```

```
                    .get();
    }
...
```

入站流的配置在本质上与出站流的配置相反。在这里，也可以从直接信道的定义开始。然后，使用 Spring Integration 的 DSL 配置 IntegrationFlow。通俗地讲，就是"当从 RabbitMQ 的回复队列中接收到请求时，取出消息并放在回复信道中"。

主步骤的最后一部分是配置步骤和作业。提醒一下，在使用远程分区时，应该将作业配置在主配置(master profile)中，这是因为：如果把主步骤和从步骤放在相同的 Spring Boot 超级 Jar 中(这也是我们建议的方式)，那么如果不把作业专门放在主配置中，从应用和主应用中的作业都会被 Spring Boot 自动执行。代码清单 11-8 展示了 Partitioner 的创建以及执行远程分区作业所需的步骤。

代码清单 11-18 Partitioner 的创建以及执行远程分区作业所需的步骤

```
...
    @Bean
    @StepScope
    public MultiResourcePartitioner partitioner(
            @Value("#{jobParameters['inputFiles']}") Resource[] resources) {
        MultiResourcePartitioner partitioner = new MultiResourcePartitioner();

        partitioner.setKeyName("file");
        partitioner.setResources(resources);

        return partitioner;
    }

    @Bean
    public Step masterStep() {
        return this.masterStepBuilderFactory.get("masterStep")
                    .partitioner("workerStep", partitioner(null))
                    .outputChannel(requests())
                    .inputChannel(replies())
                    .build();
    }

    @Bean
    public Job remotePartitioningJob() {
        return this.jobBuilderFactory.get("remotePartitioningJob")
                    .start(masterStep())
                    .build();
    }
}
```

代码清单 11-18 从 MultiResourcePartitioner 的定义开始。由于数据并没有改变，因此划分的方式也不会变。接下来定义了远程的分区步骤。我们首先从工厂中获取构建器，然后传入分区器、输出信道(requests)和输入信道(replies)。配置完这些组件后，Spring Batch 就会装入所有的必要组件，发送出站请求，接收回复，最后聚合成主步骤的单个结果。

在配置了主步骤之后，下面看看从步骤的配置。同样，我们将使用一个新的配置类 WorkerConfiguration。实际上，从步骤的配置与主步骤十分相似，也会配置入站流和出站流。不过，这里不会配置作业,而只需要配置从步骤使用的读取器和写入器。代码清单 11-19 展示了 WorkerConfiguration 配置类。

代码清单 11-19　WorkerConfiguration 配置类

```
...
@Configuration
@Profile("!master")
@EnableBatchIntegration
public class WorkerConfiguration {

    private final RemotePartitioningWorkerStepBuilderFactory
        workerStepBuilderFactory;
    public WorkerConfiguration(
        RemotePartitioningWorkerStepBuilderFactory workerStepBuilderFactory) {

        this.workerStepBuilderFactory = workerStepBuilderFactory;
    }

    /*
     * Configure inbound flow (requests coming from the master)
     */
    @Bean
    public DirectChannel requests() {
        return new DirectChannel();
    }

    @Bean
    public IntegrationFlow inboundFlow(ConnectionFactory connectionFactory) {
        return IntegrationFlows
                .from(Amqp.inboundAdapter(connectionFactory, "requests"))
                .channel(requests())
                .get();
    }

    @Bean
    public DirectChannel replies() {
        return new DirectChannel();
    }
}
...
```

与代码清单 11-16 中的 MasterConfiguration 配置类相似,代码清单 11-19 的开头也是 @Configuration、@Profile 和@EnableBatchIntegration 注解。为了防止代码重复,也可以把这些配置放在主类中。

然后,WorkerConfiguration 配置类以另一个新的构建器工厂开始,也就是 RemotePartitioningWorkerStepBuilderFactory,以与 RemotePartitioningMasterStepBuilderFactory 对应。这个构建器工厂提供了一个步骤构建器,从而构建远程的分区步骤所需的组件,包括 StepExecutionRequestHandler 和其他组件。StepExecutionRequestHander 负责从主步骤中接收消息,并在远程的 JVM 中执行主步骤。接下来是入站信道 requests 的定义,以及用于定义请求路由的集成流 (integration flow)。通俗地讲,inboundFlow 能够接收来自 AMQP 队列(名为 requests)的消息,并将其传给 requests 信道。从步骤的最后一部分是 replies 信道的定义,这个信道充当每个分区结果的返回路由。

在构建了入站管道之后，从步骤可使用新的工厂构建器进行构建，详见代码清单 11-20。

代码清单 11-20　从步骤的配置

```
...
public Step workerStep() {
        return this.workerStepBuilderFactory.get("workerStep")
                        .inputChannel(requests())
                        .outputChannel(replies())
                        .<Transaction, Transaction>chunk(100)
                        .reader(fileTransactionReader(null))
                        .writer(writer(null))
                        .build();
}
...
```

与主步骤类似，从步骤也设置了名称、输入信道和输出信道。不过，接下来的配置与其他任何 Spring Batch 步骤的配置方式完全相同。在本例中，首先设置块大小、读取器和写入器，然后调用 build 方法。这个作业的读取器和写入器与基于线程的分区示例相同。

现在，我们在和多个 JVM 打交道。在执行作业之前，需要运行消息中间件。本例使用的是 RabbitMQ，所以需要确保 RabbitMQ 在本地运行。对于 OS X 用户，可通过安装 RabbitMQ 和执行命令 rabbitmq-server 来运行。在 RabbitMQ 启动后，就可以启动作业和 JVM。在本例中，我们将使用三个 Worker JVM。为此，在三个不同的 shell 窗口中执行相同的命令 java -jar target/partitioned-demo-0.0.1-SNAPSHOT.jar --spring.profiles.active=worker。在每个应用都启动后，你将看到它们仅仅在静静地等待。由于从配置中没有作业的 Bean，因此 Spring Boot 不会像其他示例那样自动地启动处理。为了启动作业，可使用命令 java -jar target/partitioned-demo-0.0.1-SNAPSHOT.jar --spring.profiles.active=master 启动主配置。这将会启动主应用，并且 Spring Boot 将启动作业。如果监视从步骤的日志，你将看到每个从步骤都会从队列中取出请求并进行处理，就像是在主步骤中完成一样。在干净的数据库中，数据库输出应该与 TaskExecutorPartitionHandler 示例的输出一样。

到目前为止，我们已经了解了 Spring Batch 处理分区工作负载时的两种不同方式。第一种是在单个 JVM 中使用线程；第二种是静态地在从步骤中等待工作负载，这是经典的消息工作负载风格。然而，这两种方式都不能很好地发挥云计算的优势。下面将介绍第三种方式，目的就是希望当在 Spring Batch 中使用分区时能够利用云端的动态资源。

3. DeployerPartitionHandler

云计算的最大驱动力就是对按需计算的需求。如果在 80%的时间里需要两台服务器，而在另外 20%的时间里需要 100 台服务器，那么使用云计算就可以轻易地实现，并且无须在 100%的时间支付 100 台服务器所需的费用。这种按需扩展的能力能为许多工作负载带来巨大的好处。批处理可能是最理想的示例。大多数批处理都是按照某种计划进行的，在有限的时间内执行，然后直到下一个时间窗口才需要它们。在传统的 Java 部署中，这些应用将在应用容器中处于空闲状态，浪费资源。在云计算中，情况不一定是这样。

DeployerPartitionHandler 是 Spring Portfolio 最新提供的能够开箱即用的一个 PartitionHandler。这个 PartitionHandler 利用了另一个 TaskLauncher 抽象，按需执行从步骤。流程类似于下面这样：在执行主应用时，没有任何从应用在运行；当分区步骤开始时，主步骤确定将有多少分区；确定之后，这个 PartitionHandler 将在平台上启动新的应用实例来执行分区,并将尽可能多地启动分区(取决于最大配置)以完成工作。例如,如果主步骤确定了 10 个分区,并且 DeployerPartitionHandler 被配置为最多 4 个 Worker,那么 DeployerPartitionHandler 将使用 4 个 Worker,直到执行了所有分区。

通过使用这种方法,就能够同时获得 Spring Batch 的分区处理好处(可重启、更高的吞吐量等)和云端的动态伸缩性。现在,为了让作业能够工作,需要同时支持运行平台和 Spring Cloud Task 的 TaskLauncher。在写作本书时,CloudFoundry、Kubernetes 和本地环境都有这样的 TaskLauncher。代码清单 11-21 展示了在本地运行这个示例时需要引入的 Maven 依赖项。

代码清单 11-21　Spring Cloud Task 和 Local Deployer 所需的 Maven 依赖项

```
...
<dependency>
    <groupId>org.springframework.cloud</groupId>
    <artifactId>spring-cloud-starter-task</artifactId>
</dependency>
<dependency>
    <groupId>org.springframework.cloud</groupId>
    <artifactId>spring-cloud-deployer-local</artifactId>
</dependency>
...
```

为了研究 DeployerPartitionHandler 的实现,我们将继续使用前面的作业。总体配置就像是对前面两种配置的结合。由于存在多个独立的 JVM,因此有两个配置(Profile),它们各司其职(一个用于运行主应用,另一个用于运行从应用)。不过,这里不包括 Spring Integration 的配置,因为主应用和从应用之间并不需要任何消息中间件。代码清单 11-22 展示了作业的配置。

代码清单 11-22　BatchConfiguration

```
...
@Configuration
public class BatchConfiguration {

    @Autowired
    private JobBuilderFactory jobBuilderFactory;

    @Autowired
    private StepBuilderFactory stepBuilderFactory;

    @Autowired
    private JobRepository jobRepository;

    @Autowired
    private ConfigurableApplicationContext context;

    @Bean
    @Profile("master")
    public DeployerPartitionHandler partitionHandler(TaskLauncher taskLauncher,
                JobExplorer jobExplorer,
                ApplicationContext context,
                Environment environment) {
        Resource resource =
            context.getResource("file:///path-to-jar/partitioned-demo-0.0.1-
```

```
                    SNAPSHOT.jar");
            DeployerPartitionHandler partitionHandler =
                    new DeployerPartitionHandler(taskLauncher, jobExplorer, resource,
                    "step1");

            List<String> commandLineArgs = new ArrayList<>(3);
            commandLineArgs.add("--spring.profiles.active=worker");
            commandLineArgs.add("--spring.cloud.task.initialize.enable=false");
            commandLineArgs.add("--spring.batch.initializer.enabled=false");
            commandLineArgs.add("--spring.datasource.initialize=false");
            partitionHandler.setCommandLineArgsProvider(
                    new PassThroughCommandLineArgsProvider(commandLineArgs));
            partitionHandler.setEnvironmentVariablesProvider(
                    new SimpleEnvironmentVariablesProvider(environment));
            partitionHandler.setMaxWorkers(3);
            partitionHandler.setApplicationName("PartitionedBatchJobTask");

            return partitionHandler;
        }

        @Bean
        @Profile("worker")
        public DeployerStepExecutionHandler stepExecutionHandler(JobExplorer jobExplorer) {
            return new DeployerStepExecutionHandler(this.context, jobExplorer, this.
            jobRepository);
        }
...
```

代码清单 11-22 的开头是配置类 BatchConfiguration，我们为它使用了 @Configuration 注解。之后，我们装配了通常的 Spring Batch 步骤和作业构建器工厂。由于这些步骤并无特别之处，因此这里并没有使用特殊的构建器来创建步骤。我们还装配了 JobRepository 和当前应用的上下文，因为从配置(worker profile)会使用它们。我们很快就会讲到它们的用途。

这里配置的第一个 Bean 是 DeployerPartitionHandler。内容有点多，所以在逐一配置之前，我们先对接下来的事情做一些分解。DeployerPartitionHandler 将在给定的平台上启动应用的一个新实例。考虑一下将应用部署到 Kubernetes 时的情况：首先创建 Docker 镜像并发布到 Docker 注册中心(Docker Registry)，然后使用 Kubernetes 工具下载 Docker 镜像并推到 Kubernetes 集群中。DeployerPartitionHandler 做了同样的事情：在创建 Docker 镜像之后，将其发布到 Docker 注册中心，然后使用 DeployerPartitionHandler 下载并自动推到 Kubernetes 集群中(假定这里使用的是 Kubernetes)。现在，回到刚才的配置。

在 DeployerPartitionHandler 的配置中，首先是获取从应用执行的工件(artifact)。如果底层使用的是 Kubernetes，那么指向的是 Docker 资源。由于这里使用的是默认的本地部署器，因此我们将资源指向 Spring Boot 的超级 Jar。当启动从应用时，DeployerPartitionHandler 将委托 TaskLauncher 创建用于执行的命令(当在本地时，将生成可从 shell 中执行的 java -jar 命令)。

在得到运行的 Resource(资源) 引用之后，我们将创建 DeployerPartitionHandler 实例。DeployerPartitionHandler 的构造函数有三个参数：taskLauncher、jobExplorer 以及刚刚获得的资源。taskLauncher 与平台相关，它知道如何在平台上启动应用；在写作本书时，Spring 提供了三个选项——Local(本地)、CloudFoundry 和 Kubernetes。jobExplorer 用于轮询作业存储库，以查看作业是否完成。

最后，step1 代表从应用在运行后将要启动的步骤。

在创建了 DeployerPartitionHandler 之后，还需要配置两个与之相关的抽象。首先是 CommandLineArgsProvider。它是一个策略接口，允许用户自定义传给超级 Jar 的命令行参数。在本例中，我们将使用 Spring 提供的 PassThroughCommandLineArgsProvider，用它传递一个参数列表。其中，传入的参数如下。

- --spring.profiles.active=worker：这个参数用于把应用作为从应用启动。
- --spring.cloud.task.initializer.enable=false：让 Spring Cloud Task 不要为从应用启动数据库初始化代码。因为数据库会在主应用启动时(或以其他方式)初始化，所以我们不想让从应用再次初始化数据库。
- --spring-batch.initializer.enabled=false：与前一个参数的概念相同，针对 Spring Batch 而不是 Spring Cloud Task 中的数据表。
- --spring.datasource.initialize=false：与前两个参数的概念相同，针对的是超级 Jar 中声明的其他数据库脚本。

在定义了 CommandLineArgsProvider 之后，另一个与 PartitionHandler 相关的抽象是 EnvironmentVariablesProvider。它也是一个策略接口，用于设置从应用运行的脚本中的任何环境变量。在本例中，可以使用 SimpleEnvironmentVariablesProvider 把当前的环境变量复制到从应用中。

最后两个需要配置的部分是允许同时运行的从应用的最大数量以及应用的名称。配置完最大数量后，Spring 就不会启动未知数量的从应用了。如果不设置这个值，Spring 会为每个分区启动一个从应用。如果使用分区器划分了 1000 个分区，那么将会启动 1000 个从应用，这并不是你期望的结果。将最大数量设置为 3 之后，如果分区数量小于 3，那么 PartitionHandler 将为每个分区启动一个从应用。如果分区数量超过 3，那么 PartitionHandler 将会启动 3 个从应用，然后随着每个从应用的完成，将会接着启动最多 3 个从应用，直到所有分区都完成。应用的名称将由 Spring Cloud Task 使用，供任务存储库用来跟踪从应用的执行。

我们现在已经配置了 DeployerPartitionHandler，下面在从应用中使用某种机制来启动请求的步骤。当使用 MessageChannelPartitionHandler 时，会有监听器用于监视信道中的请求，并在请求到来时启动它们(详见 StepExecutionRequestHandler)。当使用 DeployerPartitionHandler 时，需要另一种机制来启动从步骤(详见 DeployerStepExecutionHandler)。处理器并不会通过信道中的消息来获取步骤，而是通过已知的属性从环境中拉取作业执行 id、步骤执行 id 和步骤名称。然后执行步骤，就像前面示例中的 StepExecutionRequestHandler 一样。为了配置 DeployerStepExecutionHandler，只需要提供上下文(用来让处理器获取将要执行的步骤)、jobExplorer(用于从作业存储库中获取 StepExecution 以执行)以及作业存储库(JobRepository，用于在步骤自身未能处理故障时更新步骤的执行)。

这里的内容有点多，但却展示了与前面示例的几乎所有区别(除了如何执行)。在前面的示例中，我们必须首先启动 RabbitMQ，然后手动启动从应用。在这种方式下，我们只需要关心主应用的启动。主应用将处理从应用的启动。因此在构建应用之后，只需要执行 java -jar target/partitioned-demo-0.0.1-SNAPSHOT.jar --spring.profiles.active=master 命令即可。尽管我们正在作业存储库中监视执行的过程，但结果并没有说明其他的从应用正在启动。当它们启动时，你就会在主应用的日志中看到日志文件的位置，参见代码清单 11-23。

第 11 章　伸缩和调优

代码清单 11-23　启动 Worker JVM

```
2019-01-05 10:34:16.533 INFO 67745 --- [main] o.s.c.t.b.l.TaskBatchExecutionListener:
The job execution id 1 was run within the task execution 1
2019-01-05 10:34:16.562 INFO 67745 --- [main] o.s.batch.core.job.SimpleStepHandler:
Executing step: [step1]
2019-01-05 10:34:16.640 DEBUG 67745 --- [main] o.s.c.t.b.p.DeployerPartitionHandler:
3 partitions were returned
2019-01-05 10:34:16.684 INFO 67745 --- [main] o.s.c.d.spi.local.LocalTaskLauncher:
launching task PartitionedBatchJobTask-a5e75fd0-0c90-49b3-9e2b-428d5182765c
Logs will be in /var/folders/6s/2mwfrcbx5tg1mxr251bbl44m0000gn/T/spring-clouddataflow-
9037584903989022167/PartitionedBatchJobTask-1546706056645/PartitionedBatchJob
Taska5e75fd0-0c90-49b3-9e2b-428d5182765c
2019-01-05 10:34:16.697 INFO 67745 --- [main] o.s.c.d.spi.local.LocalTaskLauncher:
launching task PartitionedBatchJobTask-4a4c7152-8f3c-48ce-84d3-cac0919d4385
Logs will be in /var/folders/6s/2mwfrcbx5tg1mxr251bbl44m0000gn/T/spring-clouddataflow-
9037584903989022167/PartitionedBatchJobTask-1546706056689/PartitionedBatchJob
Task-4a4c7152-8f3c-48ce-84d3-cac0919d4385
2019-01-05 10:34:16.709 INFO 67745 --- [main] o.s.c.d.spi.local.LocalTaskLauncher:
launching task PartitionedBatchJobTask-705cf4dd-b708-491f-8191-f6f520cc018c
Logs will be in /var/folders/6s/2mwfrcbx5tg1mxr251bbl44m0000gn/T/spring-clouddataflow-
9037584903989022167/PartitionedBatchJobTask-1546706056699/PartitionedBatchJob
Task-705cf4dd-b708-491f-8191-f6f520cc018c
```

如代码清单 11-23 所示，当使用默认的 LocalTaskLauncher 时，以上日志就会出现日志文件中。打开其中的一个目录，就会看到 stdout.log 文件和 stderr.log 文件，它们分别代表标准输出和标准错误。stdout.log 文件与其他的 Spring Boot 日志文件类似，里面是正常情况下的日志。这里还有一种方式用来表明额外的 JVM 正在启动，就是通过 jps 命令查看是否有额外的 JVM 正在执行。jps 命令与 UNIX 中的 ps 命令很像，但是 jps 命令仅仅列出了正在运行的 Java 虚拟机。代码清单 11-24 展示了当执行作业时 jps 命令的输出。

代码清单 11-24　jps 命令的输出

```
➜  ~ jps
68944 RemoteMavenServer
42899
88027 partitioned-demo-0.0.1-SNAPSHOT.jar
88045 partitioned-demo-0.0.1-SNAPSHOT.jar
88044 partitioned-demo-0.0.1-SNAPSHOT.jar
88047 Jps
88046 partitioned-demo-0.0.1-SNAPSHOT.jar
```

你在 jps 命令的输出中可以看到四个 Java 进程：1 个是主应用，3 个是从应用。与所有其他示例一样，输出如代码清单 11-25 所示。

代码清单 11-25　分区作业的输出

```
mysql> select step_name, status, commit_count, read_count, write_count from
SCALING.BATCH_STEP_EXECUTION;
+------------------+-----------+--------------+------------+-------------+
| step_name        | status    | commit_count | read_count | write_count |
+------------------+-----------+--------------+------------+-------------+
| step1            | COMPLETED | 303          | 30000      | 30000       |
| step1:partition1 | COMPLETED | 101          | 10000      | 10000       |
```

```
| step1:partition0 | COMPLETED | 101      | 10000      | 10000      |
| step1:partition2 | COMPLETED | 101      | 10000      | 10000      |
+------------------+-----------+----------+------------+------------+
4 rows in set (0.01 sec)
```

对工作负载进行分区十分有用,无论是在单个 JVM 中,还是在集群中。对于 I/O 密集型作业,这将对作业的性能产生深远的影响。

然而,并非所有进程都受 I/O 约束。如果处理器是瓶颈,并且单个 JVM 无法进行处理,那么远程分块可能是正确的伸缩方案。

11.2.5 远程分块

在分布式计算中,把数据处理转移到集群是实现性能提升的一种常见模式。其中一个更为极端的示例是 Berkley 开发的 BOINC 系统。这个系统最初是为 SETI@Home 项目开发的,该项目通过让未使用的个人计算机处理射电望远镜记录的无线电信号,来寻找外星智能存在的证据。BOINC 后来从 SETI@Home 项目剥离出来,成为一个独立的框架。围绕 BOINC 的想法实际上非常简单。BOINC 由一台主命令服务器组成,这台主命令服务器接收要处理的数据请求,并对请求工作者(requesting worker)处理的输入进行应答。请求工作者下载输入,执行所需的处理,然后上传结果。

远程分块类似于 BOINC,只不过使用的是推模型而不是拉模型。与通过网络将元数据发送给工作者的远程分区不同,远程分块(如 BOINC)通过网络发送要处理的真正数据。主应用读取数据,将它们发送给从应用进行处理,然后由从应用写到输出。这种处于单个 JVM 之外的扩展只有在数据处理成为瓶颈时才有用。如果输入或输出是瓶颈,那么这种扩展只会使情况变得更糟。在使用远程分块作为扩展批处理过程的方法之前,需要考虑如下一些事情。

- 处理是瓶颈:由于读写是在主 JVM 中完成的,为了从远程分块中获益,将数据发送到从应用进行处理的成本必须小于并行处理带来的好处。
- 有保障的交付是必需的:由于 Spring Batch 并不维护任何类型的与"谁在处理什么"相关的信息,因此如果有从应用在处理期间宕机,Spring Batch 将无法知道处理的是什么数据。为此,使用持久化通信形式(通常是一种基于持久消息的解决方案)是必要的。

为了让作业使用远程分块,我们首先配置一个普通的作业,其中包含一个将在远程执行的步骤。Spring Batch 允许在添加远程分块这一功能时,不改变作业本身的配置。作为替代,你可以劫持步骤的 ItemProcessor,并插入一个 ChunkHandler 实例(由 Spring Integration 提供)。org.springframework.batch.integration.chunk.ChunkHandler 接口只有一个方法 handleChunk,这个方法的工作方式与 ItemProcessor 接口很类似。不过,ChunkHandler 接口的实现并不是处理给定的数据,而是发送要处理的数据并监听响应。当数据返回时,正常情况下会由本地的 ItemWriter 写入。图 11-16 展示了使用远程分块的步骤的结构。

如图 11-16 所示,作业中的任何步骤都可以配置为通过远程分块完成处理。当配置一个给定的步骤时,它的 ItemProcessor 将由 ChunkHandler 代替。ChunkHandler 接口的实现将使用一个特殊的写入器(org.springframework.batch.integration.chunk.ChunkMessageChannelItemWriter)将数据写入队列。从步骤仅仅是消息驱动的 POJO 并处理业务逻辑。当处理完成时,ItemProcessor 的输出可通过 ItemWriter 进行持久化。

第 11 章 伸缩和调优

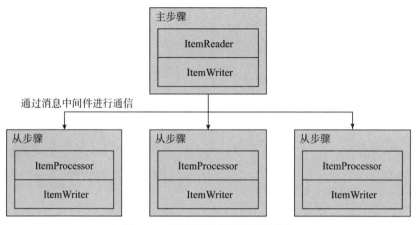

图 11-16 使用了远程分块的步骤的结构

对于本例，我们仅仅演示处理过程中的数据流，并执行与前面的分区示例相同的导入作业。就像分区作业使用 MessageChannelPartitionHandler 那样，我们也会使用两个配置文件(profile)和 Spring Integration 来配置主节点和从节点之间的通信。下面首先看看主配置。代码清单 11-26 展示了主步骤的配置。

代码清单 11-26　远程分块的主步骤

```
...
@EnableBatchIntegration
@Configuration
public class BatchConfiguration {
@Configuration
@Profile("!worker")
public static class MasterConfiguration {

    @Autowired
    private JobBuilderFactory jobBuilderFactory;

    @Autowired
    private RemoteChunkingMasterStepBuilderFactory
        remoteChunkingMasterStepBuilderFactory;

    @Bean
    public DirectChannel requests() {
        return new DirectChannel();
    }

    @Bean
    public IntegrationFlow outboundFlow(AmqpTemplate amqpTemplate) {
        return IntegrationFlows.from(requests())
                    .handle(Amqp.outboundAdapter(amqpTemplate)
                            .routingKey("requests"))
                    .get();
    }

    @Bean
    public QueueChannel replies() {
        return new QueueChannel();
```

361

```java
    }

    @Bean
    public IntegrationFlow inboundFlow(
        ConnectionFactory connectionFactory) {

        return IntegrationFlows
                    .from(Amqp.inboundAdapter(connectionFactory,
                            "replies"))
                    .channel(replies())
                    .get();
    }

    @Bean
    @StepScope
    public FlatFileItemReader<Transaction> fileTransactionReader(
        @Value("#{jobParameters['inputFlatFile']}") Resource resource) {

        return new FlatFileItemReaderBuilder<Transaction>()
                    .saveState(false)
                    .resource(resource)
                    .delimited()
                    .names(new String[] {"account",
                                "amount",
                                "timestamp"})
                    .fieldSetMapper(fieldSet -> {
                        Transaction transaction = new Transaction();

                        transaction.setAccount(
                            fieldSet.readString("account"));
                        transaction.setAmount(
                            fieldSet.readBigDecimal("amount"));
                        transaction.setTimestamp(
                            fieldSet.readDate("timestamp",
                                    "yyyy-mm-dd hh:mm:ss"));
                        return transaction;
                    })
                    .build();
    }
    @Bean
    public TaskletStep masterStep() {
        return this.remoteChunkingMasterStepBuilderFactory.get("masterStep")
                    .<Transaction, Transaction>chunk(100)
                    .reader(fileTransactionReader(null))
                    .outputChannel(requests())
                    .inputChannel(replies())
                    .build();
    }

    @Bean
    public Job remoteChunkingJob() {
        return this.jobBuilderFactory.get("remoteChunkingJob")
                    .start(masterStep())
                    .build();
    }
}
...
```

第 11 章 伸缩和调优

与本章前面的分区示例类似，这里的远程分块配置以 @EnableBatchIntegration 注解开始。Spring Batch 提供了特殊的构造器工厂来构建使用远程分块的步骤，而这个注解则用于启用这个功能。在 MasterConfiguration 类中，我们使用内部类划分了主配置和从配置。

主配置首先装入了 jobBuilderFactory(这与 Spring Batch 的所有配置类一样)和新的构建器工厂 RemoteChunkingMasterStepBuilderFactory。就像 RemotePartitioningMasterStepBuilderFactory 提供特殊的构建器来创建用于远程分区的主步骤一样，RemoteChunkingMasterStepBuilderFactory 也提供了一个构建器用于创建远程分块步骤中的主步骤。

在自动装入两个工厂对象之后，我们为出站流创建了一条信道。这条信道将用于给 RabbitMQ 发送数据，并由从节点进行处理。接下来是使用这条信道的流。注意这里的出站流与远程分区示例中使用的出站流是相同的。也就是说，Spring Integration DSL 声明了从请求信道取得消息，并将它们传给 AMQP 出站适配器。后者再将它们发送到名为 requests 的 RabbitMQ 队列。

在配置了出站流之后，接下来配置入站流。现在你可能想知道，由于写入操作发生在步骤的从节点一边，那么我们期望得到什么回复呢？来自从节点的回复实际上是节点要应用的 StepContribution。这样即使是分布式作业，作业存储库中的状态也是准确的。为了配置入站流，我们配置了一条直接信道和 Spring Integration 入站流。这个流读取 RabbitMQ 中的回复队列，取得消息，然后将它们放入回复信道。

主配置中的下一个组件是 FlatFileItemReader，它用来读取交易输入文件，并且与前面配置的读取器相同。我们在这里强调它的原因是，它是在主节点一边配置的分块步骤的唯一部分。ItemProcessor 和 ItemWriter 都配置在应用的从节点一边。

在配置了步骤的组件之后，就可以配置步骤本身。masterStep 的配置从读取开始，并且与其他基于块的步骤类似。我们指定了名称、块大小和读取器，但是并没有配置 ItemProcessor 和 ItemWriter。这里还指定了输入信道和输出信道，它们用于配置步骤，从而给从节点发送消息以及从节点接收消息。主配置中的最后一个 Bean 是作业的配置。我们将这个 Bean 放在了 master 配置文件(master profile)中，这样 Spring Boot 将仅仅在主节点上执行它，而不是在每个 JVM 上都执行一次。

主步骤有 7 个 Bean，即使不使用远程分块扩展步骤，其中 3 个也是必需的。我们下面看看从步骤，参见代码清单 11-27。

代码清单 11-27　远程分块的从步骤

```
...
    @Configuration
    @Profile("worker")
    public static class WorkerConfiguration {

        @Autowired
        private RemoteChunkingWorkerBuilder<Transaction, Transaction> workerBuilder;

        @Bean
        public DirectChannel requests() {
            return new DirectChannel();
        }

        @Bean
        public DirectChannel replies() {
```

```java
                return new DirectChannel();
        }

        @Bean
        public IntegrationFlow inboundFlow(ConnectionFactory connectionFactory) {
                return IntegrationFlows
                                .from(Amqp.inboundAdapter(connectionFactory,
                                        "requests"))
                                .channel(requests())
                                .get();
        }

        @Bean
        public IntegrationFlow outboundFlow(AmqpTemplate template) {
                return IntegrationFlows.from(replies())
                                .handle(Amqp.outboundAdapter(template)
                                        .routingKey("replies"))
                                .get();
        }

        @Bean
        public IntegrationFlow integrationFlow() {
                return this.workerBuilder
                                .itemProcessor(processor())
                                .itemWriter(writer(null))
                                .inputChannel(requests())
                                .outputChannel(replies())
                                .build();
        }

        @Bean
        public ItemProcessor<Transaction, Transaction> processor() {
                return transaction -> {
                        System.out.println("processing transaction = " + transaction);
                        return transaction;
                };
        }

        @Bean
        public JdbcBatchItemWriter<Transaction> writer(DataSource dataSource) {
                return new JdbcBatchItemWriterBuilder<Transaction>()
                                .dataSource(dataSource)
                                .beanMapped()
                                .sql("INSERT INTO TRANSACTION(ACCOUNT, AMOUNT, TIMESTAMP)" +
                                        "VALUES (:account, :amount, :timestamp)")
                                .build();
        }
    }
}
```

同样，这里使用内部类对 Bean 进行分组，从而使它们更容易合理化。WorkerConfiguration 配置类首先自动装入了 RemoteChunkingWorkerBuilder。注意与其他配置不同，这里并不需要工厂 Bean。在使用其他构建器时，工厂 Bean 会在幕后施展一些"魔法"；但对于这个构建器而言，这里并不需要魔法，所以我们使用构建器本身。

在接下来的配置中，出现了一个入站流和一个出站流，它们要做的事情与主步骤中的相同。不一样的地方是，这里有三个集成流，而在主配置中有两个流。在从节点一边，我们并没有配置步骤，而是在使用远程分块时配置了一个集成流。在幕后，这个构建器将创建一条处理链，用于接收传入的请求，并传给服务激活器(service activator)以进行相关的批处理，并且把结果返回给出站流。为了配置这个流，我们将使用新的构建器 RemoteChunkingWorkerBuilder 来配置 ItemProcessor、ItemWriter 以及输入信道和输出信道。

以上配置的其余部分是一个 ItemProcessor(一个简单的 lambda 表达式，用于将交易传递给 System.out)和一个用于将数据保存到数据表中的 ItemWriter。

完成配置后，就可以构建项目并执行作业。作业的启动机制与使用 RabbitMQ 的远程分区示例相同。首先通过命令 rabbitmq-server 启动 RabbitMQ(如果没有运行的话)，然后通过命令 java -jar target/chunking-demo-0.0.1-SNAPSHOT.jar 启动从应用，最后使用命令 java -jar target/chunking-demo-0.0.1-SNAPSHOT.jar --spring.profiles.active=master 启动主应用。

在运行了所有的组件之后，就可以在数据中查看作业的输出，参见代码清单 11-28。

代码清单 11-28　远程分块作业的输出

```
mysql> select step_name, status, commit_count, read_count, write_count from SCALING.BATCH_STEP_EXECUTION;
+-----------------+-----------+--------------+------------+-------------+
| step_name       | status    | commit_count | read_count | write_count |
+-----------------+-----------+--------------+------------+-------------+
| step1           | COMPLETED | 303          | 30000      | 30000       |
+-----------------+-----------+--------------+------------+-------------+
1 rows in set (0.01 sec)
```

11.3　本章小结

我们选择使用 Spring Batch 的主要原因之一是为了能够在不对原有代码库产生较大影响的前提下进行伸缩。尽管你可以自己编写这些特性，但是实现起来并不容易，而且"重新发明轮子"没有意义。Spring Batch 提供了一组很好的方法，可以使用许多选项来伸缩作业的规模。

本章讨论了如何对作业进行性能分析，以获得瓶颈的位置信息。然后通过示例，逐一介绍了 Spring Batch 提供的四种伸缩方法：并行步骤、多线程步骤、分区以及远程分块。

第 12 章
云原生的批处理

批处理已经存在了很长的时间。自从自动化计算出现以来，收集数据、在其上进行处理并从中生成输出一直是批处理的基础部分。随着企业转向云端，把批处理也迁移到那里是很自然的事情。

不过，是什么让应用准备好转向云或"元原生"呢？我们是否可以捡起 Spring Batch 应用，然后将它抛向云端？简而言之，答案可能是否定的。对于在云端运行程序，就算能"运行"，也还是令人担忧。"12 要素应用"的概念正是为此目的而设计的。"12 要素应用"解决了在云端运行应用的额外担忧。

Spring Cloud 是构建在 Spring Boot 之上并支持云原生开发的项目组合。诸如断路器、服务发现、配置管理、作业编排等都属于此类项目。不知道这些是什么？别担心，本章将逐一介绍。

在本章，我们将迭代一个非常简单的 Spring Batch 应用，将它从一个传统的基于 Spring Boot 的 Spring Batch 应用转换为云原生的应用。在每一次迭代中，我们都向它添加新的特性以利用云原生特性，直到我们有了一个能利用 Spring Cloud 特性的应用。本章主要介绍如下内容：

- 介绍"12 要素应用"包括什么，以及如何将其应用到批处理。
- 介绍一个非常简单的 Spring Batch 应用，以及如何将其迁移到云原生的应用。
- 通过使用断路器模式来处理与发生故障的 REST API 的交互，从而为批处理应用增加额外的弹性。
- 使用 Spring Cloud Config Server 和 Spring Cloud Eureka 把配置移到应用之外。
- 使用 Spring Cloud Data Flow 编排批处理作业。

在开始查看细节之前，让我们先看看什么是"12 要素应用"。

12.1 "12 要素应用"

"12 要素应用"的概念来自 Heroku 及其在云计算方面所做的工作，目标是开发出一定的模式，用于开发应用即服务(Applications as Services)，这 12 个要素如下。

1) 代码库：一份代码库，多份部署。
2) 依赖：显式地声明和隔离依赖。
3) 配置：在环境中存储配置。

4) 支持服务(Backing Service)：把支持服务当作附加资源。

5) 构建、发布、运行：严格分离构建和运行阶段。

6) 进程：以一个或多个无状态进程运行应用。

7) 端口绑定：通过端口绑定提供服务。

8) 并发：通过进程模型进行伸缩。

9) 可丢弃性(Disposability)：通过快速启动和优雅终止最大化健壮性。

10) 开发环境与线上环境的等价：保持开发、预发布和生产环境尽可能相同。

11) 日志：把日志当作事件流。

12) 管理进程：把管理任务当作一次性进程运行。

下面逐一定义它们的真正含义，并讨论如何在批处理中应用它们。

12.1.1 代码库

应用的代码库只应存在于单一的版本控制仓库中。理念是：如果需要把代码库拆分成多个仓库，那么它们可能不是应用，而应该是分布式系统。多个"12 要素应用"可以组成分布式系统，但是其中的每个应用都是自包含和独立的。从批处理的角度看，在本书的大部分场景中，我们都遵循了上述模型，让每个应用都包含单个批处理作业。但是，遗留环境中很少有这样的代码库，而是有一个单一的 WAR 或 EAR 文件，其中包含了多个批处理作业。在云原生的环境中，你可能期望将其拆分成多个应用。

12.1.2 依赖

依赖管理是 Java 开发中的一部分。使用构建系统(Maven 或 Gradle)下载依赖，并将其包含在 Spring Boot 的超级 Jar 中，这是本书一直在使用的方式。理念是：不要让应用依赖外部的东西，所有的依赖都应该通过某种机制封装在应用中。

12.1.3 配置

在"12 要素应用"中，配置必须与代码分离，为什么呢？因为一些工件(artifact)需要用于多个环境，而它们必须独立于这些环境。尽管 Spring 通过配置文件(profile)提供了独立配置环境的机制，但是伸缩性比不上使用环境变量或集中地配置服务器。在批处理的世界里，我们需要配置大量的东西。无论是作业存储库中的数据库，还是访问用于输入输出的其他系统，Spring Batch 应用都需要遵循以上原则。

12.1.4 支持服务

支持服务是指应用依赖的任何服务，比如 RDBMS(关系数据库管理系统)、SMTP 服务器、S3、第三方 API 等。其中的关键是：所有这些都应该能够通过配置 URL 或其他的定位器(locator)进行引用。在代码中，不应该对特定的服务实例进行直接引用。例如，无论使用本例的 MySQL 还是云端的 Amazon RDS，代码都不应该有改动。唯一需要改动的地方是配置(修改 URL、用户名、密码等)。在批处理应用中，这意味着只有分开代码的组织和配置，应用才能够以这种方式进行配置。在 Spring Boot 中，可以通过使用良好的开发实践加深这些理念。

12.1.5 构建、发布、运行

"12 要素应用"中的 12 个要素严格地划分了构建、发布和运行应用的过程。构建应用就是编译和测试应用的代码。发布工件不仅仅是创建工件，还要提供唯一的版本标识符，并将其存储在不能修改的地方(例如，Maven 存储库)。运行应用就是获取发布的工件并在某个环境中执行。使用 Spring Batch 时，需要配置诸如 Jenkins 或 Concourse 的持续集成系统来运行构建管道。在此基础上，Spinnaker 之类的工具能够以非常健壮的方式处理发布和运行，从而在部署失败时提供回滚等功能。

12.1.6 进程

"12 要素应用"是无状态的，没有共享任何数据。这意味着如果只执行一次应用，那么我们并不期望文件系统或内存中有任何预先存在的数据。在 Web 世界中，有悖于以上原则的典型示例是构建需要粘滞会话(sticky session)的应用。这种设计意味着有些数据会从一个请求持续到另一个请求，因此是有状态的，并且伸缩性很差。在批处理领域，Spring 批处理作业从一开始就被设计为无状态的。你一定在想：等等，批处理作业是有状态的，它们可以重新启动，并且状态位于 ExecutionContext 中。没错，但状态是在作业存储库(一种云友好的关系数据库)中维护的。你可以在一个节点上执行一个正确设计的批处理作业，让它失败，然后在另一个完全不同的节点上重新启动它，除了需要连接到作业存储库中的相同数据库之外，不需要定义其他预先存在的状态。

12.1.7 端口绑定

"12 要素应用"不需要在运行时添加某种服务器以公开为服务，这种应用是自包含的，如果需要执行为之设计的功能，可将其绑定到端口本身。例如，在"非 12 要素应用"中，将把 WAR 文件部署到 Tomcat 上；而在"12 要素应用"中，由于应用是完全自包含的，因此 Tomcat 被嵌入应用(就像在 Spring Boot 中所做的那样)。在 Spring 批处理作业或其他的批处理作业中，这不是什么大的问题，因为批处理作业的定义是自包含的。也就是说，Spring 为你提供了处理这种场景的能力，可出于某种原因让 Spring 批处理作业打开一个端口，将自己暴露出来。

12.1.8 并发

"12 要素应用"必须能够通过进程进行伸缩，但这不是指禁止使用线程或者不鼓励使用线程。JVM 中的线程是扩展应用的好工具，而且应该在正确的场景中使用。然而，JVM 的增长是有极限的，所以必须使用多个实例。在扩展 Spring Batch 应用时，要么通过单个 JVM 中的多个线程，要么对外使用远程分块和分区。

12.1.9 可丢弃性

在遗留系统中，进程的宕机是一件可怕的事情。在应用必须保持正常运行或客户因宕机受到影响的世界中，换页往往让开发人员脊背发凉。然而，在云原生的世界中，进程是可以随意丢弃的。它们在停止和重启时并不需要发出通知，并且需要以这种理念进行架构。进程应该能够尽可能快地启动，并且在收到停止请求时优雅地关闭。Spring Batch 很自然地采用了这种方式。基于作业存储库的运作原理，进程能够按需关闭和重启。也就是说，进程本身是有状态的，并且在重启时，状态确实需要恢复。Spring Batch 优化了这一点，例如在重启时会跳过已经处理过的记录。

12.1.10　开发环境与线上环境的等价

云计算的目标之一是提高业务敏捷性，消除诸如服务器机架安装和数据库配置方面的时延。从业务角度看，速度是计算的关键驱动因素。速度的最终目标是持续部署。从每季度或每月部署一次，到每天部署多次，只有在各个环境之间等价时，才会发生这种情况。如果正在基于 MySQL 进行开发，但是在生产环境中部署基于 Oracle，那么在生产环境中可能出现一些尚未发现的问题。让环境之间等价是最小化这些差异可能暴露的问题的关键。Spring Batch 并没有解决这种类型的非功能性需求，但是这样的做法会让发布和部署过程更加顺畅。

12.1.11　日志

前面说过，进程必须是可丢弃和无状态的，这也包括它们的日志。理想情况下，云原生的应用会把所有的日志写入标准输出，以便开发人员在工作时能通过终端窗口查看，或者由日志收集系统(例如 Splunk)或相关系统使用。这种做法将允许把多个实例的日志聚合到一个地方，以便更好地理解系统中发生了什么。Spring 与主流的 Java 日志框架实现了很好的集成，因而能够很好地处理上述需求。

12.1.12　管理进程

管理进程是指应该只运行一次的进程。"12 要素应用"的这一要素的有趣之处在于，批处理过程本身常常成为管理进程。

以上就是云原生应用开发的 12 个要素。但是，从战术的角度看，如何将它们应用到我们的 Spring 批处理应用中呢？接下来将进行探索，看看如何把一个基本的 Spring Batch 应用演进为云原生的应用。

12.2　一个简单的批处理作业

在本章的剩余部分，我们将使用 Spring 提供的功能，把一个简单的 Spring Batch 应用演进为云原生的应用。我们的关注点不在于领域模型，而在于每一次迭代中增加的功能。这个简单的 Spring Batch 应用是一个单步骤的批处理作业，作用是把来自 Amazon S3 的文件下载到本地目录中，按行读取文件并调用 REST API 来充实数据，然后把结果存储到数据库中。

在创建批处理作业之前，我们将从 Spring Initializr 中创建项目。除了添加 Batch、JDBC 和 MySQL 依赖项，我们还会添加 AWS 的依赖项，以便能够与 AWS 的 S3 存储进行交互。我们还将在添加新特性时添加依赖项。不过，对于我们的基础应用，这些已经足够了。

对于这个批处理作业，我们首先从主类(main class)开始。它只是一个普通的 Spring Boot 主类，与本书其他示例使用的主类相似。代码清单 12-1 展示了这个主类。

代码清单 12-1　CloudNativeBatchApplication.java

```
...
@EnableBatchProcessing
@SpringBootApplication
public class CloudNativeBatchApplication {
```

```java
    public static void main(String[] args) {
        SpringApplication.run(CloudNativeBatchApplication.class, args);
    }
}
```

这个主类从两个普通的注解开始：@EnableBatchProcessing 用于启用所有的 Spring Batch 基础设施；@SpringBootApplication 用于启用类路径扫描和自动配置。这个主类的 main 方法只是简单指明了类路径扫描的开始位置。

这会将我们引导至 JobConfiguration 类，该类包含了执行 Spring Batch 作业所需的所有组件的配置。代码清单 12-2 展示了 JobConfiguration 类的代码。

代码清单 12-2　JobConfiguration.java

```java
...
@Configuration
public class JobConfiguration {

    @Autowired
    private StepBuilderFactory stepBuilderFactory;

    @Autowired
    private JobBuilderFactory jobBuilderFactory;

    @Bean
    public DownloadingJobExecutionListener downloadingStepExecutionListener() {
        return new DownloadingJobExecutionListener();
    }

    @Bean
    @StepScope
    public MultiResourceItemReader reader(
        @Value("#{jobExecutionContext['localFiles']}")String paths) throws Exception {

        System.out.println(">> paths = " + paths);
        MultiResourceItemReader<Foo> reader = new MultiResourceItemReader<>();

        reader.setName("multiReader");
        reader.setDelegate(delegate());

        String [] parsedPaths = paths.split(",");
        System.out.println(">> parsedPaths = " + parsedPaths.length);
        List<Resource> resources = new ArrayList<>(parsedPaths.length);

        for (String parsedPath : parsedPaths) {
            Resource resource = new FileSystemResource(parsedPath);
            System.out.println(">> resource = " + resource.getURI());
            resources.add(resource);
        }
        reader.setResources(resources.toArray(new Resource[resources.size()]));

        return reader;
    }
```

```java
@Bean
@StepScope
public FlatFileItemReader<Foo> delegate() throws Exception {
    FlatFileItemReader<Foo> reader = new FlatFileItemReaderBuilder<Foo>()
                    .name("fooReader")
                    .delimited()
                    .names(new String[] {"first", "second", "third"})
                    .targetType(Foo.class)
                    .build();
    return reader;
}
@Bean
@StepScope
public EnrichmentProcessor processor() {
    return new EnrichmentProcessor();
}

@Bean
public JdbcBatchItemWriter<Foo> writer(DataSource dataSource) {
    return new JdbcBatchItemWriterBuilder<Foo>()
                    .dataSource(dataSource)
                    .beanMapped()
                    .sql("INSERT INTO FOO VALUES (:first, :second, :third, :message)")
                    .build();
}

@Bean
public Step load() throws Exception {
    return this.stepBuilderFactory.get("load")
                    .<Foo, Foo>chunk(20)
                    .reader(reader(null))
                    .processor(processor())
                    .writer(writer(null))
                    .build();
}

@Bean
public Job job(JobExecutionListener jobExecutionListener) throws Exception {
    return this.jobBuilderFactory.get("s3jdbc")
                    .listener(jobExecutionListener)
                    .start(load())
                    .build();
}

@Bean
public RestTemplate restTemplate() {
    return new RestTemplate();
}
}
```

这段代码有点多，但是其中大部分代码你应该很熟悉。在@Configuration 注解和类定义之后，我们装入了两个字段，它们是用于创建步骤和作业的构建器。

JobConfiguration 类定义的第一个 Bean 是一个 DownloadingJobExecutionListener。稍后我们会实现

这个监听器，它负责从 S3 存储桶中下载文件，以便导入数据库。下载是在 beforeJob 方法中完成的。这种设计允许我们遵从"12 要素应用"中的"进程"要素。如果作业失败，并且在云端的新节点或容器中重启，那么 beforeJob(JobExecution jobExecution)方法将会重新执行，文件会重新下载。

由于处理了多个文件，JobConfiguration 类定义的下一个 Bean 是 MultiResourceItemReader。这个条目读取器可通过前面的监听器查看下载目录，并配置为读取目录中的所有文件。注意这里通过 @StepScope 将这个 Bean 标记为使用步骤作用域，这是由于监听器将下载的文件列表放到了作业的 ExecutionContext 中以便拉取。

在读取时，MultiResourceItemReader 需要把读取委托给另一个 ItemReader。于是，配置中的下一个 Bean 就是委托。我们提供了一个配置过的 FlatFileItemReader 来读取 CSV 文件，其中的每条记录都有三个值。这些值的名称分别是 first、second 和 third(非常别出心裁)。其中每条记录的值被映射到名为 Foo 的领域模型，Spring 会自动调用这个领域模型的 setFirst(int value)、setSecond(int value)和 setThird(String value)方法。

在定义了条目读取器之后，接下来就是 ItemProcessor 的定义，它被用于调用 REST API。创建 ItemProcessor 的工厂方法会返回一个新的 EnrichmentItemProcessor 实例。我们稍后会介绍这个自定义的 ItemProcessor，它要做的事情是调用一个返回调用次数的 REST API。EnrichmentItemProcessor 将使用返回结果设置 message 字段。

这个步骤的最后一个组件是 ItemWriter。为此，我们使用 JdbcBatchItemWriter 来写入充实后的数据。使用构建器创建这个 ItemWriter 时，我们配置了数据源(DataSource)，并通过调用 beanMapped 方法来使用 BeanPropertyItemSqlParameterSourceProvider，我们还提供了 SQL 插入语句，然后调用了 build 方法。

上面就是加载数据所需的所有组件。下面构建这个步骤。我们使用自动装配的 StepBuilderFactory 来获得 StepBuilderFactory，然后指明使用基于块的步骤，传入前面配置的 MultiResourceItemReader、EnrichmentItemProcessor 以及 JdbcBatchItemWriter。

为了使用步骤，还需要配置作业。通过使用自动装入的 JobBuilderFactory，我们把 DownloadingJobExecutionListener 配置为作业的 JobExecutionListener，指定作业从加载步骤开始，然后调用构建器的 build 方法。

我们将为这个作业设定两个自定义类：DownloadingJobExecutionListner 和 EnrichmentProcessor。下面首先看看 DownloadingJobExecutionListner。在 DownloadingJobExecutionListener 中，我们使用了 Spring AWS 项目的 S3 资源处理功能，在配置过的 S3 存储桶(bucket)中遍历其中的资源，然后在 beforeJob(JobExecution jobExecution)方法中下载。代码清单 12-3 展示了这个监听器。

代码清单 12-3　DownloadingJobExecutionListner.java

```
...
public class DownloadingJobExecutionListener extends JobExecutionListenerSupport {

    @Autowired
    private ResourcePatternResolver resourcePatternResolver;

    @Value("${job.resource-path}")
    private String path;
```

```
    @Override
    public void beforeJob(JobExecution jobExecution) {

        try {
            Resource[] resources =
                this.resourcePatternResolver.getResources(this.path);

            StringBuilder paths = new StringBuilder();
            for (Resource resource : resources) {

                File file = File.createTempFile("input", ".csv");

                StreamUtils.copy(resource.getInputStream(),
                    new FileOutputStream(file));

                paths.append(file.getAbsolutePath() + ",");
                System.out.println(">> downloaded file : " +
                    file.getAbsolutePath());
            }

            jobExecution.getExecutionContext()
                    .put("localFiles",
                        paths.substring(0, paths.length() - 1));
        }
        catch (IOException e) {
            e.printStackTrace();
        }
    }
}
```

DownloadingJobExecutionListener 扩展了 JobExecutionListenerSuppor，后者提供了 JobExecutionListener 的无操作实现，从而让我们只覆盖自己关心的方法。在本例中，我们只需要实现 beforeJob 方法。这个方法首先获取 S3 存储桶中所有资源的列表，其中 S3 存储桶可通过应用参数 job.resource-path 进行配置。在获取了资源(文件)数组之后，我们将为其中的每个资源创建一个临时文件，然后使用 Spring 的 StreamUtils 下载文件，并保存每个下载文件的绝对路径。在下载完所有文件之后，我们将路径列表保存到了作业的 ExecutionContext 中。

在云原生环境中实现上述功能的关键是：在每次运行作业时，监听器都将重新执行，并重新下载所需的每个文件；然后覆盖作业上一次的 ExecutionContext 中的值，从而防止作业在新容器中运行时查找那些很可能不存在的文件。

这个示例作业的最后一部分是 EnrichmentProcessor，用于对 REST API 执行简单的 GET 调用，这里使用了 JobConfiguration 配置中的 RestTemplate。代码清单 12-4 展示了 EnrichmentProcessor。

代码清单 12-4　EnrichmentProcessor.java

```
...
public class EnrichmentProcessor implements ItemProcessor<Foo, Foo> {

    @Autowired
    private RestTemplate restTemplate;

    @Override
    public Foo process(Foo foo) throws Exception {
        ResponseEntity<String> responseEntity =
```

```
            this.restTemplate.exchange(
                    "http://localhost:8080/enrich",
                    HttpMethod.GET,
                    null,
                    String.class);
        foo.setMessage(responseEntity.getBody());

        return foo;
    }
}
```

当接收到来自 RestTemplate 调用的响应时,就将响应保存到数据条目中。在本例中,响应仅仅是一条消息 Enriched X,其中 X 代表 REST API 中的控制器被调用的次数。

在定义了作业的所有代码之后,我们需要对作业进行一些配置。首先,我们使用 Spring 的 application.yml 配置应用,如代码清单 12-5 所示。

代码清单 12-5　application.yml

```yml
spring:
  datasource:
    driverClassName: org.mariadb.jdbc.Driver
    url: jdbc:mysql://localhost:3306/cloud_native_batch
    username: 'root'
    password: 'password'
    schema: schema-mysql.sql
job:
  resource-path: s3://def-guide-spring-batch/inputs/*.csv
cloud:
  aws:
    credentials:
      accessKey: 'OPAR8O2SSRDI9NIGDBWA'
      secretKey: 'SDKEjF9IqN0IjTKIJVaE0G9UwI+=DOEFTjOkS2B4'
    region:
      static: us-east-1
      auto: false
```

■ **注意**　在这里,你需要配置自己的数据库 URL、用户名、密码、job.resource-path、AWS 凭据和区域。以上这些都是作为示例提供的,它们并不适合你。

在 application.yml 中,我们首先使用 Spring Boot 所需的普通值来配置数据源。这里唯一增加的部分是 schema 的值,从而指向用于 Foo 数据条目的数据表。配置完数据库后,job.resource-path 用于指向输入文件所在的 S3 存储桶。最后一部分是关于 AWS 凭据和区域的配置。当在 AWS 上运行代码时,Spring 会自动把区域配置为代码正在运行的区域。例如,如果代码运行在 US East 1 区域,那么 Spring 会自动将 S3 存储桶配置为这一区域。由于代码并没有配置为运行在 AWS 上,因此我们关闭了区域的自动配置。

如果现在尝试编译和运行,你会得到异常,因为我们还没有导入 EnrichmentProcessor 将要调用的 REST API。

REST API 是通过 Spring Initializr 创建的 Web 模块。将 REST API 下载并导入项目之后,只需要添加一个控制器即可。代码清单 12-6 展示了这个控制器。

代码清单 12-6　EnrichmentController.java

```
...
@RestController
public class EnrichmentController {

    private int count = 0;

    @GetMapping("/enrich")
    public String enrich() {
        this.count++;

        return String.format("Enriched %s", this.count);
    }
}
```

这个控制器使用了 @RestController 注解，从而向 Spring 表明，从方法返回的值将以原始形式返回给客户端。这里唯一的字段 counter 用于跟踪控制器被调用的次数。enrich 方法使用了 @GetMapping 注解，从而能够被映射到 /enrich URL。enrich 方法要做的事情是增加计数器和格式化消息，然后将格式化之后的消息返回给调用者[1]。

如果构建并运行 REST API 和作业，你就会看到：存储在 S3 存储桶中的所有数据已成功导入数据库。这个作业和 REST API 是在本章剩余部分进行迭代的基线。

注意，尽管本章讨论了云原生应用的构建，但我们不会在任何特定的云端运行它们。考虑到撰写本书时市场上有许多不同的选择，而且目前还没有明确的答案，我们暂不考虑针对特定云提供商的部署问题。也就是说，本章的所有内容都是跨云的，它们适用于任何主要的云提供商(CloudFoundry、Kubernetes、Google Cloud Platform、Amazon Web Services 等)。

接下来你将了解使用 Spring Batch 时的第一个云原生特性，并且它也是云原生 Web 应用中的常见特性之一：断路器。

12.3　断路器

如果需要执行大量的 API 调用，那么批处理作业将高效地执行。如果需要从队列中读取消息，那么批量读取是一种提高性能的好方法。对于写入数据库来说，也是如此。然而，效率也可能带来麻烦。

假如 REST API 过载了，那么让批处理过程无情地调用它还有意义吗？爱因斯坦把精神错乱定义为一遍又一遍地做同一件事，却期望得到不同的结果。然而，这正是可能发生的情况。

相反，如果在发送另一个请求之前，给 REST API 一次赶上来的机会，那么又会怎么样呢？Netflix 通过名为 Hystrix 的框架在其微服务体系架构中推广了这一技术。理念很简单。确定一个被断路器(circuit breaker)包围的方法。当超过异常阈值时，断路器跳闸，停止对该方法的调用，并将流量路由到另一个替代方法。这个替代方法通常以不同的方式处理事情。例如，返回默认值而不是 REST API 的返回值。根据某种算法，断路器会将流量缓慢地返回到原来的方法，以测试是否恢复了在线状态。

---

[1] 为简单起见，本例忽略了使用 int 类型时的线程安全问题。

一旦恢复，那么断路器复位，流量就会恢复正常并路由到原来的方法。

后来，Netflix 弃用了 Hystrix，而改用 resilience4j，但 Spring Batch 实际上并不需要使用它们来实现断路器模式。Spring Batch 依赖于一个名为 Spring Retry 的库。这个鲜为人知的库实际上在 Spring Portfolio 中得到了大量使用，它能在 Spring Batch 的容错步骤中提供容错功能。然而，出于我们的目的，我们将使用 Spring Batch 在最新版本中添加的功能：一个基本的断路器。

我们的用例将以下面这样的方式工作：配置 REST API 以返回随机异常，在断路器中封装 EnrichmentProcessor#process 方法，并通过替代方法将数据条目的 message 字段设置为 error。

考虑到 Spring Batch 的容错功能，你可能想知道为什么要这样做，而不是直接使用这种能力。实际上，使用断路器而不是容错功能的原因有两个。首先是性能。当重试某个操作时，如前所述，Spring Batch 框架会回滚事务，将提交计数设置为 1，然后在自己的事务中重试每一个数据条目。这个操作可能是性能杀手。如果可以将数据条目标记为错误，并在以后重新运行它们，那么这将是一种更有效的错误处理机制。

其次是示例本身。Spring Batch 允许重试数据条目，但却无法减轻问题代码的调用压力。如果服务需要一段时间才能恢复，那么 Spring Batch 并没有很好的内置工具用来处理这种情况。下面介绍 Spring Retry 的断路器。

Spring Retry 的断路器的关键组件是两个注解。第一个是@CircuitBreaker，这并不奇怪。这个方法级的注解表明应该在断路器中封装某些东西。默认情况下，断路器将关闭，直到配置的三种类型的异常(默认情况下是所有异常)在 5 秒内从方法抛出为止。默认情况下，断路器一旦跳闸，就将保持打开 20 秒，然后试图进入主路径。所有这些都可以通过表 12-1 所示的注解属性进行配置。

表 12-1 注解属性

属性	描述	默认值
exclude	所要排除的异常数组，在排除异常的特定子类时很有用	空(如果 include 属性也为空，那么包括所有异常)
include	进行重试的异常数组	空(如果 exclude 属性也为空，那么包括所有异常)
label	用于断路器报告的唯一标签用于	用于声明注解的方法签名
maxAttempts	在打开断路器之前重试的最大尝试次数(包括第一次失败)	3
openTimeout	断路器跳闸前必须达到最大尝试次数(maxAttempts)的时间(单位为毫秒)	5000(5 秒)
resetTimeout	再次重试主路径的时间(单位为毫秒)	20000(20 秒)
value	进行重试的异常数组	空(包括所有异常)

Spring Retry 使用的另一个注解是@Recover。@Recover 也是方法级的注解，用于指示在可重试(retryable)方法失败或者断路器翻转时要调用的方法。使用@Recover 注解的方法必须和与之关联并使用@CircuitBreaker 注解的方法的签名一致。

用于将断路器功能添加到 ItemProcessor 的最后一部分是@EnableRetry 注解。这个注解启用了 Spring 提供的一种用于代理可重试方法调用的机制。在与@EnableBatchProcessing 一起添加到主类之后,就有了我们需要的基础设施。代码清单 12-7 展示了更新后的代码。

代码清单 12-7　使用了断路器的 EnrichmentProcessor

```
...
public class EnrichmentProcessor implements ItemProcessor<Foo, Foo> {

    @Autowired
    private RestTemplate restTemplate;

    @Recover
    public Foo fallback(Foo foo) {
        foo.setMessage("error");
        return foo;
    }

    @CircuitBreaker(maxAttempts = 1)
    @Override
    public Foo process(Foo foo) {
        ResponseEntity<String> responseEntity =
            this.restTemplate.exchange(
                    "http://localhost:8080/enrich",
                    HttpMethod.GET,
                    null,
                    String.class);
        foo.setMessage(responseEntity.getBody());

        return foo;
    }
}
```

与代码清单 12-6 所示的原始版本相比,这里有两处不同:首先是为 process 方法添加了 @CircuitBreaker 注解,其次是添加了使用@Recover 注解的 fallback 方法。fallback 方法与 process 方法的签名一致。不过,在 fallback 方法中没有执行远程调用,而是设置了默认显示的消息。我们为 process 方法的@CircuitBreaker 注解做了如下配置:在 5 秒内超过一次尝试就会跳闸,并在 20 秒后重置断路器。

为了测试以上配置,我们将更新 EnrichmentController,使其以 50%的概率抛出异常。这模拟了系统中的不稳定情况,并且将导致断路器跳闸。代码清单 12-8 展示了更新后的控制器代码。

代码清单 12-8　抛出随机异常的 EnrichmentController

```
...
@RestController
public class EnrichmentController {

    private int count = 0;

    @GetMapping("/enrich")
    public String enrich() {
        if(Math.random() > .5) {
                throw new RuntimeException("I screwed up");
        }
        else {
                this.count++;
```

```
            return String.format("Enriched %s", this.count);
        }
    }
}
```

在更新完控制器之后，如果再次运行作业，你就会看到导入记录的一些消息是 Enriched X(其中的 X 表示控制器未抛出异常的调用次数)。不过，你在一些消息中还会看到 error。如果在数据库中比较错误消息的数量与 REST 应用日志中栈跟踪信息的数量，就会看到错误消息比栈跟踪信息多。这可以证实断路器跳闸了，而且调用了回退(fallback)方法，但没有尝试调用 REST API。

现在已经为应用添加了一些额外的弹性，接下来我们将外部化应用的配置。有两种方式：第一种是使用 Spring Cloud Config Server(配置服务器)，第二种是使用 Eureka 提供的服务发现。

12.4 外部化配置

到目前为止，本书中的每个 Spring Boot 应用都是使用 application.properties 或 application.yml 进行配置的。然而，这在云原生环境中会导致一个问题。因为配置与 Jar 文件中的应用绑定在了一起，所以当从一个环境转向另一个环境时，很难对其进行修改。目前的方法还存在安全问题，因为我们在工件中是以纯文本的形式存储机密信息的(比如数据库的用户名和密码，以及 Amazon 凭据等)，这些机密信息通常会被发布到某种公共存储库中。

一定还有更好的办法。多亏了 Spring Cloud 项目，我们有了外部化配置的两种机制。第一种是使用 Spring Cloud Config Server 提供和保护当前存储在 application.yml 文件中的值。我们还将使用 Spring Cloud CLI 提供的加密工具来保护它们。第二种是使用服务绑定，从而允许批处理作业定位和访问 REST API，而不是直接在应用中硬编码或配置 URL、端口等。

12.4.1 Spring Cloud Config

Spring Cloud Config 是配置服务器，用于为存储在 Git 存储库中或数据库后端的配置提供服务。为了使用配置，可在应用中使用 Spring Cloud 客户端。Spring Cloud 客户端将调用配置服务器，从配置服务器获取配置属性，并使用这些值填充 Spring Environment。因此，在使用 application.yml 时，Spring Boot 所做的一切(例如注入的所有普通属性)依然有效。

在我们的应用中，这是通过在 pom.xml 中添加 spring-cloud-starter-config 依赖项来实现的。代码清单 12-9 展示了所需的依赖项。

代码清单 12-9　Spring Cloud Config Client 依赖项

```
...
<dependency>
        <groupId>org.springframework.cloud</groupId>
        <artifactId>spring-cloud-starter-config</artifactId>
</dependency>
...
```

在添加了依赖项之后，对应用要做的其他修改是将 application.yml 文件中特定于应用的配置替换为客户端所需的配置。为了达到我们的目的(在本地机器中运行一切)，我们只需要配置两个属性：

spring.application.name 和 spring.cloud.config.failFast。客户端使用第一个属性 spring.application.name 向服务器询问正确的配置，可以将属性值设定为 cloud-native-batch。Spring Cloud Config 支持基于 Spring 属性的所有普通功能，例如配置文件。另一个属性 spring.cloud.config.failFast 会让客户端在无法从配置服务器获取配置时抛出异常，同时阻止应用启动。默认情况下，当无法获取配置时，客户端会忽略并使用本地的配置；但是在本例中，我们想要确保从配置服务器中读取配置。在批处理应用中配置了上面这两个属性后，就可以删除所有其他配置。

在配置了客户端之后，需要向配置服务器提供配置。最简单的启动和运行配置服务器的方法是使用 Spring Cloud CLI。Spring Cloud CLI 提供了一行命令用于启动各种 Spring Cloud 服务器组件，还提供了一些有用的实用程序，详见项目页面[1]。

在安装了 Spring Cloud CLI 之后，需要对配置服务器进行一些配置。我们将使用 Git 存储库来存储配置。为了让配置服务器使用 Git 存储库，我们需要告诉它将要使用的 Git 存储库的位置。为此，可在 ~/.spring-cloud 目录下的 configserver.yml 文件中进行配置，如代码清单 12-10 所示。

代码清单 12-10　configserver.yml

```
spring:
  profiles:
    active: git
  cloud:
    config:
      server:
        git:
          uri: file:///Users/mminella/.spring-cloud/config/
```

有了代码清单 12-10 中的配置之后，在 ~/.spring-cloud/config 目录下创建一个 Git 存储库，它是存储配置的地方。把之前的 application.yml 文件复制到这个目录，再把它提交到一个新的 Git 存储库(可使用命令 git init、git add 和 git commit)。不过，因为我们还将保护这个文件中的机密信息，所以不必担心攻击者获取这个文件。为此，我们使用了 Spring Cloud CLI 的加密功能。

Spring Cloud CLI 支持使用字符串密钥或密钥文件(例如，RSA 公钥)进行加密。为了简单起见，我们使用字符串密钥。代码清单 12-11 展示了使用 Spring 加密工具后的结果。

代码清单 12-11　加密与解密

```
➜ config git:(master) ✗ spring encrypt mysecret --key foo
ea48c11ca890b7cb7ffb37de912c4603d97be9d9b1ec05c7dbd3d2183a1da8ee
➜ config git:(master) ✗ spring decrypt --key foo
ea48c11ca890b7cb7ffb37de912c4603d97be9d9b1ec05c7dbd3d2183a1da8ee
mysecret
```

有了这项技术后，就可以加密所有的机密信息，把加密后的值粘贴到配置文件中，由配置服务器进行管理。机密信息的粘贴格式是在值的前面加上 {cipher}，并将整个值放在单引号中。代码清单 12-12 显示了完整的 cloud-native-batch.yml 文件，其中包含了加密后的值。

---

[1] https://cloud.spring.io/spring-cloud-cli/。

代码清单 12-12　cloud-native-batch.yml

```
spring:
  datasource:
    driverClassName: org.mariadb.jdbc.Driver
    url: jdbc:mysql://localhost:3306/cloud_native_batch
    username: '{cipher}19775a12b552cd22e1530f745a7b842c90d903e60f8a934b072c21454321de17'
    password: '{cipher}abcdefa44d2db148cd788507068e770fa7b64c4d1980ef6ab86cdefabc118def'
    schema: schema-mysql.sql
  batch:
    initalizr:
      enabled: false
job:
  resource-path: s3://def-guide-spring-batch/inputs/*.csv
cloud:
  aws:
    credentials:
      accessKey: '{cipher}a7201398734bcd468f5efab785c2b6714042d62844e93f4a436bc4fd2e95fa4bcd
26e8fab459c99807d2ef08a212018b'
      secretKey: '{cipher}40a1bc039598defa78b3129c878afa0d36e1ea55f4849c1c7b92e809416737
de05dc45b7eafce3c2bc184811f514e2a9ad5f0a8bb3e503282158b577d27937'
region:
  static: us-east-1
  auto: false
```

在创建了配置文件并将配置服务器指向正确的位置之后，就可以启动配置服务器并用它测试批处理作业。首先，可通过前面使用过的标准的 Spring Boot 命令启动 REST API：java -jar rest-service/target/rest-service-0.0.1-SNAPSHOT.jar。在运行了 REST API 之后，就可以使用命令 spring cloud configserver 启动配置服务器。配置服务器会指向我们刚刚提交到 Git 存储库的配置文件。一旦配置服务器开始运行，我们就可以启动批处理作业了。有了这些更新，你应该不会看到作业的输出有任何不同，因为行为是相同的。唯一的区别是获取作业配置的机制不同。

外部化配置的另一部分是通过 Eureka 进行服务绑定，将批处理作业连接到 REST API。

12.4.2　通过 Eureka 进行服务绑定

Spring Cloud Netflix 提供了 Eureka——一款由 Netflix 开发的服务发现工具。Eureka 为服务提供了注册能力，这样它们就可以被其他服务动态地发现。在本小节中，我们将了解如何通过 Eureka 使用服务发现，以允许在不进行显式配置时连接到 REST API。

启用服务发现的方式与云配置的方式类似。Eureka 也有客户端和服务器。应用在服务中心注册，以表明是可发现的。我们的 REST API 已将自己注册为可发现的。然后，我们的批处理作业将在启动时从 Eureka 获取如何与 REST API 通信的相关信息。这意味着需要为作业提供 Eureka 的位置以及需要访问的服务，Eureka 将处理其余配置。为了实现这一点，只需要修改几行代码即可。

我们首先从需要的依赖项开始。我们需要将客户端依赖项添加到 REST API 和批处理作业中。代码清单 12-13 展示了添加的依赖项。

代码清单 12-13　Eureka 依赖项

```
...
<dependency>
```

```xml
        <groupId>org.springframework.cloud</groupId>
        <artifactId>spring-cloud-starter-eureka</artifactId>
</dependency>
...
```

在把依赖项添加到批处理作业和 REST API 后，我们就能够对它们进行适当的配置了。比如，配置 REST API，使其在启动时把自身注册到 Eureka。为此，我们将进行两个小的修改。首先是为主类添加@EnableDiscoveryClient 注解，如代码清单 12-14 所示。

代码清单 12-14　添加@EnableDiscoveryClient 注解

```java
...
@EnableDiscoveryClient
@SpringBootApplication
public class RestServiceApplication {

    public static void main(String[] args) {
        SpringApplication.run(RestServiceApplication.class, args);
    }
}
```

在添加了这个注解之后，REST API 就会自动使用本地(localhost)的 Eureka 进行注册。在生产环境中，可以通过标准属性配置远程实例，甚至使用 Spring Cloud Config 指定位置。

其次是在项目中添加 boostrap.yml 文件。这个文件的格式与 application.yml 相同，区别在于 application.yml 会在加载 ApplicationContext 之后加载。但是，对于一些 Spring Cloud 特性来说，这已经太晚了，所以 Spring Cloud 创建了引导应用上下文(bootstrap ApplicationContext)，以充当应用上下文的父上下文。正是引导应用上下文加载了 bootstrap.yml 文件。对于本例来说，还需要在 bootstrap.yml 中配置应用的名称。代码清单 12-15 展示了 REST API 的 bootstrap.yml 文件。

代码清单 12-15　bootstrap.yml 文件

```yaml
spring:
  application:
    name: rest-service
```

有了这些更改，就可以通过 Eureka 使用 REST API 了。接下来修改批处理作业。为了通过 Eureka 提供的配置使用 REST API，需要对批处理作业进行四个改动。第一个是添加客户端依赖项，这与代码清单 12-13 提供的依赖项相同。

在添加了依赖项之后，就可以更新主类了。同样，给主类添加@EnableDiscoveryClient 注解。不过这里有一个小小的修改。我们希望 REST API 作为服务注册到 Eureka，但不希望注册作业，我们只想获得其他服务的配置细节。为此，将这个注解中的 autoRegister 设置为 false，如代码清单 12-16 所示。

代码清单 12-16　CloudNativeBatchApplication

```java
@EnableRetry
@EnableBatchProcessing
@SpringBootApplication
@EnableDiscoveryClient(autoRegister = false)
public class CloudNativeBatchApplication {
```

```
    public static void main(String[] args) {
        SpringApplication.run(CloudNativeBatchApplication.class, args);
    }
}
```

在批处理应用中启用了 Eureka 客户端之后，就需要配置 RestTemplate 了。这可以通过另一个名为@LoadBalanced 的注解来实现。在把这个注解添加到 RestTemplate 的 Bean 定义中之后，就会自动配置 RestTemplate，从而使用通过 Eureka 提供的配置，包括客户端负载平衡等。代码清单 12-17 展示了更新后的 RestTemplate 的 Bean 定义。

代码清单 12-17　使用了负载均衡的 RestTemplate

```
...
@Bean
@LoadBalanced
public RestTemplate restTemplate() {
    return new RestTemplate();
}
...
```

对批处理作业要做的最后一个改动是真正地通过名称引用服务。在 EnrichmentProcessor 的前一次迭代中，我们直接在代码中指定了 REST API 的主机和端口。这在云环境中显然不是理想做法，因为你可能并不知道主机名是什么。使用 Eureka 的好处是，我们只需要指定服务的名称即可，其余的都由 Spring Cloud 处理。更新后的 EnrichmentProcessor 调用了服务的名称(bootstrap.yml 中设定的 rest-service)而不是主机和端口。代码清单 12-18 展示了更新后的 EnrichmentProcessor。

代码清单 12-18　EnrichmentProcessor

```
...
public class EnrichmentProcessor implements ItemProcessor<Foo, Foo> {

    @Autowired
    private RestTemplate restTemplate;

    @Recover
    public Foo fallback(Foo foo) {
        foo.setMessage("error");
        return foo;
    }

    @CircuitBreaker
    @Override
    public Foo process(Foo foo) {
        ResponseEntity<String> responseEntity = this.restTemplate.exchange(
                        "http://rest-service/enrich",
                        HttpMethod.GET,
                        null,
                        String.class);
        foo.setMessage(responseEntity.getBody());

        return foo;
    }
}
```

这就是所有的改动。现在为了运行这些组件，需要像启动 Spring Cloud Config 那样在本地启动 Eureka。为此，可以使用 Spring Cloud CLI 和 spring cloud eureka 命令。如果要同时启动配置服务器和 Eureka，那么可以通过命令 spring cloud configserver Eureka 来简化它们的启动，并且两台服务器都将在本地启动。

Eureka 启动后，可以通过日志中提供的 URL（默认为 http://localhost:8761）导航到 Eureka 仪表盘，并查看注册了哪些服务。在启动时，注册过的服务是空的。在启动了 Eureka 之后，就可以启动 REST API 并监视 Eureka 仪表盘中的注册项。图 12-1 展示了在将 REST API 注册到 Eureka 后的情况。

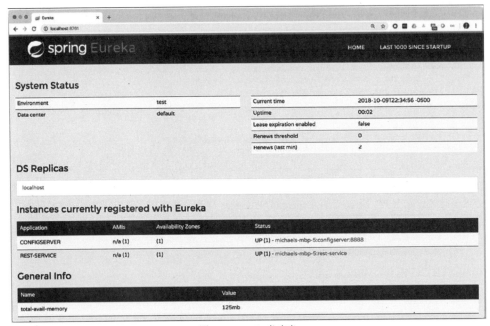

图 12-1　Eureka 仪表盘

注意在默认情况下，configServer 也被注册到了 Eureka 中。如果单击 REST API 的状态 URL，就会得到一个空白的错误页面，因为我们还没有为这个 Spring Boot 应用的/info 端点配置任何东西。但是，如果把 URL 的路径从/info 修改为/，就可以看到 REST API 的调用结果。

最后一步是运行作业。使用之前用过的 java 命令进行验证，运行结果应该与上一次相同。

外部化配置是让云原生处理过程适应新的、动态环境的最重要部分。然而，考虑到云环境的动态特性，编排机制需要与它们兼容。在 12.5 节中，我们将查看另一个 Spring Cloud 项目，从而进行批处理过程的编排。

12.5　批处理过程的编排

按照设计，Spring Batch 并不处理编排事宜。Spring Batch 框架没有用于在给定时间内启动批作业的调度器或其他机制，而是将这一职责委托给其他机制，以允许与其他编排工具进行集成，进而满足企业的具体需求。无论是像 Control-M 这样的大规模企业调度器，还是像 cron 这样简单的调度器，

Spring Batch 都可以使用它们来运行。

也就是说，Spring Portfolio 确实有一个工具用于编排数据处理应用，这个工具名为 Spring Cloud Data Flow。与大多数 Spring Portfolio 成员(其中包含了用于构建自定义应用的框架和库)不同，Spring Cloud Flow 是完全构建好的工具，可用于为流式处理或基于任务的工作负载编排应用。

12.5.1 Spring Cloud Data Flow

Spring Cloud Data Flow 的核心是编排工具。到目前为止，如果想使用 Spring Boot 的超级 Jar 启动 Spring Batch 作业，那么需要用到命令 java -jar、Jar 文件的名称以及运行作业的其他参数。在开发环境中这么做是可以的。甚至可以说，偶尔运行临时任务是没有问题的。但是，在云环境中该怎么办呢？如何将应用部署到云端？如何使用正确的参数启动应用？如何监控批处理作业？如何管理批处理作业之间的依赖关系？

这些都是 Spring Cloud Data Flow 要解决的问题。Spring Cloud Data Flow 是服务器应用，能在支持的平台上启动批处理作业。Spring Cloud Data Flow 的服务器支持三种不同的平台——CloudFoundry、Kubernetes 和本地，Spring Cloud Data Flow 在每个平台上都能部署和启动批作业。

Spring Cloud Data Flow 由一个 Spring Boot 应用组成，这个 Spring Boot 应用充当服务器，负责在给定平台上部署和启动批处理作业。可以通过交互式 shell 或基于 Web 的用户界面与服务器交互。它们都通过一组 REST API 与服务器通信，也可以直接使用这些 API。图 12-2 展示了使用 Spring Cloud Data Flow 时涉及的架构。

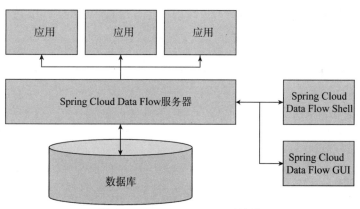

图 12-2　Spring Cloud Data Flow 的架构

注意，Spring Cloud Data Flow 中的所有东西都是基于 Spring Boot 的。比如，服务器是 Spring Boot 应用，shell 也是 Spring Boot 应用，使用 Spring Cloud Data Flow 部署的应用更是典型的 Spring Boot 应用。

Spring Cloud Data Flow 在开始使用之前，必须先通过 wget 命令下载服务器和 shell：

```
wget https://repo.spring.io/milestone/org/springframework/cloud/spring-cloud-
dataflow-server-local/1.7.0.M1/spring-cloud-dataflow-server-local-1.7.0.M1.jar
```

```
wget https://repo.spring.io/milestone/org/springframework/cloud/spring-cloud-
dataflow-shell/1.7.0.M1/spring-cloud-dataflow-shell-1.7.0.M1.jar
```

在下载了这些 Jar 文件后，需要配置 Spring Batch 的作业存储库。幸运的是，Spring Cloud Data Flow 的服务器已经预先配置了 Spring Cloud Config 客户端，因而能够使用我们为批处理作业配置的作业存储库。这一点很重要。为了有助于监控，Spring Cloud Data Flow 需要查看批处理作业写入的作业存储库表。

由于配置服务器已经在运行，因此现在要做的就是通过命令 java -jar spring-cloud-dataflow-server-local-1.7.0.M1.jar 启动 Spring Cloud Data Flow 服务器，并指向我们之前配置的数据库。一旦服务器启动并运行，就可以通过基于 Web 的用户界面或交互式 shell 连接服务器。

我们首先使用 shell。为了启动服务器，可执行命令 java -jar spring-cloud-dataflow-shell-1.7.0.M1.jar。默认情况下，将自动连接本地运行的 Spring Cloud Data Flow 服务器。运行之后，你将看到提示信息。

如前所述，Spring Cloud Data Flow 负责应用的编排。在处理基于 Spring 的应用时，由 Spring Cloud Data Flow 编排的工作负载主要有两种类型：流和任务。下面看看任务以及它们在这里的上下文中如何工作。

12.5.2　Spring Cloud Task

当提到大多数的云原生应用时，我们自然会想到 REST API 或集成应用。所有这些工作负载都有一个共同点——它们都不会结束。如果网站的 REST API 在半夜宕机，那么你可能会通过修改页面或调用来修复问题。然而，正如我们已经知道的，并不是所有的工作负载都适合这种永不结束的模型。这正是任务的用武之地。

Spring Cloud Task 是用来构建非持续微服务的框架，例如数据库迁移、批量作业以及数据科学中批量模型的培训。对于这些工作负载，我们期望它们在云端运行时更为健壮，而不仅仅是让它们在云端执行脚本并希望能够成功运行。Spring Cloud Task 提供了一系列功能特性和非功能特性，从而允许在云端运行有限的工作负载。Spring Cloud Task 提供了以下特性。

- 任务存储库：根据 Spring Batch 的作业存储库，Spring Cloud Task 提供了后端是数据库的任务存储库，用于存储任务的开始时间、结束时间、结果、传递过来的参数，以及在任务执行过程中抛出的任何异常。
- 监听器：Spring Cloud Task 为你提供了挂接到任务的各个执行点的能力，就像 Spring Batch 为你提供了挂接到作业生命周期的各个阶段的能力一样。
- 与 Spring Cloud Stream 集成：Spring Cloud Stream 是用来构建基于消息的微服务的框架。Spring Cloud Task 能与 Spring Cloud Stream 集成，这是通过为 Spring Cloud Task 和 Spring Batch 实现一些监听器做到的。这些监听器会通过中间件发出一些提示性消息，例如开始或完成的任务、开始或完成的工作、开始或完成的步骤，等等。
- 与 Spring Batch 集成：Spring Cloud Task 提供了与 Spring Batch 集成的两个主要集成点。前面已经提到第一个集成点：Spring Cloud Task 可以启动信息性消息。第二个集成点是第 11 章讨论过的 DeployerPartitionHandler。

为了让 Spring Cloud Data Flow 访问任务，任务必须是 Spring Cloud Task。为了实现这一点，你需要修改应用的两个小的地方。首先是添加 Spring Cloud Task 启动包的依赖项，参见代码清单 12-19。

代码清单 12-19　添加 Spring Cloud Task 启动包的依赖项

```
<dependency>
    <groupId>org.springframework.cloud</groupId>
    <artifactId>spring-cloud-starter-task</artifactId>
</dependency>
```

在将 Spring Cloud Task 添加到项目中之后，还需要启用一些功能。@EnableTask 注解可用来引导 Spring Cloud Task 的功能，以便与 Spring Cloud Data Flow 进行交互，如代码清单 12-20 所示。

代码清单 12-20　Spring Cloud Task 的 @EnableTask 注解

```
@EnableTask
@EnableRetry
@EnableBatchProcessing
@EnableDiscoveryClient(autoRegister = false)
public class CloudNativeBatchApplication {

    public static void main(String[] args) {
        SpringApplication.run(CloudNativeBatchApplication.class, args);
    }
}
```

现在，应用变成了任务，可以使用 Spring Cloud Data Flow 进行注册了。

12.5.3　注册和运行任务

在 Spring Cloud Data Flow 能够编排应用之前，你还需要知道应用在哪里。为此，将应用注册到 Spring Cloud Data Flow，并提供名称和可执行文件所在的坐标。可执行文件既可以是某个地方的 Jar 文件(比如通过 HTTP 托管的 Maven 存储库)，也可以是位于 Docker 注册中心的 Docker 镜像。在本例中，我们将为创建的 Jar 文件指定 Maven 坐标，然后把应用注册到 Spring Cloud Data Flow。为此，我们使用 app register 命令。该命令接收三个参数：name 是应用的名称，type 是应用的类型(源、处理器、接收器或任务)，URI 则指定了应用的位置。在本例中，我们将使用应用的 Maven 坐标进行注册，因为我们可以方便地将之安装到本地的 Maven 存储库中。所以，完整的命令是 app register --name fileImport --type task --uri "maven://io.spring.cloud-native-batch: batch-job:0.0.1-SNAPSHOT"。如果在运行以上命令后执行 app list，就会看到已经添加了 fileImport 任务。

在注册了应用之后，还需要添加任务的定义。任务的定义就像启动任务的模板，其中包括任务的名称，以及任务在运行时需要的所有属性。任务的定义使用了管道和过滤器的语法，它们与标准的 UNIX shell 很像，所以你应该很熟悉。我们也使用 shell 来创建任务的定义。为此，可以使用命令 task create myFileImport --definition "fileImport"。如果 fileImport 任务需要设置参数，那么可以在任务的定义中使用类似于 fileImport --foo=bar 的格式。不过，由于我们使用配置服务器来获取配置，因此可以极大地简化配置。

使用 Spring Cloud Data Flow 启动作业的最后一步是确定如何启动作业。这里有几种不同的机制。可以按需通过 shell、GUI 或 REST API 启动任务。如果是在支持 Spring Cloud Data Flow 的平台上运行任务，就可以对任务进行调度。也可以采用基于事件的方法，定义在特定事件(例如，下载文件)发生时用来启动任务的流。在本例中，我们只是临时地启动任务。用来启动任务的命令是 task launch myFileImport。

如果在运行时有额外的命令行参数，那么可以通过--arguments 追加参数(例如 task launch myFileImport –arguements foo=bar)，也可通过--properties 追加属性。

在运行作业后，就可以通过作业存储库中的数据进行监控。这也可以通过 shell、REST API 或 GUI 来完成。不过，GUI 能够更好地呈现监控情况。在浏览器中打开 http://localhost:9393/dashboard 即可打开 GUI。页面的左侧是一组选项卡，它们用于展示 Spring Cloud Data Flow 的各个特性。其中，Apps(应用)选项卡展示了所有已经在系统中注册的应用的列表，可以在这里注册更多应用。Runtime(运行时)选项卡展示了通过 Spring Cloud Data Flow 部署的所有应用的状态。Streams(流)选项卡为你提供了定义和启动基于消息的微服务的能力，这些微服务是基于 Spring Cloud Stream 的流。Tasks(任务)选项卡为你提供了定义和执行任务的能力，还提供了任务存储库的视图，供 Spring Cloud Task 使用。Jobs(作业)选项卡是对 Task 选项卡的扩展，为你提供了浏览 Spring 批处理作业的能力。在 Analytics(分析)选项卡中，可以使用 Spring Cloud Data Flow 内置的分析功能进行基本的可视化工作。最后，在 Audit Records(审计记录)选项卡中可以查看 Spring Cloud Data Flow 提供的用于安全性和合规性的审计流。

我们先看一下 Tasks 选项卡，其中还包含两个子选项卡：Task(任务)和 Executions(执行)。Task 子选项卡列出了所有任务的定义，并且为你提供了创建、启动、调度和删除任务定义的能力。不过，因为已经通过 shell 启动了任务，所以我们直接查看 Executions 子选项卡。在这里我们可以找到任务存储库中的所有任务。在本例中，有一个条目对应刚刚运行的 myFileImport 任务。图 12-3 展示了任务执行数据。

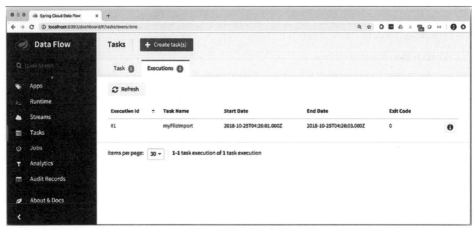

图 12-3　任务执行数据

在图 12-3 中，你可以看到任务运行后的名称、开始和结束时间以及退出码。单击#1(任务的执行 id)，就可以看到任务运行的细节，包括传入的任何参数、外部执行 id(底层系统的 id)、是否包含批处理作业、作业执行链接 id 以及任务执行列表里的其他数据。如果作业在运行时发生异常，那么栈跟踪将出现在 exit message(退出消息)字段中。图 12-4 展示了任务执行详情。

第 12 章 云原生的批处理

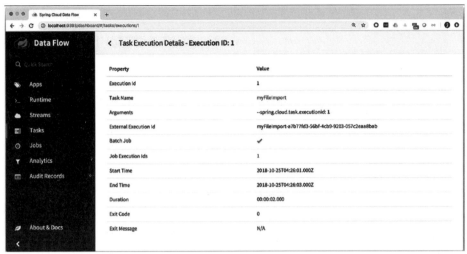

图 12-4 任务执行详情

在本例中,我们不仅关心任务的执行,还关心批处理作业中发生的事情。你可以单击页面左侧的 Jobs 选项卡,也可在前面打开的任务执行详情页面上单击作业执行 id 的链接,从而打开作业执行详情页面。在这个页面上,可以看到 BATCH_JOB_EXECUTION 表中的所有字段和作业参数,这个页面还展示了在这次作业执行中每个步骤的摘要视图,如图 12-5 所示。

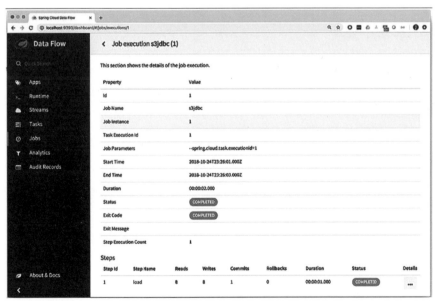

图 12-5 作业执行详情

在监视批处理作业时,我们感兴趣的最后一个页面是步骤执行详情页面。如果单击 Step Name(步骤名称)旁边的 load(加载)链接,就会看到步骤执行时的所有详情以及相关的步骤执行上下文。图 12-6 展示了步骤执行详情页面。

389

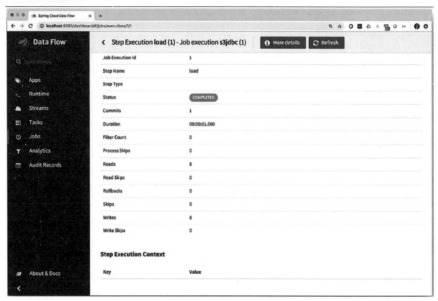

图 12-6　步骤执行详情

Spring Cloud Data Flow 为云平台上的批处理作业提供了健壮的编排和监控方案，它是云原生的批处理的重要组成部分。

12.6　本章小结

相比云计算，批处理可能没有那么强的吸引力。然而，正如我们在本章中讨论的，在现代云平台上运行批处理应用不仅是可能的，而且好处不少。在第 13 章，我们将继续讨论对批处理应用进行伸缩的好处。

第 13 章

批处理的测试

测试是编程中最不令人愉悦的环节。但有趣的是，就像生活中的大多数事情一样，一旦掌握，测试实际上是很有趣的。测试能让开发变得效率更高，能为尝试新事物提供安全的环境。程序化的测试还为你提供了尝试新技术的测试平台(大多数公司并不介意你在测试中尝试一些新技术，但是如果你在即将投入生产的代码中尝试新技术，则会非常介意)。在前 10 章，你虽然已经花费了大量时间来编写代码，但却无法证明其中任何一个作业是有效的。本章将介绍如何以多种方式执行代码，这样不仅可以证明代码在按设计工作，而且可以在更改代码时提供安全保障。

本章主要介绍如下主题。

- 使用 JUnit 和 Mockito 进行单元测试：我们首先对 JUnit 和 Mockito 框架做了高层次概述。虽然在本章的后面，你会跳过 JUnit 的基本功能，但是 Spring 中的测试设施是基于 JUnit 约定的，因此了解这些概念有助于你理解在更高级的测试中发生了什么。我们还将介绍 mock 对象的框架 Mockito，看看 Mockito 如何有助于开发批处理组件的单元测试。
- 使用 Spring Batch 的实用工具进行集成测试：批处理过程有自己特定的执行需求。我们将介绍如何使用 Spring Batch 提供的一些实用工具来测试批处理过程。

因为最基本的测试通常是单元测试，所以下面从单元测试开始讨论。

13.1 使用 JUnit 和 Mockito 进行单元测试

单元测试可能是最容易编写且最有价值，但也最容易被忽视的测试类型。尽管由于许多原因，本书没有采用测试驱动开发方式，但我们还是鼓励你在开发过程中这样做。作为一种经过验证的有效方式，测试驱动开发不仅可以提高软件的质量，还可以提高任何单个开发人员和整支团队的生产力，测试中的代码可以产生一些十分有价值的信息。本节将介绍如何使用 JUnit 和 Mockito 对批处理组件进行单元测试。

什么是单元测试？单元测试是对单个、独立组件执行的可重复测试。下面分解一下上述定义。

- 单个：单元测试旨在测试应用的最小构建块。
- 独立：依赖项会严重影响系统的测试。然而，所有的系统都有依赖项。单元测试的目标不是测试组件与这些依赖项之间的集成，而是测试组件本身的工作方式。

- 可重复：当启动浏览器并单击应用时，这不是可重复的练习。可以每次输入不同的数据，还可以按略微不同的顺序单击按钮。单元测试应该能够一次又一次地重复完全相同的场景，从而允许在对系统进行更改时使用它们进行回归测试。

能够以可重复方式对组件执行独立测试的框架有 JUnit、Mockito 和 Spring。前两个是通用的多用途框架，对创建单元测试非常有用。Spring 的实用测试工具有助于测试更广泛的问题，包括不同层的集成，甚至端到端地测试作业的执行(从服务或 Spring 批处理组件到数据库，然后返回)。

13.1.1 JUnit

JUnit 被认为是 Java 中测试框架的黄金标准，它能够以标准的方式对 Java 类进行单元测试。虽然大多数框架都需要向 IDE 和构建过程附加组件，但 Maven 和大多数 Java IDE 都内置了对 JUnit 的支持，所以不需要额外的配置。下面介绍 JUnit 及其最常用特性。

在编写本书时，JUnit 的当前版本是 JUnit 5.2.0。尽管每一次修订都包含一些细微的改进和漏洞修复，但 JUnit 框架的最后一次主要修订是从 JUnit 4 迁移到 JUnit 5，这次修订对测试用例 API 进行了重大修改。什么是测试用例？让我们先回顾一下 JUnit 测试的构建方式。

JUnit 测试的生命周期

JUnit 测试由多个所谓的测试用例组成。每个测试用例都旨在测试特定的功能块，大部分测试在类级别进行划分。通常的做法是每个类至少有一个测试用例。测试用例只不过是使用 JUnit 注解进行配置，并且由 JUnit 执行的 Java 类。测试用例中既存在测试方法，也存在用于设置前置条件以及在每个测试或测试组之后进行清理的方法。代码清单 13-1 展示了一个基本的 JUnit 测试用例。

代码清单 13-1　一个基本的 Junit 测试用例

```
package com.apress.springbatch.chapter13;
import org.junit.jupiter.api.Test;
import static org.junit.jupiter.api.Assertions.*;

public class StringTest {

    @Test
    public void testStringEquals() {
        String michael = "Michael";
        String michael2 = michael;
        String michael3 = new String("Michael");
        String michael4 = "Michael";
        assertTrue(michael == michael2);
        assertFalse(michael == michael3);
        assertTrue(michael.equals(michael2));
        assertTrue(michael.equals(michael3));
        assertTrue(michael == michael4);
        assertTrue(michael.equals(michael4));
    }
}
```

代码清单 13-1 中的单元测试并没有什么特别之处，所要做的只是证明在比较字符串时使用==运算符与使用.equals 方法是有区别的。现在，让我们逐一看看这个单元测试中的各个不同部分。首先，JUnit 测试用例是常规的 POJO。不需要扩展任何特定的类，JUnit 对类的唯一要求是必须有一个无参构造函数。

第 13 章 批处理的测试

在每个测试中，都有一个或多个测试方法(在本例中为一个)。每个测试方法都必须是公共的(public)，无返回值(void)，并且没有参数。为了指明 JUnit 执行的测试方法，可以使用@Test 注解。在执行给定的测试时，JUnit 会执行每一个使用了@Test 注解的方法。

StringTest 的最后一部分是在测试方法中使用断言(assert)方法。流程很简单。首先设置测试所需的条件，然后执行测试并同时使用 JUnit 的断言方法验证结果。org.junit.Assert 类用于验证给定测试场景的结果。对于代码清单 13-1 中的 StringTest，Assert 类用于验证.equals 方法比较的是字符串的内容，而==运算符比较的是两个字符串是否是相同的实例。

尽管这个测试十分有用，但是在使用 JUnit 时，你应该了解一些其他有用的注解。前两个与 JUnit 测试的生命周期相关。JUnit 允许将某些方法配置为在每个测试方法的前后运行，这样可以设置通用的先决条件，并在每次执行之后进行基本的清理。要在每个测试方法之前执行某个方法，可以使用@BeforeEach 注解；@AfterEach 注解表示应该在每个测试方法之后执行某个方法。与任何测试方法一样，使用@BeforeEach 和@AfterEach 标记的方法必须是公共的、无返回值且不带参数。通常，在标记为@BeforeEach 的方法中需要创建新的测试实例，以防止一个测试对另一个测试产生影响。图 13-1 展示了使用@BeforeEach、@Test 和@AfterEach 注解的 JUnit 测试的生命周期。代码清单 13-2 显示的测试用例使用了这三个注解。

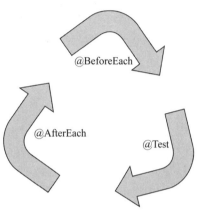

图 13-1　JUnit 测试的生命周期

代码清单 13-2　一个测试用例

```
...
public class FooTest {

    private Foo fooInstance;

    @BeforeEach
    public void setUp() {
        fooInstance = new Foo();
    }

    @Test
    public void testBar() {
        String results = fooInstance.bar();

        assertNotNull("Results were null", results);
        assertEquals("The test was not a success", "success", results);
    }

    @AfterEach
    public void tearDown() {
        fooInstance.close();
    }
}
```

JUnit 提供了这三个注解的其他变种，比如，@BeforeAll 用于对给定的测试类中的所有测试方法进行一次性设置，@Ignore 用于指示跳过的测试方法和类，@RunWith 用于设定使用类来运行测试用例而不是默认使用 JUnit。然而，这些都超出了本书的讨论范围。我们的目标是提供能够测试批处理过程所需的工具。只需要使用@BeforeEach、@Test 和@AfterEach 注解，以及 JUnit 的 Assert 类提供的断言方法，就可以测试绝大多数场景。

但这里存在一个小问题。我们在前面的单元测试的定义中讲过，单元测试是指单独地测试组件。当数据访问对象(DAO)依赖于 JDBC 和数据库时，如何使用 JUnit 进行测试？如何测试 ItemStream(其中的一些参数是 Spring 批处理组件)？mock 对象填补了这一空白，详见 13.1.2 节。

13.1.2　mock 对象

编写与前面测试的 String 对象类似的软件是非常容易的，因为没有依赖关系。然而，大多数系统是十分复杂的。批处理作业可能需要几十个类甚至更多的类，并且依赖于外部系统，包括应用服务器、消息传递中间件和数据库，而这些只是其中的一部分。所有这些都很难管理，并且涉及超出单元测试范围的交互。例如，当测试某个 ItemProcessor 的业务逻辑时，真的需要测试 Spring Batch 是否将上下文正确保存到了数据库吗？这超出了单元测试的范围。不要犯这样的错误——这确实需要测试，然而，为了测试业务逻辑，并不需要测试与生产系统交互的各种依赖项。你可以使用 mock 对象来替换测试环境中的这些依赖项，并在不受外部依赖项影响的情况下运行业务逻辑。

■ **注意**　stub 并不是 mock 对象。stub 在测试中使用了硬编码的实现，而 mock 对象是可重用的构造体，允许在运行时定义所需的行为。

下面花一点时间来说明 mock 对象不是 stub。stub 是用于替换应用的各个部分的实现。stub 包含硬编码逻辑，用于模拟执行期间的特定行为。它们不是 mock 对象(无论它们在项目中的名称是什么)！

mock 对象是如何工作的呢？大多数 mock 对象基本上采用两种不同的方法：基于代理或者基于类的重新映射。下面介绍第一种方法。

代理对象用来代替实际的对象。对于 mock 对象来说，可使用代理对象模拟代码依赖的真实对象。可首先使用 mock 框架创建代理对象，然后使用 setter 或构造函数进行设置。这指出了使用代理对象进行模拟的如下固有问题：必须能够通过外部方法设置依赖项。换句话说，不能通过在方法中调用 new MyObject()来创建依赖项，因为无法模拟通过调用 new MyObject()创建的对象。这就是像 Spring 这样的依赖注入(Dependency Injection)框架能够成功的原因之一——它们允许在不修改任何代码的情况下注入代理对象。

另一种方法是在类加载器中重新映射类文件。JMockit 是到目前为止笔者所知道的唯一能将这种功能用于 mock 对象的框架。具体的概念由 java.lang.instrument.Instumentation 接口提供，实现方式是让类加载器重新映射已加载的类文件的引用。假设存在 MyDependency 类和相应的类文件 MyDependency.class，如果希望使用 MyMock 来模拟，那么可以通过使用这种类型的 mock 对象，在类加载器中将 MyDependency 的引用重新映射到 MyMock.class。这将允许你模拟那些使用 new 操作符创建的对象。

Mockito 是一款流行的基于代理的 mock 对象框架，提供了强大的灵活性和极富表现力的语法，允许相对轻松地创建易于理解的单元测试。接下来让我们一起来看看这个框架。

13.1.3 Mockito

Mockito 允许你模拟自己关心的行为,并只验证其中重要的那些。在本小节中,你将了解 Mockito 的一些可用功能,并用它测试 Spring Batch 的组件。

JUnit 和 Mockito 都包含在 spring-boot-starter-test 依赖项中,因此在从 Spring Initializr 创建项目之后,不需要做任何事情就可以开始编写测试。

回顾第 10 章,为了验证客户是否存在而创建的 CustomerItemValidator 是使用 mock 对象的主要候选者,它依赖于外部的 JdbcTemplate。为了方便你回忆,代码清单 13-3 展示了 CustomerItemValidator 的代码。

代码清单 13-3　CustomerItemValidator

```
...
@Component
public class CustomerItemValidator
implements Validator<CustomerUpdate> {

private final NamedParameterJdbcTemplate jdbcTemplate;

public static final String FIND_CUSTOMER =
      "SELECT COUNT(*) FROM CUSTOMER WHERE customer_id = :id";

public CustomerItemValidator(NamedParameterTemplate template) {
      this.jdbcTemplate = template;
}

@Override
public void validate(CustomerUpdate customer)
      throws ValidationException {

      Map<String, Long> parameterMap =
            Collections.singletonMap("id", customer.getCustomerId());

      Long count =
            jdbcTemplate.queryForObject(FIND_CUSTOMER,parameterMap,Long.class);

      if(count == 0) {
            throw new ValidationException(
                  String.format("Customer id %s was not able to be found",
customer.getCustomerId()));
      }
}
}
```

这里要测试的方法显然是 validate。validate 方法需要一个外部依赖项——一个 NamedParameterJdbcTemplate 实例。为了测试 validate 方法,你需要两个测试方法,分别用于 validate 方法的两个执行分支(一个用于发现客户时,另一个用于未发现客户时)。

为了开始测试,下面创建测试用例并使用@BeforeEach 注解,以便构建以后使用的对象。代码清单 13-4 展示的测试用例包含一个使用了@BeforeEach 注解的 setUp 方法和两个属性。

代码清单 13-4　CustomerItemValidatorTests

```
...
public class CustomerItemValidatorTests {

    @Mock
    private NamedParameterJdbcTemplate template;

    private CustomerItemValidator validator;
    @BeforeEach
    public void setUp() {
        MockitoAnnotations.initMocks(this);
        this.validator = new CustomerItemValidator(this.template);
    }
...
}
```

测试类 CustomerItemValidatorTests 的两个属性是：将要测试的类(CustomerItemValidator)及其依赖项(NamedParameterJdbcTemplate)。@Mock 注解用于告知 Mockito 为 NamedParameterJdbcTemplate 创建 mock 对象。当执行测试时，Mockito 将创建用于测试的代理对象。

setUp 方法做了两件事情。首先，使用 Mockito 的 MockitoAnnotations.initMocks 方法初始化 mock 对象(initMocks 方法将使用 mock 对象初始化前面指定的所有对象)，这是创建将来需要使用的 mock 对象的一种快速而简单的方法。

其次，创建要测试的类的新实例，从而确保每个测试方法都包含一个要测试的类的干净实例，这可以防止来自某个方法的测试对象中的任何遗留状态对其他测试方法产生影响。在创建 CustomerItemValidator 之后，注入 mock 对象，方式与 Spring 引导应用时一样。

你已经有了要测试的类的新实例和一组新的 mock 对象，它们能满足对 Spring Batch 框架和数据库的依赖要求，现在可以编写测试方法了。第一个测试方法用于对查询到的客户进行测试，参见代码清单 13-5。

代码清单 13-5　testValidCustomer 测试方法

```
...
@Test
public void testValidCustomer() {

    // given

    CustomerUpdate customer = new CustomerUpdate(5L);

    // when
    ArgumentCaptor<Map<String, Long>> parameterMap =
            ArgumentCaptor.forClass(Map.class);
    when(this.template.queryForObject(eq(CustomerItemValidator.FIND_CUSTOMER),
                        parameterMap.capture(),
                        eq(Long.class)))
            .thenReturn(2L);

    this.validator.validate(customer);

    // then
```

```
        assertEquals(5L, (long) parameterMap.getValue().get("id"));
    }
    ...
```

这里使用了行为驱动的设计风格，注释中包含了 //given、//when、//then，它们表示给定这些输入，当这些动作发生时结果应该什么样。在本例中，给定 id 为 5 的 CustomerUpdate 对象，当进行验证时，不应该抛出任何异常。

但是，我们的代码稍微复杂一些。这里在 when 部分使用了 Mockito 的用于捕获传递给 mock 对象的值这一特性。在本例中，由于捕获了传递给 NamedParameterJdbcTemplate 的 Map，因此可以断言传递的参数是符合预期的。当执行想要测试的方法，并且给 mock 对象传递正确的参数时，Mockito 将返回 2。当没有指定返回值时，Mockito 将返回类型适当的默认值。在本例中，如果使用任何其他的参数值调用这个测试方法，将返回 null(null 是对象的默认值)，从而导致测试失败。一旦 Mockito 返回 2，我们的逻辑就会验证值不等于 0，并且不会抛出异常。

我们将要编写的第二个测试方法实际上几乎与第一个相同。主要的区别是，当没有找到提供的客户时，我们期望测试代码抛出异常。代码清单 13-6 展示了 testInvalidCustomer 测试方法的代码。

代码清单 13-6　testInvalidCustomer 测试方法

```
...
@Test
public void testInvalidCustomer() {

    // given

    CustomerUpdate customerUpdate = new CustomerUpdate(5L);

    // when
    ArgumentCaptor<Map<String, Long>> parameterMap =
            ArgumentCaptor.forClass(Map.class);
    when(this.template.queryForObject(eq(CustomerItemValidator.FIND_CUSTOMER),
                        parameterMap.capture(),
                        eq(Long.class)))
        .thenReturn(0L);

    Throwable exception = assertThrows(ValidationException.class,
                    () -> this.validator.validate(customerUpdate));

    // then

    assertEquals("Customer id 5 was not able to be found",
                exception.getMessage());
}
...
```

testInvalidCustomer 测试方法使用了 JUnit 5 中的新特性——assertThrows 方法。在以前的 JUnit 版本中，我们需要以不同的方式处理异常的断言(例如，使用 JUnit 规则、自己捕获异常等)。现在，你可以通过 org.junit.jupiter.api.Assertions.assertThrows 方法验证抛出的异常类型，这个方法的参数是期望的异常类型以及用来执行被测试代码的闭包。如果没有异常或抛出错误类型的异常，那么断言失败。

如果类型正确，那么断言通过并返回抛出的异常。在此基础上，就可以断言错误消息正是我们期望的结果。

以上两个测试方法使我们能够可靠地验证类的行为并进行重构，而不必担心会影响代码库的其他部分。单元测试是构建坚实系统的基础。它们不仅让你有了无忧地进行更改的能力，而且迫使你保持代码简洁。

13.2 使用 Spring 的实用工具进行集成测试

13.1 节讨论了单元测试。无论单元测试多么有用，它们总有局限性。集成测试将自动化测试提升到一个新的层次，本节将介绍如何使用 Spring 的集成测试工具来测试与各种 Spring Bean、数据库和批处理资源的交互。

13.2.1 使用 Spring 进行通用集成测试

集成测试就是测试不同部分之间的通信：是否正确连接了 DAO、Hibernate 映射是否正确，以便可以保存所需的数据？服务是否从给定的工厂检索正确的 Bean？在编写集成测试时，将测试这些内容以及其他情况。但是，如何在不设置所有的基础设施时做到这一点，并确保基础设施在运行这些测试的任何地方都可用呢？幸运的是，你不必这么做。

使用核心的 Spring 集成测试工具进行集成测试的两个主要用例是：测试数据库交互和测试 Spring Bean 交互(是否正确装入了服务，等等)。为此，先看看我们在前面的单元测试(CustomerItemValidator)中模拟的 NamedParameterJdbcTemplate。但是这一次，我们让 Spring 自己装入 CustomerItemValidator，并使用 HSQLDB 的内存实例作为数据库，这样就可以随时随地执行测试。HSQLDB 是百分百由 Java 实现的数据库，作为轻量级实例，HSQLDB 非常适合进行集成测试。下面看看如何配置测试环境。

配置测试环境

为了将测试的执行与外部资源需求(特定的数据库服务器，等等)隔离开来，应该配置一些东西。具体来说，应该为数据库使用测试配置，并在内存中创建 HSQLDB 实例。为此，需要更新 POM 文件以包含 HSQLDB 数据库驱动程序。需要添加的特定依赖项如代码清单 13-7 所示。

代码清单 13-7　HSQLDB 的数据库驱动依赖项

```
...
<dependency>
        <groupId>org.hsqldb</groupId>
        <artifactId>hsqldb</artifactId>
        <scope>test</scope>
</dependency>
...
```

在配置了额外的依赖项之后，就可以编写集成测试了。Spring Boot 提供了一些工具，使得编写集成测试非常容易。事实上，由于我们不需要定义任何 mock 对象，因此集成测试将比单元测试更短。

首先，为测试类使用 @ExtendWith(SpringExtension.class) 注解，这在 JUnit 5 中相当于

第 13 章 批处理的测试

@RunWith(SpringRunner.class)注解。通过使用这个注解,就会得到用于测试的所有 Spring 特性。从这里开始,我们将使用 Spring Boot 的自动配置功能来创建数据库并进行填充。像使用 Spring Boot 的大多数功能一样,这非常简单,只需要使用@JdbcTest 注解。这个注解将创建一个内存数据库,并通过 Spring Boot 通常使用的数据进行填充(通过初始化脚本)。

在 Spring Boot 自动创建和填充数据库后,你需要做的就是把与之相关的数据源(DataSource)装配到测试用例中。代码清单 13-8 展示了使用类级别注解的初始类的定义,并且展示了自动装入的数据源。

代码清单 13-8　CustomerItemValidatorIntegrationTests

```
@ExtendWith(SpringExtension.class)
@JdbcTest
public class CustomerItemValidatorIntegrationTests {

    @Autowired
    private DataSource dataSource;

    private CustomerItemValidator customerItemValidator;
...
}
```

有了数据源之后,就可以配置 setUp 方法以创建校验器,参见代码清单 13-9。

代码清单 13-9　CustomerItemValidatorIntegrationTests 的 setUp 方法

```
...
@BeforeEach
public void setUp() {
    NamedParameterJdbcTemplate template =
            new NamedParameterJdbcTemplate(this.dataSource);
    this.customerItemValidator = new CustomerItemValidator(template);
}
...
```

有了校验器之后,就可以创建测试了。但是,与需要进行模拟的单元测试不同,这里只需要执行调用并断言结果。代码清单 13-10 列出了一些功能上与单元测试相同的测试,它们就像集成测试一样。

代码清单 13-10　测试 CustomerItemValidatorIntegrationTests

```
...
@Test
public void testNoCustomers() {
    CustomerUpdate customerUpdate = new CustomerUpdate(-5L);
    ValidationException exception =
                assertThrows(ValidationException.class,
                    () -> this.customerItemValidator.validate(customerUpdate));

    assertEquals("Customer id -5 was not able to be found",
                exception.getMessage());
}

@Test
```

```java
public void testCustomers() {
    CustomerUpdate customerUpdate = new CustomerUpdate(5L);
    this.customerItemValidator.validate(customerUpdate);
}
...
```

这一次，对于没有客户的测试，我们创建了一个 CustomerUpdate 对象，它的 id 不在测试数据中。然后，捕获抛出的异常并断言结果，就像在单元测试中所做的那样。在验证客户存在的测试中，我们创建了一个 id 在测试数据中的 CustomerUpdate 对象，并且调用了校验器。如果没有抛出异常，那么测试通过。

在开发系统时，类似于 CustomerItemValidatorIntegrationTests 中的集成测试非常有价值。在处理复杂系统时，通过判断组件是否正确装入、SQL 是否正确以及系统组件之间的操作顺序是否正确等，可以提供相当高的安全性。

使用 Spring Batch 进行测试的最后一部分是测试 Spring Batch 组件本身。可以使用 Spring 提供的工具测试 ItemReader、步骤甚至整个作业。

13.2.2　测试 Spring Batch

尽管在处理健壮的批处理作业时，测试诸如 DAO 或服务的组件的能力是绝对需要的，但是 Spring Batch 框架引入了一些额外的复杂性。为了构建健壮的测试套件，这些复杂性是需要解决的。本小节介绍如何测试 Spring Batch 的特定组件，包括依赖于自定义范围(custom scope)的元素、Spring Batch 步骤甚至完整的作业。

1. 测试步骤范围和作业范围的 Bean

正如你在本书的许多示例中看到的那样，Spring Batch 定义的步骤范围(StepScoped)和作业范围(JobScoped)对 Spring Bean 来说是非常有用的工具。然而，当为使用步骤范围的组件编写集成测试时，就会遇到如下问题：如果在步骤范围之外执行这些组件，那么如何解决这些依赖项呢？下面介绍 Spring Batch 提供的两种方法以模拟在步骤范围内执行的 Bean。

你在前面已经看到，步骤范围允许 Spring Batch 使用作业和/或步骤上下文把运行时的值注入 Bean 中。前面的示例包括输入或输出文件名的注入，以及涉及特定数据库的查询条件。在这些示例中，Spring Batch 将从 JobExecution 或 StepExecution 中获取值。如果不在作业中运行步骤，那就不会有这两个执行对象。Spring Batch 提供了两种不同的方法来模拟步骤中的执行对象，以便注入这些值。第一种方法是使用 TestExecutionListener。

TestExecutionListener 允许定义测试方法前后发生的事情。与使用 JUnit 的@BeforeEach 和@AfterEach 注解不同，TestExecutionListener 允许以一种更加可重用的方式将行为注入测试用例的所有方法中。尽管 Spring 提供了三个有用的 TestExecutionListener 实现(DependencyInjectionTestExecutionListener、DirtiesContextTestExecutionListener 和 TransactionalTestExecutionListener)，但是 Spring Batch 只提供了两个接口：StepScopeTestExecutionListener 和 JobScopeTestExecutionListener。它们的功能相同，因此我们接下来选择使用其中更为常用的 StepScopeTestExecutionListener。

StepScopeTestExecutionListener 提供了你所需要的两个特性。首先，可以使用来自测试用例的工厂方法获得 StepExecution，并把返回的实例作为当前测试方法的上下文。其次，可以为每个测试方法的生命周期提供 StepContext。图 13-2 显示了使用 StepScopeTestExecutionListener 执行测试的流程。

第 13 章 批处理的测试

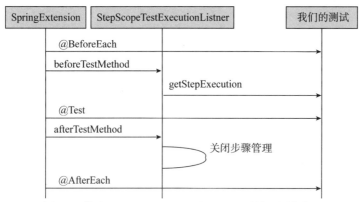

图 13-2 使用 StepScopeTestExecutionListener 执行测试的流程

如你所见,在调用每个测试方法之前,调用测试用例的工厂方法(getStepExecution)以获取一个新的 StepExecution。如果没有工厂方法,那么 Spring Batch 将使用默认的 StepExecution。

为了进行测试,需要配置 FlatFileItemReader,以获取从 jobParameters 读取的文件的位置。这个条目读取器来自读取客户更新文件的样例应用。代码清单 13-11 展示了这个条目读取器的配置。

代码清单 13-11　ImportJobConfiguration#customerUpdateItemReader

```
...
@Bean
@StepScope
public FlatFileItemReader<CustomerUpdate> customerUpdateItemReader(
        @Value("#{jobParameters['customerUpdateFile']}") Resource inputFile)
        throws Exception {
    return new FlatFileItemReaderBuilder<CustomerUpdate>()
                .name("customerUpdateItemReader")
                .resource(inputFile)
                .lineTokenizer(customerUpdatesLineTokenizer())
                .fieldSetMapper(customerUpdateFieldSetMapper())
                .build();
}
...
```

我们将使用这个条目读取器来读取一个包含三条记录的测试文件,每条记录对应测试文件中可能存在的三种记录格式之一。然后,我们将断言这个条目读取器返回的类型都是正确的。代码清单 13-12 展示了一个测试文件。

代码清单 13-12　测试文件 customerUpdateFile.csv

```
2,5,,,Montgomery,Alabama,36134
3,5,,,,316-510-9138,2
1,5,Rozelle,Heda,Farnill
```

代码清单 13-11 中定义的 customerUpdateItemReader 需要一些额外的依赖项,但是我们无须担心它们的配置。我们实际上会在测试中使用来自应用的原始配置,其中提供了测试所有组件是否按预期一起工作的最准确方式。这需要在测试中添加一些基础设施,所以我们首先看看所需的这些基础设施,如代码清单 13-13 所示。

代码清单 13-13　FlatFileItemReaderTests 的基础设施

```
...
@ExtendWith(SpringExtension.class)
@ContextConfiguration(classes = {ImportJobConfiguration.class,
                                 CustomerItemValidator.class,
                                 AccountItemProcessor.class})
@JdbcTest
@EnableBatchProcessing
@SpringBatchTest
public class FlatFileItemReaderTests {

        @Autowired
        private FlatFileItemReader<CustomerUpdate> customerUpdateItemReader;
...
}
```

你应该已经熟悉代码清单 13-13 中的第一个注解，因为之前在集成测试中用过。@ExtendWith(SpringExtension.class)用于触发 Spring 的测试功能。注意这个测试再次使用了@JdbcTest 注解，原因在于：虽然我们在单元测试中不会使用创建的数据库，但是现在需要重新使用应用的配置，并且需要提供数据库。下一个引人关注的注解是@ContextConfiguration，这个注解用于指定构建 ApplicationContext 的类(如果在使用 XML 配置，那么就是资源)。在本例中，我们提供了三个类。ImportJobConfiguration 类是用@Configuration 注解的，我们在其中完成了所有@Bean 风格的配置。但是，我们也确实定义了其他两个组件，分别是 CustomerItemValidator 和 AccountItemProcessor，它们可由 ImportJobConfiguration 中的 Bean 使用。这两个类都使用了@Component 注解，因此也需要包含在此处提供的 classes 数组中。

到目前为止，所有的注解都来自 Spring 框架或 Spring Boot。我们要介绍的最后两个注解是特定于 Spring Batch 的。第一个注解是本书的每个示例都在使用的@EnableBatchProcessing。因为我们需要重新使用应用的配置，所以应用上下文(ApplicationContext)将包含作业和步骤(不要担心，它们不会执行)。Spring Batch 需要将它们连接到 JobRepository，这就是使用这个注解的原因。

第二个注解是来自 Spring Batch 4.1 的@SpringBatchTest。这个注解为自动测试 ApplicationContext 提供了许多实用工具。具体来说，添加了四个 Bean：

- 一个 JobLauncherTestUtils 实例，用于启动作业或步骤。
- 一个 JobRepositoryTestUtils，可用于从 JobRepository 中创建 JobExecutions。
- 一个 StepScopeTestExecutionListner 和一个 JobScopeTextExecutionListner，它们允许测试处在步骤范围和作业范围内的 Bean。

我们现在感兴趣的是最后两个 Bean，因为我们的读取器处在步骤范围内。为了使用 StepScopeTestExecutionListener 处理步骤范围内的依赖项，我们需要创建一个工厂方法，从而提供一个 StepExecution 以填充我们需要的内容。在本例中，我们测试的读取器需要提供一个名为 customerUpdateFile 的作业参数，以指向要读取的文件。这个工厂方法如代码清单 13-14 所示。

代码清单 13-14　FlatFileItemReaderTests#getStepExecution

```
...
public StepExecution getStepExecution() {
    JobParameters jobParameters = new JobParametersBuilder()
```

```
                .addString("customerUpdateFile", "classpath:customerUpdateFile.csv")
                .toJobParameters();
        return MetaDataInstanceFactory.createStepExecution(jobParameters);
    }
...
```

代码清单 13-14 中的 getStepExecution 方法非常直观：首先创建一个 JobParameters 对象，其中的 customerUpdateFile 参数指向 customerFile.csv 文件的测试版本；然后使用 MetaDataInstanceFactory 创建一个 StepExecution。MetaDataInstanceFactory 是一个用于创建步骤和作业执行实例的实用工具类，它与 JobRepositoryTestUtils 的区别在于：从 MetaDataInstanceFactory 产生的步骤或作业执行不会持久化到作业存储库中，而从 JobRepositoryTestUtils 获得的步骤或作业执行则会持久化到作业存储库中。

有了创建 StepExecution 的工厂方法之后，现在需要做的就是编写测试。由于我们严重依赖 Spring Batch 的正常行为，因此测试不是很复杂。我们将打开已经注入测试中的读取器，然后读取测试文件中的三条记录，并验证它们是否返回正确的类型，如代码清单 13-15 所示。

代码清单 13-15　FlatFileItemReaderTests#testTypeConversion

```
...
@Test
public void testTypeConversion() throws Exception {
    this.customerUpdateItemReader.open(new ExecutionContext());

    assertTrue(this.customerUpdateItemReader.read() instanceof CustomerAddressUpdate);
    assertTrue(this.customerUpdateItemReader.read() instanceof CustomerContactUpdate);
    assertTrue(this.customerUpdateItemReader.read() instanceof CustomerNameUpdate);
}
...
```

这种性质的集成测试对测试自定义组件(例如自定义的 ItemReader 和 ItemWriter 或者相关组件)非常有用。但是，如你所见，测试 Spring Batch 自身组件的价值是微不足道的。相反，通过执行整个步骤来测试批处理作业可能更有用。

2. 步骤的测试

作业被分成多个步骤，其中的每个步骤都是独立的功能块，几乎可以在不影响其他步骤的情况下执行。由于步骤与批处理作业之间本身就是解耦的，因此步骤成为测试的主要候选对象。下面介绍如何完整地测试 Spring Batch 作业的步骤。

在基于步骤范围的示例中，我们测试了用于读取文件的 ItemReader。下面看看如何在步骤上下文中进行同样的测试。我们将执行步骤(按预期读取文件并更新数据库)，然后查看结果，以验证步骤是否正确执行。

我们首先配置测试的基础设施，如代码清单 13-16 所示，与基于步骤范围的测试的基础设施非常相似。

代码清单 13-16　ImportCustomerUpdatesTests 的基础设施

```
...
@ExtendWith(SpringExtension.class)
@JdbcTest
@ContextConfiguration(classes = {ImportJobConfiguration.class,
```

```
                                        CustomerItemValidator.class,
                                        AccountItemProcessor.class,
                                        BatchAutoConfiguration.class})
@SpringBatchTest
@Transactional(propagation = Propagation.NOT_SUPPORTED)
public class ImportCustomerUpdatesTests {

        @Autowired
        private JobLauncherTestUtils jobLauncherTestUtils;

        @Autowired
        private DataSource dataSource;

        private JdbcOperations jdbcTemplate;

        @BeforeEach
        public void setUp() {
                this.jdbcTemplate = new JdbcTemplate(this.dataSource);
        }
...
}
```

开头使用的注解与基于步骤范围的测试使用的注解一样。

- @ExtendWith(SpringExtension.class)：在 JUnit 5 中启用所有的 Spring 特性。
- @JdbcTest：为进行数据库测试提供包括内存数据库在内的基础设施。
- @ContextConfiguration：提供用于构建 ApplicationContext 的类。
- @SpringBatchTest：提供用于测试 Spring Batch 作业的实用工具。在本例中，我们特别关心的是 JobLauncherTestUtils。
- @Transactional(propagation = Propagation.NOT_SUPPORTED)：默认情况下，@JdbcTest 会把每个测试方法封装到一个事务中，并在完成时回滚。在正常的单元测试场景中，这是有意义的。但是，在我们的示例中，Spring Batch 负责管理事务，如果再使用另一个事务进行封装的话，实际上会导致错误。@Transaction 注解关闭了 @JdbcTest 注解的事务行为。

在应用了这些注解之后，就可以自动装配一些可用的基础设施。在本例中，它们是 JobLauncherTestUtils 和 DataSource。你还可以创建 setUp 方法，从而使用提供的数据源创建 JdbcTemplate。

接下来我们看看要在测试中使用的数据。表 13-1 展示了每个数据表字段以及它们在测试运行前后的期望值。

表 13-1 测试值

数据表字段	初始值	最终值
customer_id	5	5
first_name	Danette	Rozelle
middle_name	null	Heda
last_name	Langelay	Farnill
address1	36 Ronald Regan Terrace	36 Ronald Regan Terrace
address2	P.O. Box 33	P.O. Box 33

第 13 章　批处理的测试

(续表)

数据表字段	初始值	最终值
city	Gaithersburg	Montgomery
state	Maryland	Alabama
postal_code	99790	36134
ssn	832-86-3661	832-86-3661
email_address	tlangelay4@mac.com	tlangelay4@mac.com
home_phone	240-906-7652	240-906-7652
cell_phone	907-709-2649	907-709-2649
work_phone	null	316-510-9138
notification_pref	3	2

初始值已经存在于数据库中，它们可通过脚本进行加载。最终值是使用代码清单 13-17 所示的文件作为输入执行测试后的结果。

代码清单 13-17　输入文件

```
2,5,,,Montgomery,Alabama,36134
3,5,,,,316-510-9138,2
1,5,Rozelle,Heda,Farnill
```

在确定了输入输出后，就可以编写测试方法了。令人惊讶的是，测试方法中的大部分代码实际上用于在数据库中进行查询以获得结果。测试方法首先定义了运行步骤所需的作业参数，在本例中，这些作业参数与上一个测试使用的作业参数相同。一旦定义了参数，就可以通过调用 this.jobLauncherTestUtils.launchStep("importCustomerUpdates", jobParameters)来执行步骤。这个调用将查找名为 importCustomerUpdates 的步骤，并使用传入的作业参数运行该步骤。这个测试的最后一部分操作是断言数据库中的数据就是我们期望的值。我们使用 JdbcTemplate 执行查询，将结果映射到 Map，然后断言每个结果都是我们期望的值。代码清单 13-18 展示了完整的测试方法。

代码清单 13-18　ImportCustomerUpdatesTests#test

```
...
    @Test
    public void test() {
        JobParameters jobParameters = new JobParametersBuilder()
                .addString("customerUpdateFile", "classpath:customerFile.csv")
                .toJobParameters();

        JobExecution jobExecution =
                this.jobLauncherTestUtils.launchStep("importCustomerUpdates",
                jobParameters);
        assertEquals(BatchStatus.COMPLETED,
                            jobExecution.getStatus());
        List<Map<String, String>> results =
                this.jdbcTemplate.query("select * from customer where customer_id = 5",
                (rs, rowNum) -> {
                    Map<String, String> item = new HashMap<>();
```

```java
                    item.put("customer_id", rs.getString("customer_id"));
                    item.put("first_name", rs.getString("first_name"));
                    item.put("middle_name", rs.getString("middle_name"));
                    item.put("last_name", rs.getString("last_name"));
                    item.put("address1", rs.getString("address1"));
                    item.put("address2", rs.getString("address2"));
                    item.put("city", rs.getString("city"));
                    item.put("state", rs.getString("state"));
                    item.put("postal_code", rs.getString("postal_code"));
                    item.put("ssn", rs.getString("ssn"));
                    item.put("email_address", rs.getString("email_address"));
                    item.put("home_phone", rs.getString("home_phone"));
                    item.put("cell_phone", rs.getString("cell_phone"));
                    item.put("work_phone", rs.getString("work_phone"));
                    item.put("notification_pref", rs.getString("notification_pref"));

                    return item;
                });

        Map<String, String> result = results.get(0);

        assertEquals("5", result.get("customer_id"));
        assertEquals("Rozelle", result.get("first_name"));
        assertEquals("Heda", result.get("middle_name"));
        assertEquals("Farnill", result.get("last_name"));
        assertEquals("36 Ronald Regan Terrace", result.get("address1"));
        assertEquals("P.O. Box 33", result.get("address2"));
        assertEquals("Montgomery", result.get("city"));
        assertEquals("Alabama", result.get("state"));
        assertEquals("36134", result.get("postal_code"));
        assertEquals("832-86-3661", result.get("ssn"));
        assertEquals("tlangelay4@mac.com", result.get("email_address"));
        assertEquals("240-906-7652", result.get("home_phone"));
        assertEquals("907-709-2649", result.get("cell_phone"));
        assertEquals("316-510-9138", result.get("work_phone"));
        assertEquals("2", result.get("notification_pref"));
    }
}
```

你可能需要进行测试的最后一部分是整个作业。下面将介绍真正的功能测试,并端到端地测试批处理作业。

3. 作业的测试

测试整个作业可能是一项艰巨的任务。如你所见,有些作业可能非常复杂,并且很难完成配置。然而,自动化地执行作业和验证结果的好处是不容忽视的。因此,强烈建议你尽可能尝试在这个级别进行自动化测试。下面将介绍如何使用 JobLauncherTestUtils 执行整个作业以进行测试。你很快就会发现,这实际上与单独地执行步骤非常类似。

为了进行测试,我们将使用一个比银行对账单作业更容易测试的作业。在本例中,我们将使用一个单步作业,用于从列表中读取数据并将它们写入 System.out。尽管这个作业可能不怎么花哨,但它可以让我们关注作业的测试部分而不是作业本身。

与完成集成测试一样,我们首先从基础设施开始。这个测试将使用与其他集成测试相同的基础设

第 13 章 批处理的测试

施，如代码清单 13-19 所示。

代码清单 13-19　JobTests 的基础设施

```
...
@ExtendWith(SpringExtension.class)
@SpringBatchTest
@ContextConfiguration(classes = {JobTests.BatchConfiguration.class,
BatchAutoConfiguration.class})

    @Autowired
    private JobLauncherTestUtils jobLauncherTestUtils;
...
```

在完成基础设施后(包括添加注解并在测试类中注入 JobLauncherTestUtils)，我们需要定义作业。对于这个示例，我们将在测试用例的静态类中定义作业。代码清单 13-2 展示了测试作业的代码。

代码清单 13-20　测试作业

```
...
@Configuration
@EnableBatchProcessing
public static class BatchConfiguration {

        @Autowired
        private JobBuilderFactory jobBuilderFactory;

        @Autowired
        private StepBuilderFactory stepBuilderFactory;

        @Bean
        public ListItemReader<String> itemReader() {
            return new ListItemReader<>(Arrays.asList("foo", "bar", "baz"));
        }

        @Bean
        public ItemWriter<String> itemWriter() {
            return (list -> {
                list.forEach(System.out::println);
            });
        }

        @Bean
        public Step step1() {
            return this.stepBuilderFactory.get("step1")
                        .<String, String>chunk(10)
                        .reader(itemReader())
                        .writer(itemWriter())
                        .build();
        }

        @Bean
        public Job job() {
            return this.jobBuilderFactory.get("job")
                        .start(step1())
                        .build();
```

```
        }

        @Bean
        public DataSource dataSource() {
                return new EmbeddedDatabaseBuilder().build();
        }
}
...
```

代码清单 13-20 中没有什么太复杂的内容。BatchConfiguration 是一个基本的 Spring 配置类，使用了我们在整本书中创建步骤和作业时一直在使用的构建器。ItemReader 使用的是 Spring Batch 提供的 ListItemReader，在调用 read 方法时，将返回 foo、bar 和 baz。ItemWriter 使用的是 lambda 表达式，由于这里把写入数据的操作委托给了 System.out.println，因此会把每一个数据都写到标准输出。最后配置的是步骤(一个基于块的步骤，块大小为 10，并且使用了前面定义的条目读取器和条目写入器)以及由步骤组成的作业。

最后一部分是测试作业。为此，使用 JobLauncherTestUtils 启动作业。在本例中，没有任何作业参数，因此可以使用 JobLauncherTestUtils.launchJob()方法执行作业。由于上下文中只有一个作业，因此不需要指定，实用工具会自动装配这个作业。这个方法调用将返回一个 JobExecution，可以验证 BatchStatus 为 COMPLETED。我们还可以检查 StepExecution，验证 BatchStatus 为 COMPLETED 以及读取和写入的数据量。代码清单 13-21 展示了这个测试方法。

代码清单 13-21　JobTests#test

```
...
@Test
public void test() throws Exception {
        JobExecution jobExecution =
                this.jobLauncherTestUtils.launchJob();

        assertEquals(BatchStatus.COMPLETED,
                jobExecution.getStatus());

        StepExecution stepExecution =
                jobExecution.getStepExecutions().iterator().next();

        assertEquals(BatchStatus.COMPLETED, stepExecution.getStatus());
        assertEquals(3, stepExecution.getReadCount());
        assertEquals(3, stepExecution.getWriteCount());
}
...
```

13.3　本章小结

从对系统中的任何组件的单个方法进行单元测试，直到以编程方式执行批处理作业，你已经看到了批处理编程人员可能遇到的绝大多数测试场景。本章首先概述了用于单元测试的 JUnit 测试框架和 Mockito 对象模拟框架；然后探索了如何使用 Spring 提供的类和注释进行集成测试，包括在事务中执行测试；最后介绍了特定于 Spring Batch 的测试，包括测试步骤范围内定义的组件以及作业中的各个步骤和整个作业。